Towards Holistic Medicine

Bing Yuan

HK Modern Chinese Medicine R&D Center
Hong Kong
China

CRC Press is an imprint of the
Taylor & Francis Group, an **informa** business
A SCIENCE PUBLISHERS BOOK

Cover credit: Image provided by the author.

First edition published 2024
by CRC Press
2385 NW Executive Center Drive, Suite 320, Boca Raton FL 33431

and by CRC Press
4 Park Square, Milton Park, Abingdon, Oxon, OX14 4RN

CRC Press is an imprint of Taylor & Francis Group, LLC

Library of Congress Cataloging-in-Publication Data (applied for)

ISBN: 978-0-367-14534-7 (hbk)
ISBN: 978-1-032-48035-0 (pbk)
ISBN: 978-0-429-03219-6 (ebk)

DOI: 10.1201/9780429032196

Typeset in Times New Roman
by Prime Publishing Services

Preface

This book was written in response to an invitation from CRC Press, the world's leading scientific publishing house. In 2018, CRC Press included 'precision medicine', which is at the forefront of medicine, as a key option of the publishing house, and was seeking authors and topics related to this field. Seeing my previous publications on methodological research on precision medicine, Raju Primlani from the publishing house contacted me, hoping that I could write a monograph related to this field, with no less than 125,000 words in English.

At that time when precision medicine was hyped to be 'in full swing', through multi-angled methodological research on the background, essence, and significance of precision medicine, I clearly pointed out the problems in the development of precision medicine and its scientific limitations in my published papers. It was clear to me that precision medicine could only be a 'flash in the pan' concept in the long history of scientific development due to the inability to solve the problem of medical integration which is as important as personalization. After precision medicine and systems biology, modern medicine and even life sciences will face another more profound revolution. In my reply to the publisher, I expressed the hope that the scope of this book would not be limited to the field of precision medicine. My idea was recognized by the publishing house. From this, there is this book that shows the current crisis, development trend and future medical model of modern medicine under the broad perspective of the method evolution of Eastern and Western medicine, biology, and even natural science in the past few hundred years.

"Viewed horizontally as a ridge, and side viewed as a peak:

It looks different from a distance, near, high, and low.

I don't know the true face of Mount Lu,

Only because I am in this mountain."

This is a popular poem by the famous Chinese poet Su Shi in the Northern Song Dynasty more than a thousand years ago. The words are simple, but the meaning is profound. What kind of method humans adopt to establish a medical theory system, and the future direction of medical development cannot be completely determined by medicine itself. Future medical research methods and medical systems will inherit the advantages of past medicine in dealing with human diseases and health problems, overcome their defects, and make up for their shortcomings. At the same time, they will inevitably be branded with the prevailing methodology of human research and understanding of natural objects (including the human body) in this era. There is an old saying in China: "Those who do not have long-term considerations are not enough to seek for a while; those who do not have overall considerations are not enough to seek for a single domain." Following the evolution of natural science methods and mainstream medical methods, this book provides an in-depth comparative analysis of the effects and features of past medical systems (including traditional medicine and modern medicine) in dealing with diseases and health problems. Based on contemporary science's deepest understanding of nature (including the human body) in depth and breadth and the most advanced methodology of scientific research, this book proposes a complete solution carrying the past into the future for future medical methods and medical systems from a broader perspective of biology and even natural sciences.

In recent years, in the field of natural sciences, the rise of complexity science has triggered changes in scientific concepts and ways of thinking. People no longer attribute the problems of complex systems to problems of simple systems through analysis and simplification, but start to face complexity directly and conduct research in accordance with the true colors of complex systems. In the field of biology, the birth of systems biology marked the beginning of a holistic study of organisms. In the field of medicine, on one hand, the concept of integrated medicine has been widely recognized, and medical scientists are striving to overcome the limitations of analysis methods to realize the overall synthesis of disease treatment. On the other hand, modern medicine has gradually got rid of the fetters of disease medicine. Precision medicine, which was pioneered in the United States, has begun an attempt to formulate treatment plans based on the individual characteristics of patients.

However, after more than 30 years of fruitful exploration, systems biology, which aims at holistic research, seems to have encountered insurmountable obstacles when facing of the complexity and adaptability of organisms. Currently, the concepts and methods employed in precision medicine research are still largely based on reductionism. With the establishment of the biomarker system and the continuous discovery of new target medications, medical scientists are still facing the problems of the complexity of the human body and the overall synthesis of disease treatments.

Over the past few hundred years, every breakthrough in natural science methods and technology has rapidly expanded into the field of life sciences, propelling the advancement of biology and medicine. In recent years, as natural sciences have evolved from simple science to complexity science, the method of dealing with complexity has also transcended the 'analysis-refactoring' approach typically used in the initial stage of system science to integrate simple systems. Instead, the holistic approach to directly confront complexity has gradually matured. This seems to indicate that if biology aims to achieve a holistic study of the human body, it will eventually have to abandon the 'analysis-reconstruction' method and embrace the era of holistic biology utilizing holistic methods for research.

When we look at traditional Chinese medicine (TCM) which has a history of more than 2000 years in China and created countless medical miracles, from the perspective of complexity science, we find that: the methods used by ancient TCM scholars to establish medical theories are surprisingly consistent with the methods used by complexity science to deal with complex systems; the overall concept as the core concept of TCM is exactly the same as the integrity pursued by systems biology; the unique syndrome differentiation and treatment of TCM and the individualization embodied in precision medicine are also completely consistent in concept. Obviously, modern life sciences have begun the process of returning to TCM in terms of concepts and methods, and this process is completely consistent with the directions indicated by complexity science for life sciences and medicine.

TCM embodies the concept and methodology of complexity science, and perfectly solves the integration problems faced by biology and medicine. The description and regulation-control of the human body state in its treatment system based on syndromes are all personalized, which reflects the pursuit of individualization in precision medicine. In TCM, the adaptability that makes systems biology helpless is no longer an obstacle to scientifically controlling the complexity of life, but a powerful assistant for mankind to treat diseases and care for health more effectively and simply.

The human body model and the human body state description system (SDS) are established based on the TCM method consistent with the holistic method of complexity science; the regulation-control of diseases are based on TCM syndrome differentiation system, which is consistent with the state regulation-control methods commonly used in natural sciences. As a result, a new medical system will be formed that inherits the traditions of TCM and conforms to scientific norms. Since this new medical system is in line with the holistic concept of TCM and the holistic method of complexity science, we call it holistic medicine.

Holistic medicine is medicine that embodies the holistic concept and the principle of individualization and is based on empirical evidence. On the basis of the functional model, state identification and state regulation-control based on the SDS is the mainstream regulation-control mode of this medicine. We call this mode the state medicine mode. However, holistic medicine is not limited to the comprehensive regulation-control of the human body based on the model of state medicine. It also integrates the model of disease medicine based on the cause, location, and pathological changes, so as to more comprehensively and accurately grasp the patient's personalized health state during the disease process from different perspectives, making the treatment more comprehensive and holistic.

Holistic medicine is a medical system that integrates state medicine and disease medicine. It will still follow the methods of TCM and complexity science, using metaphors and analogies to build functional models that reflect human physiological and pathological activities. But with the help of modern scientific structure analysis concepts and methods, it will give the model a rigorous logical structure. Moreover, with the help of big data analysis and artificial intelligence technology, the model can be realized through computer simulation with empirical support. Holistic medicine will still use TCM treatment based on syndrome. However, through the introduction of modern scientific concepts and methods to improve and optimize the completeness and independence of state variables, the SDS will conform to the structured and standardized scientific norms. And deredundancy and optimization of the syndrome system based on the principle of 'the simplest and applicable' will further simplify the SDS. The state medicine of holistic medicine is holistic and comprehensive in grasping and regulation-controlling the state of the human body. However, by introducing detection methods developed in disease medicine and precision medicine, as well as detection indicators that are valuable for its state description, the SDS can be refined and objectified. Therefore, the human body model and SDS will be established on the basis of evidence that can be verified by observation and trial.

The disease medicine system of holistic medicine includes the traditional disease medicine system of TCM and modern medicine, as also the medical system that is being establishing by precision medicine based on the new disease classification. The current development trend of disease classification systems (DCS) in modern medicine and TCM is getting more and more refined, and there are more and more types of diseases. Of course, holistic medicine does not rule out the refinement of the DCS when necessary. However, in the current situation where the modern medical DCS is already very large, the development trend of disease medicine in holistic medicine is just the opposite. Based on the principle of 'the simplest and applicable', on the one hand, it is necessary to remove the redundancy of the SDS, and on the other hand, it is necessary to greatly simplify the existing DCS under the premise of satisfying the distinction between different treatment methods. In other words, although the disease medicine of holistic medicine covers the traditional disease medical system of traditional Chinese and Western medicine, as well as the new disease system of precision medicine, it will contain far fewer types of diseases after eliminating the redundancy.

In holistic medicine, the integration of TCM treatment based on syndrome differentiation and modern medicine's causal treatment and symptomatic treatment will reach a higher level. The integration of traditional Chinese and Western medicine based on the concept of reductionism is being carried out in mainland China for more than 50 years, which has changed TCM treatment based on syndrome to an appendage of differentiated disease medicine through the implementation of 'Western medicine diagnosis, TCM classification'. In contrast, the integration with treatment methods from traditional Chinese and Western medicine in holistic medicine is based on the holistic concept and is dominated by the state medicine based on TCM syndrome differentiation.

The theoretical model of holistic medicine, the SDS, and the corresponding state regulation-control and disease treatment system are laid on an empirical basis. However, the randomized controlled trial and its corresponding clinical efficacy evaluation system were designed based on the reductionist scientific concept of 'change only one variable at a time'. Simply applying this method is obviously not suitable for studying the full range of actions of medications on various parts of the human body under various conditions. As a result, research on medications and treatment methods in holistic medicine will gradually move towards an evaluation system that combines traditional randomized controlled trials and 'real world research'.

In reality, it is common for patients to suffer from multiple diseases and multiple syndromes at the same time, and some of the curative effects are long-term effects that require long-term follow-up observation. Real-world research can include complex patients with multiple diseases, can allocate treatment in a non-random way, and requires a large number of cases. Therefore, the real-world research approach adopted in medication research of holistic medicine is obviously more appropriate. It's just that today's real-world research can no longer be the same as that of ancient times. It will be based on rigorous empirical and statistical analysis. To realize this kind of large-scale real-world research, the standardization of the SDS is an indispensable prerequisite. Therefore, the standardized system of symptoms, signs, and detection indicators established under the overall medical framework should be a unified system suitable for the state description and the classification of diseases.

To organically combine and apply the treatment methods based on the development of state medicine and disease medicine in clinical practice, the clinical guidelines of modern medicine provide an effective mode of operation. However, the clinical guidelines of holistic medicine are quite different from the static clinical guidelines currently used in modern medicine. It not only will reflect the relationship between syndromes, and the relationship among syndromes and diseases, but also will reflect the dynamic process of the evolution of the human body's state and the occurrence, development and outcome of diseases. It does not independently issue a treatment plan for each disease of the patient, but comprehensively considers all the patient's abnormal state variables and the severity and importance of all diseases and the relationship among them, and seeks for the best overall treatment plan. In the process of establishing and improving clinical guidelines, artificial intelligence and big data analysis technology will play a vital role.

Holistic medicine is a disciplinary system that integrates methods of complexity science and TCM methods, and integrates knowledge systems and diagnosis and treatment technology of TCM and modern medicine. Today, among the tentacles of the advancement of medicine, precision medicine, integrated medicine, and evidence-based medicine show human's hope for the future medical model from different perspectives: towards individualization, towards overall integration, and laying on an empirical basis. However, what is needed is medicine with these three characteristics at the same time, rather than three independent medicines. Holistic medicine is precisely such a medical system that perfectly embodies the concept of holistic integration and the principle of personalization, and is based on evidence.

The holistic medicine system based on the principle of 'the simplest and applicable' will end the medical development trend of the continuous explosive increase in the total amount of knowledge. In its development process, just like in today's physics, there will be continuous development by absorbing of new knowledge, experience and technology related to disease diagnosis and treatment, state identification and regulation-control; there will also be continuous simplification and updating of the knowledge system through structural merging, integration of knowledge, and elimination of useless knowledge. In the era of holistic medicine, it is not ruled out that the treatment of diseases will still be divided into different departments, but it will not continue to be more and more detailed, and limit the medical practitioners to increasingly narrow fields. Simple and practical clinical pathways and clinical guidelines will greatly expand the special fields that medical practitioners can master. State identification and state regulation-control based on theoretical models and SDS are the basic skills that medical practitioners trained in this discipline should master. That is to say, no matter which clinical specialty area the medical practitioner will focus on in the future, the skills of state recognition and state regulation-control at the holistic level are indispensable. For a SDS with only a two-digit scale for state variables (syndrome), mastering the state identification and regulation-control skills based on it will no longer require five years of study at the university of TCM. The realization of artificial intelligence based on the SDS, as a decision support for medical practitioners in clinical diagnosis and treatment, will effectively ensure the diagnosis and treatment level of inexperienced medical practitioners.

Today's medicine is moving towards an era of great changes. This is a revolution in which modern medicine started by systems biology and precision medicine returns to the tradition of TCM. The development of methodologically-significant complexity science, the maturity of artificial intelligence and big data analysis technology, the rapid development of disease detection and diagnosis technology in modern medicine, and the deepening of understanding of the human body and diseases based on these technologies, all provide the strong knowledge base and technical support to this kind of change. The reconstruction of medicine according to a new concept is a huge project that requires the continuous efforts of one or several generations. However, as you will see from this book, medicine is moving towards such an era.

I would like to thank Dr. Meng Fanyi for undertaking the review of the English version of this book. Dr. Meng is my friend from the Beijing University of Chinese Medicine, where we were classmates for five years. After graduating in 1983, we both stayed on at the University to work together, and from 1984 to 1986, we also studied computer software design together in a computer specialization course organized by the Ministry of Health. Dr. Meng went to the United Kingdom in 1997. Since 2004, he has been the director of the acupuncture degree program at Lincoln University and concurrently the director of the complementary medicine discipline group. Since 2005, he has served as an independent examiner in many universities, and also served as a member of the British Acupuncture Certification Committee, TCM Certification Committee and other TCM quality control organizations. In 2019, he served as a commissioner of the Quality Supervision Agency (QAA) of the National University of the United Kingdom. He is currently the Vice President of the British Association of Chinese Medicine (CMA.UK), a director of the World Federation of Chinese Medicine Societies (WFCMS), and director of a number of branches. Two overseas textbooks of the Beijing University of Chinese Medicine, 'Basic Theory of Traditional Chinese Medicine' and 'Diagnostics of Traditional Chinese Medicine', edited by Dr. Fanyi Meng, were published in 2016. In the experience of working and studying together, we often discussed the issue of medical development, and this gave great support and help to my research work. We have agreed that after the publication of this book, we will further cooperate in the construction of the holistic medicine system based on the concepts and methods of this book.

I would also like to thank my son, Kevin Yuan, who is currently a fourth-year student at Cardiff University School of Medicine in the United Kingdom, for his great help in the process of writing and proofreading the book.

Nov. 18, 2023, in HK — Bing Yuan

Contents

Part II: Towards Holistic Medicine

Part I

The Human Body and Medical Tradition of Holism from the Perspective of Complexity Science

Chapter 1
System, Model and State Description

Just two decades ago, concepts such as system and model were little known as fashionable terms for emerging disciplines at the frontiers of science. Today, whether in China or the West, the terms system, model, function and structure are so deeply-rooted in people's minds and they can be seen everywhere in textbooks from elementary school to middle school. They have become the basic scientific concepts that elementary and middle school students must master when learning scientific methods.

1.1 System

System is a concept widely used in science, engineering and many fields. It involves a set of interrelated and interacting elements that operate according to certain rules to form a unified whole and exhibit a complex range of behaviors and properties. A system, surrounded and influenced by its environment, described by its boundaries, structure and purpose, and expressed through its functionality. In the modern world we can find various types of systems in nature, social systems, technological applications and many other fields.

1.1.1 The concept of system

A system is usually defined as a set of interrelated, interdependent, and mutually restrictive elements (components) that are combined according to certain rules to form a whole with a certain structure and function. The definition includes the four concepts of system, element (part), structure, and function, and it reveals the relationship between element and element, element and system, and system and environment.

We can understand the concept of the system from three aspects (Zhang Lianming 2023):

1. *The system is composed of two or more elements (parts).* These elements may be individuals, components, parts, or they may themselves be a system (or subsystem). For example, the computer, the controller, the memory, and the input/output devices form the hardware system of the computer. These hardware components with different functions, as a subsystem of the computer system, are composed of low-level components and integrated circuits.
2. *The system has a certain structure.* A system is a collection of its constituent elements, which are interrelated, interdependent and mutually restrictive. There are relatively stable contact methods and organizational order among the elements inside the system, between elements and the whole, and between the whole and the external environment, which is the structure of the system. For example, clocks and watches are assembled by gears, springs, hands and other components in a certain way, but gears, springs and hands which are randomly stacked together cannot constitute a clock. The human body is composed of various organs, but the various organs are simply put together and cannot be a person with the ability to act.

3. *The system has certain functions*. The function of the system refers to the nature and capabilities shown in the interconnection and interaction between the system and the external environment. For example, the function of the clock system is to accurately indicate the time; the function of the human respiratory system is to complete the process of inhaling oxygen and exhaling carbon dioxide. If it is merely a simple stack of elements without any function, it cannot be considered a system.

Systems are ubiquitous, and everything in the world can be regarded as a system. From elementary particles to extragalactic galaxies, from human society to human thinking, from inorganic to organic, and from natural science to social science, systems are omnipresent. The vastness of the universe, the microscopic elementary particles, a seed, a swarm of bees, a machine, a factory, an association... are all systems. The whole world is a collection of systems.

At the same time, systems and elements are relative. Elements belonging to a system often also constitute a system (or a subsystem of this system), and a system is often an element or subsystem of a larger system. For example,

University: University → College → Department → Class → Group → Student
Human body: human body → system → organ → cell → molecule
Substance: molecule → atom → elementary particle

Among them, each level is an element that constitutes the system of the previous level, while at the same time, it is a system composed of the elements of the next level.

A system can be concrete, referred to as a real system, or it can be abstract, also known as a conceptual system or theoretical model. The atoms, cells, and human body systems that we encounter in natural science exist in the real world as actual systems, regardless of how people perceive them. However, the descriptions of these systems in textbooks are, in fact, conceptual systems or theoretical models abstracted from real systems.

When understanding the concept of a system, it is necessary to understand the relationship between the system and the environment:

The whole of everything related to it outside a system constitutes the environment of the system. Usually, in the process of studying the system, people think that the state of the environment is known. People only care about the action of the environment on the system, not the composition, nature and changes of the environment itself, nor the impact of the system on the environment. In order to simplify the problem, people often classify the parts that are not concerned or difficult to understand as the environment in the research process.

The division of the system and the environment is relative, and people usually divide according to actual needs and the convenience of handling problems. For the same problem, you can choose different systems and environments. The interaction between the system and the environment is achieved through the exchange of matter, energy and information, and the result of the interaction may change the nature and function of the system. Any system is generated in a certain environment and develops and evolves in the environment. Therefore, the research system must study its interaction with the environment, especially the impact of the environment on the development of the system.

1.1.2 The structure and function of the system

For any system to perform certain functions, it must have a corresponding structure. A chair can sit because of the supporting leg, seat plate and corresponding backrest structure; the bird can fly because of its structure with wings; the fish can breathe in the water because of its gill structure. This is true in nature and living things, as well as in the human body. Lungs can breathe because of the existence of alveoli, trachea, chest cavity and other structures; cells can exchange energy and substances with the external environment because of the special permeability structure of the cell membrane. It is precisely because of this dependence between structure and function that, when constructing a theoretical model based on a real system, we always have to conceive certain elements and structures so that phenomena occurring in the real system can be fully explained by the theoretical model.

(1) The structure of the system

A component of a system can be a single element that cannot be further subdivided, or it can be a system, also known as a subsystem. The interaction of components to form a unified whole of the system. The way in which components are related and interacted with each other is called the structure of the system, including the quantity ratio, arrangement order, combination mode and their evolution of each component. Structure is the form of existence of the system, and the hierarchy of the system is an important feature of the system structure.

Usually, people judge the complexity of a system not by the number of components it contains, but more by the number of types of components and the number of levels of the system. The more types and layers of components a system contains, the more complex the system will be. There is a close relationship between the various levels of the system, but the properties of a level are not simply summed up from the properties of the lower level. When a complex system is composed of low-level components into a high-level system, the system often produces new properties that the original level did not have. This process is called emergence in complexity science. Emergence phenomena widely appear in various fields of natural science. In recent years, its nature and characteristics have become a hot topic in complexity science research. What needs to be emphasized here is that different levels of a complex system sometimes present different characteristics and require different methods for research (Tan Lu and Jiang Lu 2009).

(2) System behavior and function

Any change that the system exhibits relative to its environment, or all changes that the system can detect from the outside, is called the system's behavior. Behavior is a manifestation of the system's own characteristics, but it is also related to the environment and reflects the action or impact of the environment on the system.

Different systems have different behaviors, and the same system has different behaviors in different situations. The various behaviors of the system include life-supporting behaviors, learning behaviors, adaptive behaviors, evolutionary behaviors, balanced behaviors, stable behaviors, dynamic behaviors, etc. Systems science leaves aside the particularities of various individual systems, and is a discipline focused on studying the general behavior and laws of various systems.

The function of the system is an important concept to describe the behavior of the system, especially the relationship between the system and the environment. In principle, any behavior of the system will have an impact on the environment. System behaviors that are conducive to the survival and development of certain things in the environment and even the entire environment is called system functions. All systems have certain functions. For example, the function of the engine is to provide power for vehicles or aircraft, and the function of the school is to cultivate talents and provide scientific research results for the society. Function is an overall characteristic of the system.

The concept of function is also commonly used in subsystems, which refers to the responsibilities and contributions of subsystems to the survival and development of the entire system. Subsystems are usually divided according to their functions in the entire system, and are related, interacted, and restricted according to their respective functions. They jointly maintain the survival and development of the entire system. We refer to the functional division of subsystems and the way they are interconnected as the functional structure of the system. To understand the properties and behaviors of a system, one must understand its functional subsystems. The function of a system is based on its structure, but the function is not solely determined by the structure; it is jointly determined by the structure and the environment. For the system to play a certain function, the environment needs to have appropriate conditions and atmosphere, which means that the system can be used for its proper function only when it is used for specific objects and given appropriate environmental conditions. That is to say, it is only when the environmental conditions are given that we can say 'structure determines function'.

1.1.3 System environment and input and output

A system's environment and input and output are key concepts in systems theory. They help us understand and describe how a system operates in a specific context. The environment of a system refers to the external conditions, factors, and other systems surrounding the system that can affect the behavior and properties of the system. The environment provides the resources required by the system and determines the feasibility and efficiency of the system to a certain extent. Inputs to a system are information, energy, or substances obtained from the environment that are necessary for the system to operate and perform its functions. Input is processed and transformed by the system, and output is produced, which is the system's response or result to the input. Outputs reflect the behavior and properties of a system, and they can affect the environment and even other systems.

By studying a system's environment, inputs, and outputs, we can better understand its functionality, behavior, and interrelationships. Whether in the field of engineering or other fields such as ecology and social sciences, understanding the environment and the input and output of a system is a critical step in solving complex problems and improving system performance.

(1) System environment

A collection of everything related to a system outside the system itself is called the environment of the system. More precisely, the environment of a system refers to everything outside the system that has a non-negligible connection with the system. This means that the environment of the system can only be determined in a relative sense. Under different research purposes, or for different researchers, the environmental division of the same system will be different.

Any system is generated, operated, continued and evolved in a certain environment, and there is no system without an environment. The structure, state, attributes, and behavior of the system are more or less related to the environment. In the real system, the same components combine in different ways to form different structures in order to adapt to different environments, and the properties of the components may also change with different environments. Generally speaking, the environment is also an important factor that determines the overall structure of the system. Under certain environmental conditions, the system can only adapt to the environment and form a stable environmental dependency relationship when it evolves a specific overall structure. As the environment changes, the system must generate a new overall structure to achieve a new environment adaptation relationship. Environmental complexity is an important source of system complexity. Therefore, studying a system must study its environment and its interaction with the environment (Xu Guozhi 2000).

Compared with the system, the interconnections between environmental components are generally weaker, less regular, and less systematic. This provides the possibility for the system to seek advantages and avoid harm, and protect and develop itself. But there is always some kind of connection between different things in the environment of the same system, and usually some kind of bigger system can be formed through the connection with the system. Any system is relatively independent of the environment, yet the division of the system and the environment is relative. Therefore, the system can be defined as a relatively independent part separated from the interrelated things according to the research concerns.

(2) System input and output

The interaction and mutual influence between the system and the environment is realized through the input of the environment to the system and the output of the system to the environment. Generally, the effect and influence of the environment on the system are known as the input of the system, and the action and influence of the system on the environment are known as the output of the system.

The input of the system can be divided into two parts, namely controllable input and uncontrollable input. To control a system is to transform the state of the system toward a specific state by changing the input of the system. But not all the inputs to the system are controllable. For example, among the input factors that affect the flight of an aircraft, the actions of the rudder, elevator, aileron and engine thrust are realized by the pilot's operation, which are controllable inputs. However, the wind, atmospheric density, and magnetic field strength, which also affect the flight of the aircraft are natural conditions that are not controlled by the pilot. These are called uncontrollable inputs, or the condition, environment or interference. In the process of medical practitioners treating patients, therapeutic medications, acupuncture stimulation, and language guidance are controllable inputs, while the invasion of pathogens, the influence of the climate environment, and the stimuli that can cause emotional changes are considered as interference to patients.

The output of the system is the result of the system reacting and processing the input, which is a function of the input. The set of system outputs reflects the behavioral effects of the system under the action of the input. In fact, whether it is input or interference, it will affect the output of the system. Interference often causes the output of the system to deviate from the set target, that is to say, it may produce a target difference between the control result and the target. The function of controllable input is reflected in two aspects: one is to make the system produce a predetermined output, and the other is to make the system try to overcome the deviation caused by the interference, and exclude those outputs that do not meet the target. The input of the system has a certain impact on the system, and it is shown as an output after being processed by the system. The processing of the input by the system depends on the structure of the system. It can be said that the output of the system is determined by the combination of the inputs and the structure of the system together.

1.1.4 Classification of the system

For the convenience of research, it is necessary to classify the various specific systems. According to different classification principles, we will see different systems (Tan Lu and Jiang Lu 2009):

1. *Classification according to the relationship between the system and the environment*:
   i. *Closed system*: There is no exchange of matter, energy, information with the outside world, that is, such a system does not have any interaction with the surrounding environment. Strictly speaking, there is no such system in nature, it is an ideal model proposed for the needs of research. When the interaction between the system and the outside world is negligible, the system can be regarded as an isolated system, such as a box of well-sealed gas.
   ii. *Open system*: A system that has both material and energy exchanges with the outside world is defined as an open system in physics. There are always inextricable connections between things in the real world, so most of the systems that exist in the real world are open systems.
2. *Classification according to the mutual relationship between the various subsystems in the system*:
   i. *Linear system*: Changes in one part of the system can cause changes in other parts to be linear, or, when the input of the system increases linearly, the output of the system also increases linearly. Such a system is called a linear system. For linear systems, modern science has very mature methods for analysis.
   ii. *Nonlinear system*: Here, the influence between the elements in the system is not linear, that is to say, the input and output of the system do not meet the principle of superposition. At present, there is no standardized solution for nonlinear systems, so it is much more difficult to deal with.
3. *Classification according to the relationship between system status and time*:
   i. *Static system*: If the state of the system does not change with time, that is, the output of the system at a certain moment is independent of the input at other moments, it is called a static

system. Studying a static system is equivalent to analyzing the properties of a fixed state of the system.

ii. *Dynamic system*: A dynamic system is a system whose system state changes with time. The output of a dynamic system at a certain moment is related to the input at other moments. To study a dynamic system, we must study the time behavior of the system and find out the law of the system state changing with time.

4. *Classification according to the evolution characteristics of the system*:
   i. *Deterministic system*: A deterministic system is a system in which external influences are determined and the evolution of the system and the interrelationships among subsystems are also fixed. Such systems can be described by deterministic equations.
   ii. *Random system*: Here, there are some uncertain factors inside the system, or the outside world imposes random disturbances on the system, so that the behavior of the system is random. When studying stochastic systems, probabilistic methods are usually used to describe their evolutionary behavior.
5. *Classification according to the complexity of the system structure*:
   i. *Simple system*: This refers to a system that contains a small number of subsystems and the interaction between the subsystems is relatively simple; or a system composed of a large number of elements with similar behaviors, such as gases in a closed container or distant galaxies. For the latter class of simple systems, simple statistical averaging methods can usually be used to study their behaviors.
   ii. *Complex system*: A complex system must have a certain scale, but it is not that the larger the system and the more elements, the more complicated it is. The complexity of the system is usually reflected in the large number of component elements and the diversification of the interaction between elements. Therefore, in complexity science, a complex system is defined as a system with a relatively large number of elements kinds and a certain number of scales, and there is a strong interaction between the elements.
6. *Classification according to the degree to which the system is understood*:
   i. *Black box system*: It cannot be opened, nor can its internal structure and state be directly observed from the outside, so its structure and parameters can only be determined through the input and output of information, such as the human brain.
   ii. *White box system*: Here, not only is the corresponding relationship between input and output clearly understood, its internal structure, laws and mechanisms of behavior occurrence are understood as well, such as a manually designed mechanical system.
   iii. *Gray box system*: An intermediate system of the black box and the white box, its internal structures can be partially observed. This usually refers to those systems whose internal structure and laws are not yet very clear and need to be further revealed, such as systems in the fields of meteorology, ecology, and the economy.

It can be seen from the above that real systems can be distinguished from different perspectives, and a system can belong to different system types at the same time.

1.1.5 Description of system status

The state is a term that people are very familiar with. It usually refers to those conditions, situations, characteristics, etc., of the system that can be observed and identified. For example, the state of human beings includes health, disease, sobriety, and sleep; the state of the economic system includes prosperity, depression, crisis, and recovery. Correctly distinguishing and describing these states is a necessary condition for regulating and controlling the system.

It should be explained here that the mathematical equations encountered in middle school are usually only suitable for describing static systems and linear dynamic systems. To describe the state of the system with equations, as long as an analytical solution to the equation is obtained, all the future states of the system can be predicted from the given initial conditions. Using this equation, all the past states of the system can also be traced back, so as to achieve a comprehensive and quantitative grasp of the behavioral characteristics of the system. Systems such as the human body, economic systems, and social systems are all nonlinear systems. There are very few cases where nonlinear dynamic equations can obtain analytical solutions. For example, celestial mechanics shows that the three-body problem cannot be solved analytically in principle, let alone multi-body problems and complex giant systems with millions of variables such as the human body. Seeking an analytical solution is not a universal method for dealing with nonlinear systems. The usual method of studying general nonlinear systems is qualitative description, that is, using qualitative means to study in the state space. From the perspective of system evolution, what is important is not to understand the quantitative properties of the system, but the qualitative properties.

The state is a concept that qualitatively describes the nature of the system, but the state can also be represented by some state quantities that quantitatively describe the characteristics of the system. For example, the state of an ideal gas system can be characterized by quantitative concepts such as temperature T, pressure P, and volume V. Human body systems, general biological systems, and social systems can all be described with appropriate state quantities.

If the quantity used to describe the state of the system can take different values, it is called a state variable. Generally, a system needs to be described by multiple state variables at the same time, and the simplest case is the system that can be described by one state variable. A set of values for given state variables is a state of a given system, and different sets of values represent different states of the system. Generally speaking, the same system can be described by different groups of state variables, and the choice of state variables has a certain degree of freedom. The selected state variables must have a specific meaning and can characterize some basic characteristics and behaviors of the system under study, so they will vary from system to system. The selection of state variables should meet the following general requirements:

1. Completeness: The number of state variables should be large enough to completely and uniquely describe the state of the system.
2. Independence: means that no state variable can be expressed as a function of other state variables.

For example, in physics, to determine the position of a particle in three-dimensional space, the state variable group we need is (x, y, z) in a rectangular coordinate system. These state variables are linearly independent in the coordinate systems. In practical application, we must give the value of three coordinates before the position state of the mass point can be determined. Figure 1.1 shows the position of a particle in Cartesian coordinate systems.

The set composed of all the states of the system is called the state space of the system, and each point of the state space is called the state point. Let the system have n independent state variables, denoted as X1, X2......, Xn. Each set of specific values of state variables (X1, X2..., Xn) represents a specific state of the system. n is the dimension of the state variable and can take any positive integer. The one-dimensional state space is a straight line or a straight line segment, and the two-dimensional state space is a plane. The state space of no more than three dimensions can be drawn, and the space above three dimensions belongs to the abstract space, which cannot be expressed intuitively.

The values of the state variables always change within a certain range (that is, the domain of the function), but they do not necessarily change with time. A system whose state variables do not change with time is called a static system. The system whose state variable x changes with time, that is, a system whose state can be expressed as a function x(t) of time t, is called a dynamic system. In principle, as long as the time scale is large enough, it can always be observed that state variables change with time, so all systems are dynamic. However, given the research objectives and

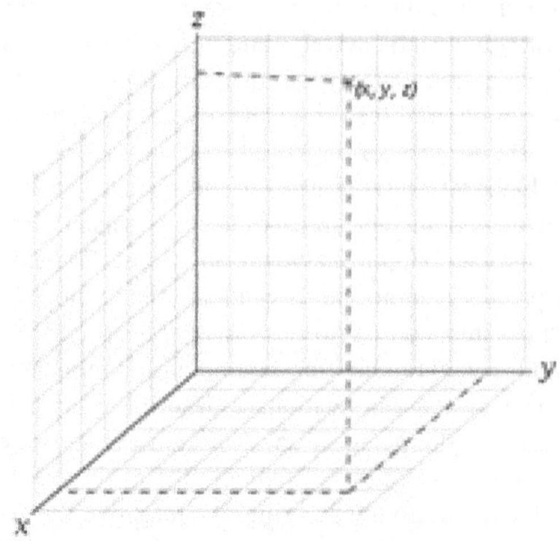

Figure 1.1. 3-dimensional space.

conditions, if the characteristic time scale of the system is much larger than the characteristic time scale of the specific problem under study, and the system state does not change significantly during the research and problem-solving, the system should be regarded as static. Doing so can greatly simplify the description of the system.

Regardless of whether systems are static or dynamic, what researchers are concerned about is the state transition of the system in the state space. Based on the way of state transition, a static system may be described as the transition from one state to another in the state space without consuming time and can be completed in an instant. Dynamic systems, however, need to spend a certain amount of time, that is, they exhibit the characteristics of state changes with time. This change shows infinite diversity in style, characteristics, degree, speed, etc., making the dynamic system theory more abundant, diverse and complex than the static system theory.

All the dynamic properties displayed by the system as a whole first come from inside the system and from the dynamic interaction between the system components. At the same time, they also come from the dynamic changes of the environment itself, and from the dynamic interaction between the system and the environment. The dynamic interaction between components and the dynamic changes of the environment must be manifested through the overall state, characteristics, and behavior of the system. The study of system dynamics shows that it is almost impossible to directly describe the kinetic interaction between components, and the feasible way is to describe the dynamic changes of the overall state, behavior, and characteristics of the system. In the following discussion, we will see that traditional Chinese medicine (TCM) grasps the disease process precisely by describing changes in the state, behavior, and characteristics of the human body during the disease process.

Generally, in the environment where the system is located, the interference and control parameters that affect the system are also variable. These changes may cause some quantitative changes in the behavior of the system, and may also lead to changes in the qualitative nature of the system. An m-dimensional space constructed with input parameters C_1, C_2, ..., C_m as coordinate axes is called an input parameter space (control space). Each point in the input parameter space corresponds to a combination of interference and control elements (Xu Guozhi 2000). We will also see later that in TCM, to describe the human body's state during the disease process, state variables are introduced, and the state space is established. To describe the pathogenic factors, input parameters compatible with the state variables are introduced, and the input parameter space is established. And to describe the properties of Chinese herbal medicines (CHM) and herbal formulas, control parameters compatible with state variables are introduced, and corresponding control spaces are established. It is just that these parameters and the parameter space they form are given different names in TCM.

Introducing the concept of state space can describe the state of the system more clearly. Within the value range of each state variable, the combination of the values of all state variables constitutes the state space of the system. The state space covers all possible states of the system. However, in reality, the possible states of a system at the next moment are not all the states in the state space except the current state. In other words, the state of the system changes according to certain rules, and the state at the latter moment usually has a certain continuity with the state at the previous moment. Thus, with the help of the state space, the evolution process of the system state can be described conveniently.

1.2 Control and feedback

Cybernetics is an emerging cross-cutting discipline that developed after the Second World War. However, looking back on the history of human development, the thought of control has existed in social production and human life from the beginning. The basic concept of cybernetics, feedback, is already deeply rooted. The eagle catching the rabbit in real life and the missile automatically seeking the target in military all run through feedback control from beginning to end. In the course of treatment, TCM physicians constantly adjust prescriptions according to changes in the patient's condition, which is also a concrete manifestation of feedback control.

1.2.1 Presentation of control issues

Since the Second World War, with the development of science and technology, the speed of aircrafts has become faster and faster, their performance has been getting better and better, and the pilots' flying technology has also been significantly improved. Due to the random nature of the pilot's movements, it is almost impossible to calculate the trajectory of the aircraft in advance, so the observation methods, calculation methods and shooting systems using old anti-aircraft guns are no longer effective. Under such circumstances, how to improve the accuracy of artillery shooting and how to organize an effective air defense and counterattack network to deal with the enemy's surprise attack became urgent problems to be solved at that time. Therefore, both sides invested a lot of manpower, financial and material resources in research. Through a series of research and experiments, many theoretical and technical problems in the control process were solved. This resulted in the emergence of missiles, an offensive weapon that can specify attack targets and precise guidance, and also led to the birth of cybernetics.

Cybernetics does not study specific material structures, movement forms and energy processes of systems, but focuses on studying the structure and behavior of abstract systems. It does not focus on studying the behavior of individual systems at a certain time, but on the possible behaviors, states and trends of all systems. Therefore, it is a science that studies the regulation and control laws of systems.

Control is the act by which the control system adjusts its behavior to achieve the goals pursued by the system by acquiring information, processing information, and using information. 'Purpose' and 'behavior' are the two earliest concepts of cybernetics. Control must have a purpose, because without a purpose, there is no need for control, so it is said that control is a purposeful behavior, or it is called a 'goal-oriented' behavior. Secondly, control is an iterative behavior process, it is usually difficult to achieve the goal of control through one control act. So, it is necessary to adjust the control behavior according to the gap between the result of the behavior and the goal, and constantly narrow the difference between behavior result and control goal.

In fact, this kind of control process of 'goal-oriented' is common in nature. The eagle soaring in the sky can not only accurately capture the fixed target on the ground, but even the rabbits and mice that are trying to evade it quickly cannot escape. Obviously, the eagle does not have the possibility to plan its own movement route to the target in advance, that is to say, the eagle does not fly according to a pre-calculated route. Immediately after the rabbit is discovered, the eagle uses his eyes to estimate the approximate distance and relative position between it and the rabbit, and then flies toward the approximate location of the rabbit. In the process, the eagle's eyes are fixed on the rabbit, and keep reporting to its brain the gap between its position and that of the rabbit. No matter how the rabbit runs, the decision made by the eagle's brain is to narrow the gap between itself and the rabbit. This decision is carried out by the wings, which change the flying direction and speed of the eagle at any time, adjust the position of the eagle, and make the gap smaller and smaller until the gap is zero, and the eagle's claws capture the rabbit. Just imagine, if the eagle aimed at the rabbit, then closed its eyes and flew straight towards the rabbit's current position—even if it was aimed very accurately, it would not be possible to catch the moving rabbit.

1.2.2 Feedback

The so-called feedback is the influence of the output of the system on the input. The concept of feedback in cybernetics refers to the process of returning the output of the system to the input and changing the input in some way, thereby affecting the function of the system. More specifically, it is the process of changing the input of the system by returning the output detected by an appropriate detector to the input and comparing it with the input, as shown in Figure 1.2. The concept of feedback in cybernetics is so prominent that some people even think that cybernetics is a discipline specializing in feedback theory.

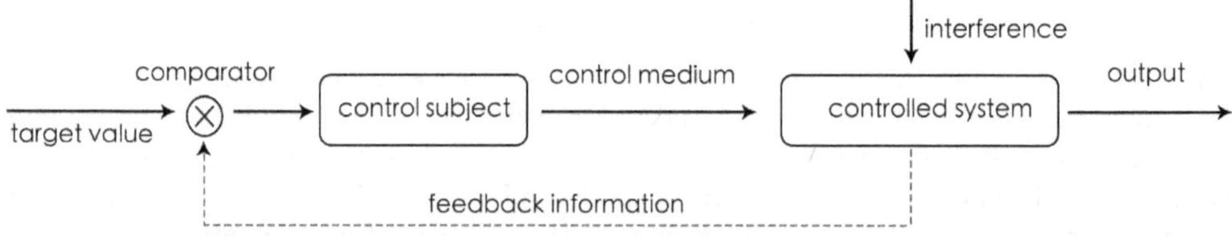

Figure 1.2. Feedback control system.

In the feedback control system, the control part regulates and controls the activities of the controlled part in a specific way, and the state information of the controlled part is fed back to the control system in time to influence and determine the corresponding regulation and control function of the control system. The feedback control system is a closed-loop system, that is to say, the entire control process does not require external intervention, and is completely automatic. Feedback control is divided into negative feedback and positive feedback (Jin Guantao and Hua Guofan 2005):

(1) Negative feedback

The information fed back to the control system by the controlled part affects the corresponding adjustment and control functions of the control system, so that the state of the controlled part eventually moves in the direction where the deviation from a certain state becomes smaller and smaller. This is negative feedback regulation. Obviously, negative feedback is a behavior toward the goal, in other words, it is the process of continuously reducing the target difference. The control system continuously compares the control effect with the target, so that the target difference continuously decreases during the continuous adjustment and control process, and finally achieves the purpose of regulation and control. The process of the eagle catching the rabbit we talked about earlier is actually a typical case of negative feedback regulation, as shown in Figure 1.3.

Let us analyze this process carefully. In fact, this control system is mainly composed of three parts: eyes, brain, and wings. While staring at the rabbit, the eagle's eyes also notice its own position and compare the two. The difference between the position of the eagle and the position of the rabbit is usually called the target difference. The eagle's eye mainly receives the information of target difference and transmits it to the brain. The brain directs the wings to change the position of the eagle, so that it moves in the direction where the target difference decreases. Repeating this control

Figure 1.3. Negative feedback regulation of eagle catching rabbit.

constitutes a continuous movement of the eagle catching the rabbit. The most important point here is that the decision of the brain always changes the position of the eagle to reduce the target difference.

The essence of negative feedback control is to form a mechanism of continuously reducing the target difference. The control system continuously compares the control result with the target, so that the target difference is gradually reduced in the continuous control process, and finally achieves the purpose of control. As a general negative feedback adjustment mechanism, obviously, there must be two links:

1. Once the target difference in the system reaches a certain level, it will automatically react to reduce the target difference.
2. The adjustment to reduce the target difference will continue to work, so that the approach to the target can be accumulated.

If these two conditions are not fully satisfied, it cannot be called perfect negative feedback regulation. For example, in the fuse device in the transmission line, if the current value deviates from the control target and increases beyond a certain limit, the fuse will blow and the power supply will be interrupted. Similarly, if the pressure of the pressure cooker deviates too much from the allowable limit, the alloy plug on the lid will melt and vent. These are the adjustment mechanisms for the system to reduce the target difference when the target difference occurs. But they are not completely negative feedback adjustments, because the second condition is not met, that is, the reduction of target difference cannot be accumulated through continuous adjustment. Although this type of semi-feedback regulation is widely used in control, it is not as complete and accurate as negative feedback regulation.

In the process of the eagle catching the rabbit, we must first regard the eagle's movement as a series of continuous dives, and each dive to the target can be regarded as the control of its own position. The eagle's control ability is limited, and it is impossible to capture the target with just one dive. It can only observe the target difference between its position and the rabbit's position at any time during the process of driving toward the target, adjust its wings at any time to adjust the direction and speed of the next dive flight, and gradually approach the target.

It is the same process when a person wants to take a cup of water placed on the table. He must first see the distance between his hand and the cup, and then determine the direction of his hand movement. When the hand starts to move towards the cup, the human eye keeps observing the distance between the hand and the cup (this distance is the target difference), and the role of the human brain (controller) is to constantly control the movement of the hand to eliminate this difference until the cup is held in the hand. As can be seen from the above example, continuously obtaining the target difference is the key to accurately completing the action of taking the cup. If this difference cannot be obtained, the entire action cannot be completed. This is why a person who is blind cannot take the cup in this way.

After World War II, rocket experts developed missiles that can attack aircraft, inspired by the process of eagles catching rabbits. The engineers installed an eye (an infrared homing device) and a brain (an electronic computer) on the missile, and at the same time gave it a pair of adjustable wings (an attitude control device). In this way, the missile can move in a direction that continuously reduces the target difference until the aircraft is shot down. Of course, the rocket that puts satellites into space orbit and sends spacecraft to the moon also uses the same principle.

The reason why negative feedback control is so effective is that it can accumulate a certain limited control ability to form a greater control ability. Each feedback takes the reduced target difference as input, allowing the control mechanism to select and adjust the control action within the reduced range. Usually, to do a complicated thing that has not been done before, it is difficult for people to arrange everything properly and thoughtfully in advance. Objective things are always changing, and unexpected situations may happen at any time. Even with the best of our forethought, there will be unpredictable interruptions. Therefore, the best way is to observe while doing, adjust the next action at any time according to the effect of the previous action, and take a step-by-step approach to gradually approach the goal.

The discovery of radium by the Curies is a good example of negative feedback control. In 1898, Madam Curie accidentally discovered that pitchblende is more than four times more radioactive than pure uranium oxide. She concluded that in addition to uranium, uranium ore apparently contains a more radioactive element. How to extract it from uranium ore? The Curies thought that since this element is much more radioactive than uranium, as long as a method is found to treat the pitchblende so that the radioactivity after the treatment is more intense than before, the new element must be enriched. This is actually a negative feedback process, which controls the entire purification process in the direction of increased radioactivity. In this way, using the feedback control method, after several years of hard work, they accumulated the control ability of experiment after experiment, and reached a result that would obviously have been impossible to achieve in one experiment. Finally, one gram of a new element, radium, was extracted from tons of pitch uranium ore.

Negative feedback regulation is a goal-oriented behavior, and purposiveness is an important aspect of biological behavior. As we will see later, negative feedback regulation widely exists at all levels and aspects of the organism, and can be seen everywhere in the process of life activities.

(2) Positive feedback

The information that the controlled part feeds back to the control system affects the corresponding regulation and control function of the control system, and finally makes the state of the controlled part move in the direction of increasing deviation from a specific state. This is positive feedback. Obviously, positive feedback is a process that gradually moves away from steady-state equilibrium.

Figure 1.4 visually shows this control mechanism of two systems with positive feedback coupling. In the figure, the states of the two systems are respectively reflected by the dials with pointers. The goal of the first system is the equilibrium state, which is 0. If the first system deviates slightly from the target for some reason, the reaction of the second system is to make the next state of the first system even further away from the target. In this way, in the interaction, they each deviate farther and farther from the target.

The arms race between the United States and the Soviet Union during the Cold War is an example of positive feedback. When each side learned that the other side had invented a new weapon, it immediately developed a more powerful weapon to counter it. As a result, atomic bombs, hydrogen bombs, intercontinental ballistic missiles, supersonic bombers, multi-warhead missiles, neutron bombs...... were continuously being manufactured in this way, getting farther and farther away from the balance goal of 'moderation'.

Many times, the reputation of positive feedback is not very good, and people often associate it with the 'vicious circle'. Since it is a process that continuously expands the target gap, it often marks the destruction of the control process to reach the predetermined target. In other words, positive feedback usually represents a process that is out of control.

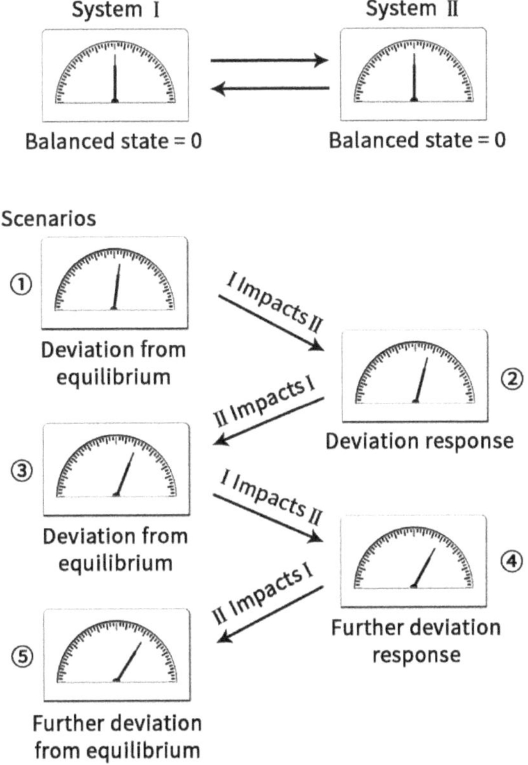

Figure 1.4. Positive feedback coupling.

In most cases of human life processes, positive feedback is a process that needs to be avoided and controlled as much as possible. Because human health is closely related to the stable state of the internal environment, pure positive feedback will take the state of the human body farther and farther from the normal stable state, so almost without exception, it will cause disease. In this regard, the understanding of traditional Chinese and Western medicine theories is consistent. For example, in modern medicine, in the case of heart failure, blood output decreases and arterial blood pressure decreases, resulting in a decrease in the supply of oxygen and nutrients to all tissues and organs of the body, and accumulation of metabolites; while the decrease in coronary blood flow, insufficient nutrition and accumulation of metabolites weaken myocardial contractility and further decrease blood output. If measures are not taken quickly to enhance cardiac function and stop this vicious cycle, the condition will deteriorate rapidly and become life-threatening. In other words, this vicious circle will eventually lead to the collapse and disintegration of the system. In TCM, the heart and the spleen are two coupling systems that interact with each other. Spleen qi deficiency will lead to loss of appetite and abnormal digestion and absorption function, blood production will be insufficient, and there will be palpitations, forgetfulness, pale complexion, weak pulse and other clinical manifestations of heart-blood deficiency. The deficiency of heart-blood leads to weakened nourishment of the spleen and further aggravates the deficiency of spleen qi. The mutual influence between the heart and the spleen forms a vicious circle of positive feedback, leading to the syndrome of heart-spleen deficiency.

Of course, the role of positive feedback regulation in human life is not always negative. In normal human physiological activities, the snowball effect of this vicious circle will prompt a certain physiological process to reach a climax quickly and exert its maximum effect. For example, in the process of urination, when the urinary center initiates urination, because urine stimulates the receptors of the posterior urethra, the latter continuously sends feedback messages to further strengthen the activity of the urinary center, so that the urination reflex is repeatedly strengthened until the urine is exhausted. Like the process of urination, the initiation of defecation, blood coagulation, ejaculation, and labor in pregnant women are all positive feedback control processes. As we will see later, the positive feedback regulation mechanism also plays a vital role in the mutual adaptation of the functions of various parts of the body and the process of life evolution.

1.3 Stability and system evolution

Using the state space method to study the system can easily distinguish different types of states of the system, describe the characteristics of various states, determine the distribution of different types of states in the state space, and clarify the connections between different states and the law of transformation from one state to another.

1.3.1 Transient and equilibrium state

There are two types of possible states for dynamic systems. The state or set of states that a system may arrive at a certain moment, but which cannot be maintained or returned without the help of external forces, is called a transient state. The set of states that a system can reach and maintain or repeatedly return to without external forces is called an equilibrium state. The equilibrium state is described by a collection of points in the state space, and the simplest equilibrium state is just a point in state space. The state space is almost entirely filled by transient points, and the equilibrium state is only a tiny part of it. Figure 1.5 is an intuitive example. The state space is the collection of all points on the curve. Among them, the three points A, B, and C represent the equilibrium state, and the other infinite points all represent the transient state. Place a small ball at any point A, B, C—as long as there is no external disturbance, the ball will always stay at that point. Place the ball at any position other than these three points, as long as there is no external force support, the ball will immediately leave these points.

The qualitative nature of a system is determined by its equilibrium states, and different equilibrium states represent different qualitative characteristics. Transient states are just accumulations of quantities that the system needs to establish a certain qualitative nature and do not represent the essential features of the system. The transition from one equilibrium state to another reflects the transformation of the system from one qualitative nature to another. The most common equilibrium states in dynamic systems include stable equilibrium states and periodic states, also known as oscillations (Xu Guozhi 2000).

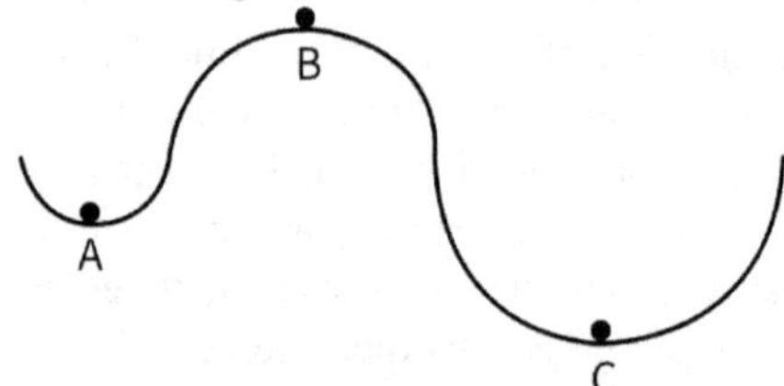

Figure 1.5. Transient and steady state.

1.3.2 Balance and oscillation

The equilibrium state is described mathematically by a fixed point. When the system is affected by multiple opposing aspects, if each aspect cancels each other out so that the state of the system does not change, it is called balance. In economics, if expenditure and income are equal, an equilibrium is reached. In chemistry, if the forward and reverse reactions of a reversible reaction are equal, an equilibrium is reached. In physics, if the forces or moments cancel each other out, then it can also create a balance. Balance can be divided into stable equilibrium and unstable equilibrium.

When an object in an equilibrium state deviate from the equilibrium position due to a slight disturbance by an external force, if that object can automatically return to its original state, such a balance is called stable equilibrium, as shown in Figure 1.6a. Unstable equilibrium is the opposite state of stable equilibrium. When an object in equilibrium deviates from the equilibrium position due to a slight disturbance of external force, if the object cannot automatically return to its original state, such a balance is called unstable equilibrium, as shown in Figure 1.6b (Xu Guozhi 2000).

Oscillation, also called periodic state, usually refers to the reciprocal change of the state of the object relative to the central value (usually the equilibrium point), or the reciprocating change between two state points/among multiple state points over time, as shown in Figure 1.7. Oscillation is a common phenomenon observed in many systems, and classic examples include the pendulum and alternating current (AC). This periodic oscillation is not only found in physical systems but also observed in various other systems, such as biology, the brain and nervous system, ecology, sociology, and economics. In these systems, oscillations can manifest as rhythmic physiological movements, neural oscillations in the brain, cyclic variations in ecological systems, and even economic business cycles.

Let's look at an example of an ecosystem. The North American fir aphid and its host balsam fir and Fraser fir form an interactive system (Figure 1.8). Fir aphids feed on the buds of these fir trees and even eat their flowers and leaves. The aphid can live as long as five years. If their populations increase, a large number of the fir plant shoots will be eaten, and fir aphids will die due to lack of food. When there are few aphids, the loss of fir trees during budding is relatively reduced, and the number of fir trees increases. In this way, the interaction of the two causes oscillations in both the populations of aphids and fir trees.

The unstable or periodic oscillation state of the system is common as a supplement to the stable state. Almost all stable systems can be transformed into instability or oscillation under certain conditions. The phenomenon of periodic fluctuations in the annual output of fruit trees, and the climatic phenomena of periodic glaciers are all related to the oscillation of the system. Symptoms of

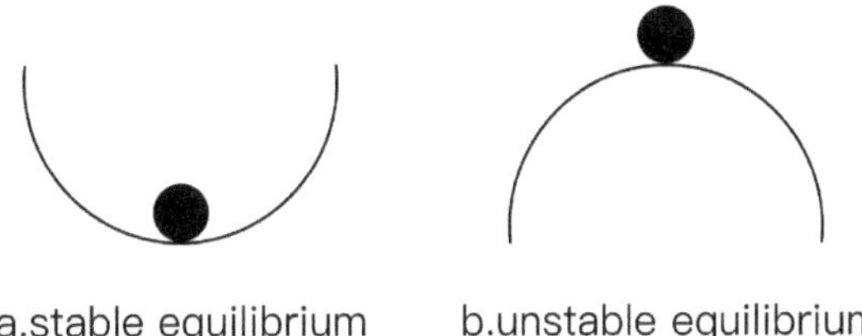

Figure 1.6. Balanced stability.

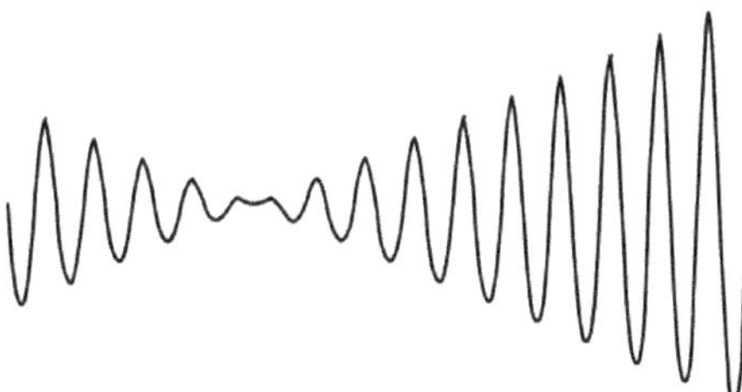

Figure 1.7. Oscillation.

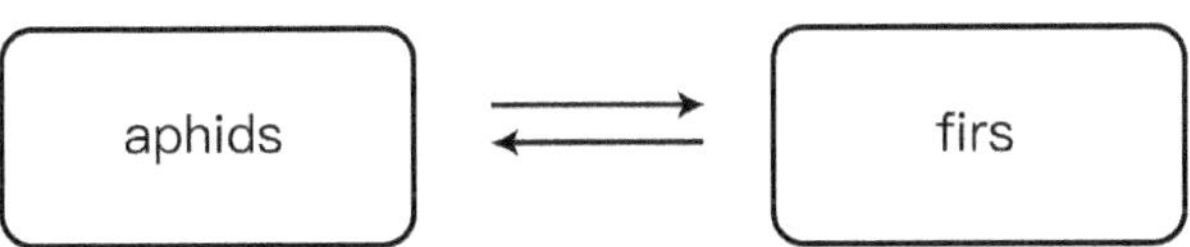

Figure 1.8. The interaction of aphids and fir.

alternating fever and chills, when malaria occurs in humans are also a reflection of the body's state of oscillation between the two states.

1.3.3 Stability

The real system will inevitably endure various disturbances from the environment or the system itself. These disturbances will generally affect the structure, state, and behavior of the system, causing the state of the system to deviate from the equilibrium state. Whether the system can restore the original equilibrium state after the state deviation occurs is the most important issue concerned by the stability research.

Stability refers to the anti-interference ability of the system's structure, state, and behavior. Generally speaking, if the deviation of the system state caused by the disturbance is small enough, the system is stable; if the deviation caused by the disturbance exceeds the allowable range, or even the deviation increases continuously, or a large-scale oscillation occurs, the system is unstable.

Stability is an important life-sustaining mechanism of the system. The better the stability, the stronger the life-sustaining ability of the system. If the state of a system is unstable, it is at most something that is fleeting in a dynamic process. If there is no stable state in the state space of a system, it must be physically unrealizable. Stability is the most important state in the process of system evolution. In general, people are only concerned with things that have stability.

From an application perspective, an unstable system cannot operate normally and cannot achieve its functional goals, so it is useless. People always strive to adopt a sufficiently stable system. But from the perspective of evolution, if all states of a system are stable under all conditions, there is no possibility of change, development, or innovation. Only when the previous state, structure, and behavior mode lose stability under certain conditions, can the system evolve to a new structure, state, and behavior mode, that is, the system has the possibility of development and innovation. Therefore, instability plays a very positive and constructive role in system evolution theory. But stability is the prerequisite for development. If the new state, new structure, and new mode are unstable and unable to preserve themselves, it will be impossible to replace the old state, old structure, and old mode. Only when the new system has a stability mechanism can it maintain the newly established structure and characteristics, preserve the accumulated information, and avoid short-lived.

The 'Deer and Wolf Story' from the United States deeply reveals the important role of stability in maintaining the ecological balance of the natural world. At the beginning of the 20th century, the Kaibab Forest in northern Arizona was still lush and vibrant. About 4000 deer lived in the forest, maintaining a dynamic balance. The ferocious wolf is the natural enemy of the deer. US President Theodore Roosevelt, in order to protect the deer in the forest and make them breed more, declared Kaibab Forest a National Hunting Reserve, and the government hired hunters to eliminate the wolves there.

Over 25 years of hunting, more than 6000 wolves were killed, and many other beasts in the forest (such as leopards) whose main prey was deer were also hunted down. The specially-protected deer became the 'darling' of the Kaibab Forest. They grew and bred freely, ate trees freely, and lived a happy life without danger and with sufficient food. Soon, the total number of deer in the forest exceeded 100,000. These deer forage for food unrestrainedly in the forest. When the bushes are exhausted, they eat the small trees. When the small trees are exhausted, they eat the bark of the big trees. All the food that can be eaten by deer will not escape bad luck. The green plants in the forest are being reduced day by day, and the withered yellow exposed on the earth is expanding day by day. The disaster finally befell the deer population. First, hunger caused a massive die-off of deer, and then a disease outbreak further decimated the population, leading to the disappearance of countless deer. Two years later, the total number of deer in the herd plummeted from one hundred thousand to forty thousand. By 1942, there were only eight thousand sickly deer left struggling to survive in the entire Kaibab Forest. President Franklin Roosevelt eventually had to order the reintroduction of wolves by the managers. Subsequently, the deer population and the forest gradually regained vitality.

In this example, deer, wolves, leopards, forests, climate, soil, etc., form a closely related ecosystem. There are many factors that affect the number of deer, such as the number of predators, wolves and leopards, how much food the forest provides for the deer, and so on. The number of deer also affects the density of the forest and the number of wolves and leopards, maintaining the density of the forest and the relative stability of the number of deer and wolves and leopards. President Theodore Roosevelt never thought that the wolves he ordered to kill were also the protector of the forest! Although the wolf eats deer, it maintains the population stability of the deer herd. This is because after the wolves eat some deer, the total number of deer in the forest can be controlled to a reasonable level, and the forest will not be ruined beyond recognition by a huge deer herd. At the same time, wolves mostly eat weak and sick deer, thus effectively controlling the threat of disease to the deer herd. As for the deer that President Roosevelt is determined to protect, once the number exceeds the limit that the forest can carry, it will destroy the stability of the forest ecosystem and bring huge ecological disasters to the forest. In other words, too many deer will become the chief culprit in destroying the forest.

Before being declared as a national hunting reserve, the number of deer in the Kaibab Forest was stable. Some small changes in the environment and climate would cause the number of deer to fluctuate. But after a period of time, it would return to its original approximate amount. Large-scale hunting of wolves and other predators broke this stable state. A system that leaves a stable state, under the action of various factors, is constantly changing until it reaches a new stable state. Of course, it is also an option to restore the previous stable state by repairing the structure that supports the previous stable state, just like the final outcome of the Kaibab Forest event (Jin Guantao and Hua Guofan 2005).

In human normal life and disease processes, stability also plays a vital role. The normal state of the human body usually has a certain degree of stability. Contracting a disease means that the body's state deviates from the normal steady-state equilibrium, while having a certain chronic and stubborn disease implies that the patient's state is in a stable pathological condition. The process of treatment involves applying artificial intervention to the links and elements that influence the stability of the pathological state in the human body, allowing the body's state to move away from the pathological steady-state equilibrium and move towards a normal steady-state equilibrium.

Stability is the most important issue in the study of dynamic systems. It involves questions such as how to determine the stability and types of stability of a system, the degree of stability of the system, whether the system has unstable points, under what conditions stable systems lose stability, how to identify the unstable points of a system, the analysis of behavioral characteristics of the system near the unstable points, and how to transition from the unstable state to a new stable state. These questions constitute essential content in the stability theory of system science and are also crucial for medical research on the occurrence, development, and evolution patterns of diseases.

1.3.4 System evolution

The changes in the structure, state, characteristics, behavior, and functions of a system over time are called system evolution. Evolution is a universal characteristic of the system. As long as it is viewed on a sufficiently large time scale, any system is in the process of rapid or slow evolution. There are two basic ways of system evolution. Evolution in the narrow sense refers only to the transformation of a system from one structure or form to another. Evolution in a broad sense includes the formation (occurrence) of the system from nothing, the development from immaturity to maturity, the transformation from one structure or form to another structure or form, the aging or degeneration of the system, and the demise from being to nothing. The survival and continuation of the system also belong to generalized evolution because during its existence, although the system may not undergo qualitative changes, variations in quantitative characteristics are inevitable.

The driving force of the system evolution may come from within the system, that is, cooperation and competition among components lead to changes in the scale of the system and the way the components are related, which in turn leads to changes in system functional characteristics. The increase in components and the increase in the scale of the system will more or less cause changes in the way components are correlated (including the correlation force). The power of system evolution may also come from the external environment. Changes in the environment, the interaction between the environment and the system, and changes in the mode of action will all lead to changes within the system in varying degrees. These changes include changes in component characteristics, structural modes, and even component metabolism, which ultimately lead to changes in the overall characteristics and functions of the system. That is to say, the system evolution is carried out under the joint promotion of internal power and external power.

There are two basic directions for system evolution: One is the evolution from low-level to high-level, from simple to complex; the other is the degradation from high-level to low-level, and from complex to simple. Real-world systems have both evolution and degradation, and the two evolutions are complementary. The general direction of system evolution is more and more complex. The key to system evolution from simple to complex is the number and distribution of potential intermediate stable forms. It is the most likely way of evolution to produce a stable intermediate form first, and then gradually produce a more complex stable form (Xu Guozhi 2000).

The natural world we face today is the product of gradual evolution in a long historical process. The evolution of the universe, the evolution of the solar system, the evolution of the earth, the evolution of life, the evolution of animals, the evolution of mankind... evolution is everywhere. Even today, we ourselves and the natural world we face are still in the process of evolution. For the human body, embryonic gestation, birth, growth and development to adulthood, as well as subsequent degradation, aging, and death, is the evolutionary process of human individual life. The process of disease occurrence, development, transformation, recovery, deterioration, and death in the human body is an evolutionary process that occurs in the human body on a smaller time scale. In the evolution of the system, stability plays an important role. After the old stability is destroyed, the system will be in an unstable state before a new stable structure is formed, and its related subsystems will be in continuous change. But as soon as it enters the range stipulated by the new structure, a new stability will be formed.

Let's look at an ecosystem composed of rats, snakes, clover and wild bees. Suppose that at the beginning, the ecosystem is in a stable state of more rats, fewer wild bees, fewer clover, and fewer snakes, the stable structure 'a' in Figure 1.9. For some reason, the number of rats reproduced suddenly increases, so this stable state is broken. The increase of rats causes a large number of honeycombs to be destroyed, thereby reducing the number of wild bees. The reduction of wild bees affects the spread of clover pollen, resulting in a corresponding decrease in the number of clovers. The reduction of clover destroys the habitat of snakes, which leads to a decrease in their number. Snakes are the natural enemies of rats, and the reduction of snakes reduces the threat to rats. As a result, the system enters an unstable state with more rats and fewer clover, wild bees and snakes. But if a large number of cats are introduced into this system at this time, because cats eat rats, changes in the number of rats will cause a series of changes in the entire system. With fewer rats, there will be fewer damaged honeycombs. More wild bees will make clover grow vigorously and lead to more snakes, and more snakes will reduce the number of rats. In this way, the entire system is constantly changing, and finally changes to a new stable structure: fewer rats, more wild bees, more clover, and more snakes, the stable structure 'b' in Figure 1.9. The new structure is also a stable structure. Even if there are fewer cats in the new structure, the number of mice will not necessarily increase. Because there are a lot of snakes, they prey on rats in large quantities, which will suppress the number of rats (Jin Guantao and Hua Guofan 2005).

It can be seen that although the interaction between the various subsystems in the system is very complicated, and the state evolution of the system when the system is unstable is also very complicated, the stable structures can be transformed into each other, and this transformation can be analyzed based on the way of interaction of subsystems. This provides feasibility for studying the evolution of complex systems: Although the changes of unstable structures in complex systems are

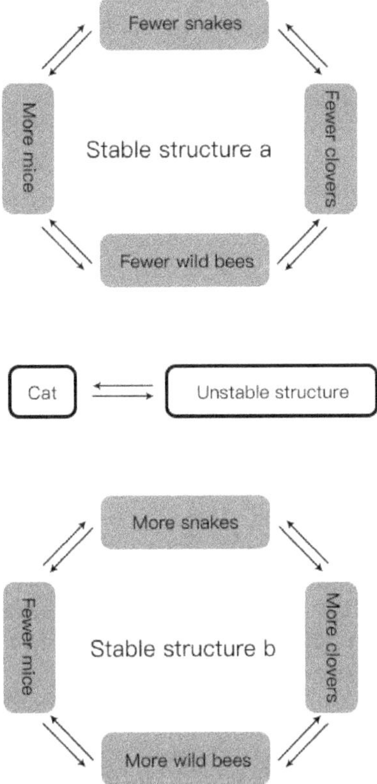

Figure 1.9. The evolution of an ecosystem.

very complex and difficult to predict, the stable structure of the system can be grasped, and people can grasp complex systems by understanding the stable structure of the system.

As we will see later, in the medical field, the disease is actually the state evolution process after the state of the human body deviates from the normal steady state. In this evolutionary process, the body's state either enters a long-term unchanging pathological steady state or enters an unstable process until death. The treatment process is to use human intervention to influence the structure of the pathological stable state, so that the body state deviates from the pathological steady state and moves toward the normal steady state; or blocks the unstable process of the body state evolving toward the worsening of the disease, and reconstructs the steady-state balance in which all parts of the human body are inter-related and restrict each other.

1.4 Model

Today, with the development of science and the renewal of scientific theories, the concept of building models has penetrated into almost every field of science. A model is an expression form of the researched system, process, and thing. It gives the object entity the necessary simplifications, and uses appropriate expressions or rules to describe its main features. Models also have structure. Model structure and prototype structure are two different things, but they are directly or indirectly related. The structural issues that must be considered in the prototype should be reflected in the model and can be described in the language of the model.

1.4.1 The concept of the model and the significance of building the model

In today's fifth grade textbook, a model is described as follows: "Models are used to explain the ideas and discoveries of scientists, and they make it easier for people to understand things that are difficult to observe directly, changes in things, and relationships between things" (Hao Jinghua and Lu Peiqi 2010). The method of inferring certain properties and laws of things through the study of models, and obtaining, expanding and deepening the knowledge of things with the help of models is commonly used in scientific research.

A model can be a physical object, a figure, or a mathematical expression. There are also conceptual models that are described purely by language or combined with language, diagrams, and expressions.

Physical model: Also known as a scale model, it is made of physical objects and requires the same or similar structure as the prototype, but the scale can be greatly different. Examples include airplane models, car models, cell models and dolls.

Appearance model: This can be some kind of graphics, such as photos, schematics or 3D graphics, such as a photo of a building or a 3D model of its appearance.

Map model: It is required to have the same topological structure as the prototype, such as topographic maps, traffic maps, and administrative division maps.

Mathematical model: Equations or inequalities composed of letters, numbers, and other mathematical symbols, or models that describe system characteristics, internal structures, and connections with the outside world using diagrams, images, block diagrams, and mathematical logic. It is an abstraction of a real system, but it must have a certain internal connection with the prototype structure. The structural problems in the prototype are described in mathematical language in the model and can be analyzed and solved by mathematical methods.

Conceptual model: There are also models that are described purely in language or are expressed through a combination of natural language, diagrams, expressions and other forms. For example, some content in physics, chemistry, and biology textbooks reveals the structure and function of the corresponding real system. Such models are usually called conceptual models.

In today's science, research systems are generally research models, and some systems can only be researched through models. The purpose of constructing a model is to study the prototype, therefore, objectivity and effectiveness are the primary requirements for modeling. All the information reflecting the essential characteristics of the prototype must be shown in the model, so that the main characteristics of the prototype can be grasped through model research.

Physical and graphical models are based on similarities in form and appearance to the prototypes represented. Topographic maps, traffic maps, and circuit diagrams are built on the basis of structural similarities to the prototypes represented. There is another kind, which may be far from the prototype in terms of form and actual structure, but it simulates or expresses the behavior and characteristics of the prototype system well. We call such a model a functional model or a behavioral model. For example, the electronic clock model with the same function as the mechanical clock, the mathematical calculation formula of practical application problems, the model of the actual scene established through computer simulation for the purpose of training or entertainment games, etc. The theories about viscera-qi-blood-body fluids of the human body, which have been gradually perfected in the development of TCM for thousands of years, are actually functional models that reflect the physiological and pathological laws of the human body from the perspective of behavior and function.

The introduction of models makes it convenient for people to recognize and understand the internal structure of things that are difficult to observe directly. The changes of things and the relations between things, and the inner laws of things can also be expressed intuitively through models. However, the characteristics of real systems are often multifaceted. A computer has a hardware structure and a corresponding software system. A person has sociological characteristics, such as temperament, character, morality, ability, intelligence and the condition of getting along with others, and also biological and medical characteristics, such as physical fitness, medical history, and physiological and pathological conditions of internal systems and organs, etc. Scientific research cannot be all-encompassing. In order to simplify the problem and facilitate understanding and research, scientists often ignore aspects that are not related to the research purpose, and focus on related aspects to build a model of the real system. Therefore, focusing on the characteristics of different aspects, there may be more than one different model for the same real system.

For example, the earth model can be a surface morphology model that indicates the topography of the earth's surface, the distribution of mountains and rivers, and the ocean, or an administrative division model that indicates the intercontinental and national divisions of the earth's surface. It can also be an international traffic model indicating the routes of sea, land and air transportation around the world. In modern medicine, human anatomy describes physical models of organs and tissue morphology; human physiology reveals physiological models of human organs, system functions and activities. The basic theory of TCM includes functional models that can describe the physiological and pathological activities of the human body on holistic level, and the state changes of the human body during various diseases.

The significance of using the modeling method to study the real system lies in (Sterman 2000):

1. *Help understand complex phenomena*: Building models can simplify and abstract complex phenomena and problems, making them easier to understand and explain.
2. *Prediction and optimization*: Models can be used to predict future trends and results, thereby providing scientific basis for decision-making. At the same time, they can be used to optimize a system or process to achieve the best results.
3. *Exploration and discovery*: Through modeling, the key factors and mechanisms of the system can be analyzed and explored, and new knowledge and laws can be discovered.
4. *Guide experiments and observations*: Modeling can explain well-designed experiments and observations, so as to obtain more data to verify and improve the model.
5. *Assisted decision-making and planning*: The model provides a scientific basis for decision-making and can make wise choices between different options.

In summary, building models is an indispensable and important step in today's scientific research and engineering applications, and they provide us with powerful tools for understanding and solving complex problems.

1.4.2 Isomorphism and homomorphism: the structural non-uniqueness of functional models

Isomorphism and homomorphism are both concepts derived from mathematics. There are two characteristics of mathematical isomorphism:

1. A one-to-one correspondence can be established between the elements of the two systems.
2. The relationship between the elements of the two systems can still remain unchanged in the respective systems after this correspondence has been established.

The isomorphic relations between different systems are equivalence relations, and equivalence relations have recurrence, symmetry, and transitivity. Therefore, with the help of mathematical isomorphism, common laws can be found in various real systems. The mathematical homomorphism between different systems is reflexive and transitive, but there is no symmetry. Therefore, mathematical homomorphism is only used for classification and model simplification, and cannot be divided into equivalence classes.

An isomorphic model is a model that has exactly the same structure and behavior as the prototype, and it accurately replicates and simulates all the properties and behaviors of the prototype. Isomorphic models are often used in place of prototype systems for experimentation, testing, or validation because they can fully reproduce the functionality and effects of the prototype. The homomorphic model is similar to the structure and behavior of the prototype, but not necessarily identical. It is a simplification and abstraction of the prototype. It preserves some key features and relationships of the archetype, but simplifies some details. Therefore, in some cases it may not be possible to accurately predict the behavior of the system and it is only suitable for understanding and analyzing the key characteristics of the system. This model is used to aid in the understanding of the prototype, providing key information without focusing on all the details.

In the research of complex systems, those systems with the same input and output and the same response to external stimuli are called isomorphic systems, and the simplified models obtained by simplifying the system through abstraction or merging are called homomorphic models. As mentioned earlier, if clocks and watches made based on mechanical or electronic principles have the same appearance, and the accuracy of time indication is exactly the same, then one can be said to be an isomorphic model of another. Suppose that two models are established to describe the structure of a school: one model with classes, teaching and research offices as the basic unit, and another with students and teachers as the basic unit. Since one element of the former model corresponds to a set of elements of the latter model, the former is a homomorphic model or simplified model of the latter. The real system is infinite in depth and breadth. Generally speaking, to study the real system through modeling, the model built can only be a system that is homomorphic with the real system.

A real system can construct different homomorphic models according to different research purposes. For systems with different structures and performances, the behavioral characteristics of their homomorphic models may have similarities in form. Therefore, the fact that there is system isomorphism or homomorphism between different subject fields and different real systems is the objective basis for the horizontal analogy of various subjects and the establishment of general system theory.

On the other hand, for the same real system, although the research purpose is the same, due to the difference in research methods and the level and phenomena of concern, models with different angles, levels, and degrees of integration can be constructed. Even at the same level and at the same angle, there is more than one model due to different selected elements and structures.

For example, in physics, regarding the nature of light, there were two models of 'particle theory' and 'wave theory'; in astronomy, there were also two models of 'big bang theory' and 'steady state theory' regarding the origin of the universe. In the medical field, the human body description system established by modern medicine based on molecules, cells, tissues, and organs, and the basic theories established by TCM based on viscera, qi-blood, and meridians are actually human models established from different levels and angles. And in TCM, which modelling based on holistic level, there are also different homomorphic models: Zangfu-qi-blood-body fluid-essence (臟腑-气血津液精) model, six meridians (六经) model, wei-qi-ying-xue (卫气营血) model, and tri-jiao (三焦) model.

The model method is commonly used in the research of each subject of modern science. Real-world systems usually have multiple characteristics and are extremely complex. Generally speaking, the more complex and detailed the model, the more comprehensive and accurate the revealing of the characteristics and internal laws of the real system. But on the other hand, the complexity and refinement of the model increase the difficulty of understanding. Overly complex models often dazzle people and lose their grasp of the whole. Therefore, in scientific research, people always build a model as simple as possible based on the purpose of the research and on the premise that the needs can be met, which is to build a model according to the principle of 'the simplest and applicable'.

1.4.3 Model building method and application

Modeling is a complex and diverse process, which requires the selection of appropriate methods and techniques in combination with specific issues and research objectives. At the same time, the establishment of the model needs continuous verification and revision to ensure the accuracy and reliability of the model. Generally, the methods and steps of modern scientific model building are as follows (Sterman 2000, Luenberger 1979):

1. *Determining the research objectives*: First of all, the problems to be studied or solved should be clarified, and the objectives and uses of the model should be determined. Clearly defining the scope and objectives of the study helps to ensure the accuracy and usefulness of the model.
2. *Data collection*: Obtaining relevant data is the basis for building a model. Data can be collected through experiments, observations, questionnaires, and literature research. The quality and accuracy of the data are critical to the reliability of the model.
3. *Select the appropriate model type*: Select the appropriate model type according to the research objectives and data characteristics. Common types of models include mathematical models, statistical models, physical models, computer models, etc. The choice of model should take into account the characteristics of the system, the complexity of the problem, the availability of data and other factors.
4. *Parameter estimation*: In the process of building a model, some unknown parameters may be involved. These parameters need to be estimated by data fitting or other methods, and the optimal parameter values should be found as much as possible, so that the model can accurately describe the real system. Artificial intelligence technology can build models through machine learning and deep learning, and learn and optimize model parameter settings from a large amount of data.
5. *Model verification and evaluation*: After the model is established, it needs to be verified. Methods such as cross-validation, prediction error, etc., can be used to evaluate the predictive ability and generalization ability of the model. By comparing with actual data, the accuracy and applicability of the model will be tested.
6. *Model analysis and interpretation*: Analyze the established model to understand the system behavior and interaction revealed by the model. In complex system modeling, it is inevitable to face various uncertainties. These uncertainties may come from incompleteness of data, simplification of models, estimation errors of parameters, etc. Therefore, it is necessary to evaluate the stability and reliability of the model through sensitivity analysis and parameter sensitivity analysis of the model.
7. *Model application*: The ultimate goal of complex system modeling is to perform prediction and control. The established models can be used to predict future trends, make decisions or optimize programs. Applying the model to practical problems, through model control, the control strategy of the system can be designed and optimized to make the system reach the desired state. In the application process, the model can be continuously optimized and improved based on actual feedback.

The purpose of model verification is to verify the correctness and reliability of the functional model, to ensure that the established model is consistent with the expected function of the actual system, so as to avoid potential problems during the system design and development phase. Here's how to do model validation (Law and Kelton 2020, Chick and Sanchez 2008):

1. *Experimental verification*: Verify the correctness of the functional model through experiments. This includes conducting experiments on real systems, or virtual experiments in simulated environments. Compare the experimental results with the expected functionality of the functional model to ensure the accuracy of the model.
2. *Expert evaluation*: Invite experts in related fields to evaluate the model to find out possible problems or missing functions in the model.
3. *Cross-validation*: Cross-validate the functional models created by different methods or tools. If multiple models produce similar results, you can increase your confidence in the model.

4. *Comparison with requirements*: Compare the model with the requirements of the system to ensure that the model covers all requirements and has no redundant or redundant functions.
5. *Validation of historical data*: If historical data are available, they can be used to verify the accuracy of the functional model. Compare the model's forecast results with historical data to assess the model's predictive power.
6. *Environmental simulation*: Carry out environmental analogy to the model, and compare different situations and conditions to ensure that the model can work normally in different situations.
7. *Logical reasoning*: Carry out logical reasoning on the functional model to ensure the logical consistency and rationality within the model.
8. *Model Demonstration*: Conduct a model demonstration for stakeholders to review and provide feedback on the model. Different scenarios and use cases can be compared during the demonstration, so that stakeholders can better understand the function and effect of the model.

Models have a wide range of applications in various fields. The following are model applications in some common fields (Sterman 2000):

1. *Natural sciences*: In fields such as physics, chemistry, and biology, models are used to explain natural phenomena, predict experimental results, and understand system behavior. For example, meteorologists use climate models to predict weather changes, and biologists use biological models to study the behavior of organisms.
2. *Engineering and Technology*: In fields such as engineering, computer science, and information technology, models are used to design and optimize systems, predict performance, and solve problems. For example, engineers use models to design building structures, and computer scientists use algorithmic models to optimize computing performance.
3. *Economics and Finance*: In the fields of economics and finance, models are used to analyze market behavior, predict economic indicators and study monetary policy. For example, economists use economic models to study inflation, and financiers use financial models to analyze investment portfolios.
4. *Medicine and health*: In fields such as medicine and epidemiology, models are used to study disease spread, evaluate treatment options, and predict health risks. For example, epidemiologists use epidemic models to predict the spread of infectious diseases, and doctors use medical models to optimize treatment options.
5. *Social Sciences*: In fields such as sociology, psychology, and pedagogy, models are used to study social behavior, psychological processes, and educational effects. For example, sociologists use social models to explain social phenomena, and psychologists use cognitive models to study thought processes.
6. *Management and decision-making*: In the fields of management, operations research, and decision science, models are used to optimize resource allocation, make decisions, and solve management problems. For example, management scientists use management models to improve organizational structures, and decision scientists use decision models to illustrate how managers make decisions.

In conclusion, models are indispensable tools in scientific research, technological development, and decision-making, and they can help people better understand and solve complex real-world problems. The application of models in various fields can help people understand the nature of problems more deeply, make scientific decisions, optimize resource utilization and solve complex real-world challenges. The establishment and application of models is the cornerstone of modern science, engineering, management and decision-making, and they promote the continuous progress and development of human society.

Chapter 2

The Rise of Complexity Science

Science Begins to Directly Face the Complexity of Life

Modern natural sciences are basically developed upon the belief of the simplicity of the world and the simplicity of natural laws. Until the 1970s, reductionist methods of analysis and decomposition occupied the leading position in scientific methods. The characteristic of this method is to reduce objects and problems into relatively simple problems through analysis and decomposition and then solve them. Those problems that cannot be simplified through analysis and decomposition are not included as objects of scientific research. Those human knowledge systems that do not conform to the reduction analysis method are regarded as pseudo-science or non-science, and are excluded from science.

However, science is progressing continuously. In the process of scientific development, many ideas and knowledge systems that were incomprehensible and had been discarded due to the limitations of past science, are being re-incorporated into the scientific system by scientists in a new way because their correctness or scientific value was confirmed.

With the deepening of human understanding of the natural world, scientists have gradually realized that the natural systems that exist around us: plants, animals, human body systems, human brain systems, geographic systems, ecosystems, galaxy systems, and social systems, all are systems with various types and numerous quantities of constituent elements, complex internal structure and interconnectedness, but difficult to decompose. Just like we cannot cut off a part of the foot to fit shoe, putting such a complex system into the framework of simplicity science, and using traditional analysis and decomposition methods for research, cannot enable us to fully grasp the behavior and internal laws of the system as a whole. Facing the ocean of complexity, if science does not want to give up its instinctive concept of revealing and using natural laws, it should introduce new methods and tools more suitable for researching and understanding complex systems, and expand its research scope and fields. As a result, complexity science came into being.

2.1 Characteristics of complex systems

'Complex' is a frequently heard and very familiar word in our daily lives, but complexity has become the object of scientific research only in recent decades. In the process of scientific development, the methods of scientific research have undergone a process of evolution from low-level to high-level. Modern scientific methods characterized by static observation and single-factor analysis are only suitable for studying the relatively simple characteristics of some relatively simple systems. In the past, when scientists had to deal with a complex system, they always decomposed it into some relatively simple systems for research. For systems that cannot be decomposed, scientists usually ignore their complexity, and only extract those aspects or characteristics that people care about

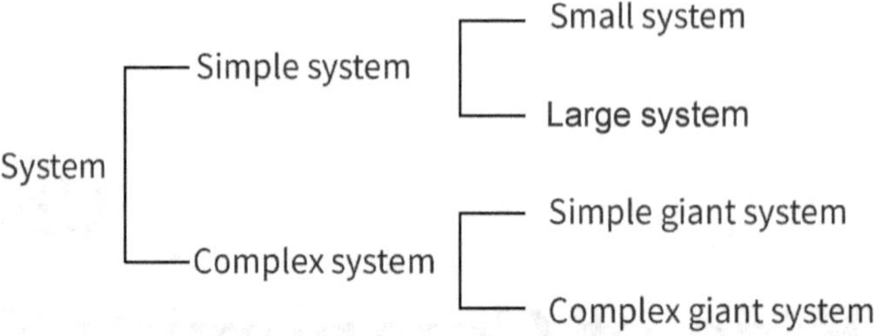

Figure 2.1. System classification diagram.

most, thus attributing the problems of complex systems to problems of relatively simple systems for research.

In recent decades, with the emergence of new disciplines of science such as information theory, cybernetics, and systems theory, and the rapid development of computer technology, the ability of science to deal with complex problems has made greater progress than before. On the other hand, the limitations of grasping the holistic performance of the system by applying the methods of dealing with simple systems to complex systems, have also prompted scientists to reflect from the perspective of methodology. As a result, a set of theories and methodologies focused on dealing with complexity was gradually formed. Scientists began to directly face the complexity of the system, and began to study the complex system as it is, instead of simply reducing it to a relatively simple system.

In the early 1990s, the well-known Chinese scientist Mr. Qian Xuesen (1911–2009) divided all systems into simple systems and giant systems based on the number of subsystems that make up the system, the number of types of subsystems, and the complexity of their relationships (Qian Xuesen et al. 1990) as shown in Figure 2.1.

2.1.1 Simple system

A simple system refers to a system that does not need to be hierarchical or has only one level. Usually, there are fewer subsystems that make up a simple system, and the relationship between them is relatively simple. Simple systems can be divided into small systems and large systems. Some non-living systems, such as a machine, are small systems. If the number of subsystems is relatively large (tens or hundreds), such as a workshop or factory, it can be called a large system. But the problems of such a large system can often be solved by decomposing the problems into problems of various subsystems. Therefore, although it is 'big', it is still relatively simple. With the advent of computers, it has become relatively easy to calculate the evolutionary behavior of systems containing hundreds of subsystems. Therefore, some systems composed of hundreds of subsystems are sometimes classified as simple systems.

From this point of view, to judge whether a system is simple, the number of subsystems included is not the most important factor. At present, to determine whether a system is a simple system, it is usually seen whether the evolution of the system can be characterized by Newtonian mechanics. The most important criterion for applying Newton's theory is to see whether it satisfies the principle of superposition.

The reality systems that exist in nature and society usually have multiple attributes, and their relationship with the environment also presents in multiple aspects. In principle, none is simple. But when we focus on the attributes and associations of a certain aspect of the system and ignore other aspects, they can usually be abstracted or simplified into a simple system for research. Therefore, a simple system usually has the following characteristics:

1. A simple system is usually an artificial system or a simplified model of a real system. It can include a wide range of systems from celestial systems to microscopic atoms.

2. The number of subsystems in a system is not the most important condition for judging a simple system. The key lies in whether analysis methods can be applied to it. If the number of subsystems reaches a scale of several hundred, but the motion state of each subsystem can be actually calculated by the computer, it can also be regarded as a simple system.
3. The correlation between subsystems and systems in a simple system satisfies the superposition principle. The motion state of all subsystems can be superimposed to get the motion state of the system; it can also be decomposed from the overall motion state of the system to get the motion state of each subsystem. The whole system is the superposition of subsystems.
4. Simple systems have no concept of hierarchy. The sum of the motion states of all subsystems is the motion state of the system. The description of the overall motion state of the system is to describe the motion state of all subsystems.
5. The state change of a simple system is completely deterministic. Given an initial state of the system, the future of the system can be accurately predicted, and it can also be accurately traced back to its past.

The 'objective world' composed entirely of simple systems is a very simple world. There is no randomness, no evolution, only cyclical movement. This is the world described by the Newtonian system. In this system, the world is unchanged. As long as 'God' 'promotes' the universe at the beginning, the entire universe will continue to move without beginning or end, and will never change. With the advancement of science, in the eyes of today's people, the world is far from the situation described by Newtonian theory. The simple system is only a simple approximation of the real system in nature at the early stage of scientific development (Xu Guozhi 2000).

2.1.2 Simple giant system

If the number of subsystems is very large (tens of thousands, tens of billions, trillions), it is called a giant system. If there are not too many types of subsystems in a giant system (several or dozens), and the relationship between them is relatively simple, it is called a simple giant system, such as a laser system. The simple giant system has many subsystems, and the interaction form is usually non-linear, so it is impossible to accurately describe the evolution of each subsystem. The overall nature of the system cannot be derived from the superposition of subsystems, but can only be described at the system level as a whole. The number of components reaches the scale of a giant system, and usually the system will produce some significantly different characteristics than a simple system.

Due to the phenomenon of 'emergence' in the transition from the subsystem level to the holistic level, new characteristics that the subsystem does not have will appear on the holistic level. Therefore, simple giant systems have different descriptions of system evolution at different levels. For example, in thermodynamic systems, physical quantities such as temperature, pressure, and volume are usually used at the system level, while physical quantities such as molecular kinetic energy and speed are used at the subsystem level. The two types of physical quantities can only describe the properties of the system at their respective levels. It is impossible to talk about temperature and pressure for the molecules in a thermodynamic system; and it is impossible to talk about kinetic energy, speed, etc., for the whole system. At present, simple giant systems are usually understood based on the following characteristics:

1. A giant system usually has a macroscopic and a microscopic level division, and the behavior characteristics of the system at these two levels are qualitatively different. This is an important feature of a giant system that is different from a simple small system and a large system.
2. Due to the large number of subsystems in this type of system, the motion state of each subsystem cannot be accurately described. The method of directly synthesizing the kinematic properties of the whole system from the interactions between subsystems is not feasible and difficult to achieve even with supercomputers.

3. The superposition principle is no longer applicable to simple giant systems, that is, the properties of the macroscopic level cannot be deduced from those of the microscopic level. The properties of different levels are usually derived through the independent analysis of each level.
4. The statistical theory of giant systems shows that phenomena that are not found in simple systems, such as self-organization, will appear in giant systems.

Since the superposition principle is still the basis of most scientific theories at present, the superposition principle is often used when quantitative analysis is needed in the theories of most simple giant systems. Since the relationship between the subsystems is relatively simple, if the relationship between the macro level and the micro level can be found, then further specific analysis can be made accordingly. In this case, scientists have achieved success by omitting the details and using statistical mechanics and thermodynamic entropy to generalize. This is the method used by Haken (Hermann Haken 1927) and Prigogine (Ilya Romanovich Prigogine 1917–2003) to establish the theory of self-organization, that is, a statistical synthesis method from micro to macro. If the relationship between them cannot be found, the characteristics and laws of system changes can only be discussed at two levels separately. In most cases, the behavior of the system can only be discussed at the macro level of the system, and not go deep into the subsystem level, and so the relationship between the motion states of the two levels also cannot be analyzed (Xu Guozhi 2000).

2.1.3 Complex giant system

In a giant system, if there are many types of subsystems (tens, hundreds, thousands or more), these subsystems have a hierarchical structure, and the way they are related is very complicated (such as non-linearity, uncertainty, fuzziness and dynamics), then such a giant system is called a complex giant system. Such systems are very complicated in terms of structure, function, behavior, and evolution, and there are hierarchical structures in time, space, and function, so there are still a lot of problems that are not clear even today. For example, due to the functions of memory, reasoning, thinking, and consciousness, the input and output response characteristics of the human brain system are extremely complex. The human brain can use memory information, reasoning information, and input information at the time to make various complex responses to the environment. From the time point of view, this response can be a real-time response, a delayed response, a leading response, a false response or no response. Therefore, human behavior is not a simple 'conditional reflection'. The human brain system has attracted many scientists for research. Its microstructure at the cellular level is gradually being recognized, but the mechanism of the extremely complex overall functions such as thinking and consciousness displayed at the macro level is still unclear.

Obviously, the methods used to deal with different types of systems are closely related to the characteristics of the system itself, and the scale and complexity of the system. The theory of large systems cannot be used to solve the problems of giant systems, and the theory of simple giant systems cannot be used to deal with complex giant systems.

If a complex giant system exchanges energy, matter and information with the environment, it is called an open complex giant system. Mr. Qian Xuesen believes that 'open' not only means that the system generally exchanges material, energy, and information with the environment, accepts environmental input and disturbance and provides output to the environment, but also has the qualities of active adaptation and evolution, which can better adapt to the environment through evolution. Open complex giant systems widely exist in the real world. The human brain system, human body system, ecological system, and social system are all such systems (Qian Xuesen et al. 1990).

Complex systems usually show different characteristics from simple systems due to the increase in the number of components and the complexity of interconnectedness:

(1) Non-linear

Linearity and non-linearity are a pair of mathematical concepts used to distinguish between two different types of relationships between different variables in mathematics. This leads to two different ways of thinking at the ontological and methodological levels: linear thinking and non-linear thinking. Linear thinking believes that the real world is essentially linear, and that non-linearity is nothing but a deviation or interference from linearity. Non-linear thinking believes that the real world is inherently non-linear, but the degree of non-linearity and manifestations vary greatly. Linear systems are just an acceptable approximate description of non-linear systems under simple circumstances.

From a methodological point of view, linear thinking believes that non-linearity can generally be reduced to linear for understanding and processing. Non-linear thinking believes that under normal circumstances, non-linearity should be treated as non-linear, and only in some simple cases can the non-linearity be reduced to linear. Therefore, non-linear actions are the root cause of the infinite diversity, unpredictability and difference of the system, and the main source of complexity. Non-linear thinking is a way of thinking that directly faces the complexity of things themselves and the complexity of their interconnections, and strives to understand and grasp the object of study according to the true colors of things. It is undeniable that when recognizing simple things, linear thinking helps improve the efficiency of recognition. But when understanding more complex things, if you only pursue simplicity, efficiency, and causality, instead of researching and understanding things as they are, the results obtained can only be an illusion of things. In fact, as human thinking changes from linear to non-linear, the understanding of the true nature of nature and society will become deeper and deeper (Yang Jisheng and Jiang Bing 2018).

(2) Uncertainty

Uncertainty is relative to certainty, and is the negation of certainty. In the development history of modern science, the classic natural sciences represented by Newtonian mechanics have painted a deterministic world picture to people, and declared that the blanks or ambiguities in this picture are areas waiting for humans to fill in gradually. However, since the 1960s, the study of chaotic phenomena has broken through the shackles of traditional scientific thinking that completely separates 'certainty' and 'uncertainty'. Chaos theory, through the study of a large number of objective phenomena and experimental results, reveals that it is precisely the mutual connection and mutual transformation of certainty and uncertainty that constitutes a rich and colorful real world (Yang Jisheng and Jiang Bing 2018).

The famous scientist Prigogine once said: "I believe that we are at an important point in the history of science. We have come to the end of the road paved by Galileo and Newton, which presented us with an image of a time-reversible, deterministic universe. We now see the erosion of determinism and the emergence of a new formulation of the laws of physics" (Prigogine and Stengers 1997). In fact, the research on 'uncertainty' in many disciplines has revealed the prevalence of uncertainty in the micro and macro world. For example, Heisenberg's uncertainty principle in quantum me chanics, Gödel's theorem in mathematical logic, Arrow's impossibility theorem in social choice theory, and fuzzy logic, etc. (Yang Jisheng and Jiang Bing 2018). Henry N. Pollack (born in 1936), a geologist at the University of Michigan in the United States, said: "Has science been debilitated by uncertainty? To the contrary, the successes of science, and indeed there are many, arise from the ways that scientists have learned to make use of uncertainty in their quests for knowledge. Far from being an impediment that stalls science, uncertainty is a stimulus that propels science forward. Science thrives on uncertainty" (Pollack 2003).

(3) Emergence

Emergence refers to the phenomenon that multiple lower-level systems interact to produce a higher-level system, and this higher-level system exhibits characteristics that the lower-level system that composes it does not have. Jeffrey Goldstein defines emergence as "the arising of novel and coherent structures, patterns and properties during the process of self-organization in complex systems" (Goldstein 1999). Complexity science calls the property that the whole has but the isolated part and its totality do not have as 'emergence'. Emergence is a structural effect stimulated by the interaction of constituent components in the way of system structure. The reason why the function of the system is often expressed as 'the whole is greater than the sum of its parts' is because of the emergence of new qualities in the system. Emergence is a transition from a low-level to a high-level, a mutation in the performance and structure of the macro system based on the evolution of the low-level object. The emergence process is the process of generating new functions and structures, and this process is the product of the interaction between the main body of the complex system and the environment.

(4) Self-organization

Self-organization refers to the phenomenon that the various subsystems within the system can form a certain structure or function according to a certain rule without external instructions (Lu Yongbo 2006). The self-organization theory is a systematic theory established and developed in the late 1960s. It mainly studies the formation and development mechanism of complex self-organizing systems (life systems, social systems), that is, under certain conditions, how the system spontaneously evolves from disorder to order, from low order to high order. Self-organization theory brings together the main ideas of dissipative structure theory, synergetics, catastrophe theory, hyper-cycle theory, fractal theory and chaos theory. It is generally believed that a system being open, and far away from equilibrium, non-linear interactions and fluctuations are the basic conditions for the formation of self-organization. The phenomenon of self-organization broadly exists both in nature and in human society. The stronger the self-organization function of a system, the stronger its ability to maintain and generate new functions. The theory of 'natural selection, survival of the fittest' proposed by Darwin can be regarded as a self-organization process of evolution and development between different species achieved by living organisms in nature through the self-regulation of the ecosystem.

2.2 The course of science from pursuing simplicity to directly facing complexity

According to the second law of thermodynamics, the entire universe will become more and more disordered, and the famous heat death theory is derived from this. However, this is obviously inconsistent with the actual situation of the generation and evolution of the universe, and contradicts the vibrant real world. Darwin's biological evolution theory further reveals that organisms have evolved from simple to complex over a long period of time, and finally formed the reality of today's orderly and colorful biological world (as shown in Figure 2.2). Obviously, the laws of physics contradict the reality of the biological world.

2.2.1 The limits of reductionism: The failure of the attempt for science to attribute entirely the world to simplicity

Reductionism is a method of reducing the high-level motion forms of matter (such as life motion) to low-level motion forms (such as mechanical motion), and replacing the laws of high-level motion forms with the laws of low-level motion forms. Reductionism believes that various phenomena can be reduced to a set of basic elements, each of which is independent of each other and does not change its essence due to external factors. Through the study of these basic elements, the essence of the overall phenomenon can be inferred. French philosopher Descartes (Rene Descartes 1596–1650)

Figure 2.2. The course of biological evolution.

is the founder of reductionism. His methodology idea has been supplemented and developed by scientists from Newton (Isaac Newton 1643–1727) to Einstein (Albert Einstein 1879–1955), and has been continuously improved after 400 years of scientific practice, forming the dominant position of reductionism in the modern scientific system.

According to the viewpoint of reductionism, both the physical world and the biological world should follow the same laws, and these laws should be simple. In the study of living organisms, we can gradually understand the parts through reduction analysis in order to grasp the whole. All life phenomena and psychological phenomena can ultimately be reduced to physical and chemical phenomena, so, in principle, they can be explained by the laws of physics and chemistry.

In the field of physics, at the beginning of the 19th century, it was still believed that atoms were the simplest building blocks of the universe. As long as the structure of the atom was clarified, the properties of matter could be grasped. However, later studies found that atoms are composed of more basic 'particles', which themselves also have a very complex structure. The birth of quantum mechanics has made people fully realize that trying to understand the properties of multi-particle systems by analyzing the properties of individual particles is not feasible. This is because the characteristics displayed by the particles in the particle system are quite different from the characteristics displayed when they exist alone, just as a hand that exists alone does not have the function of the hand of a living healthy body. This means that to study a multi-particle system composed of more than two particles, it is necessary to adopt the holistic approach to treat them as a system. Heisenberg's 'uncertainty principle' finally sounded the death knell of mechanical determinism (as shown in Figure 2.3): since it is impossible to obtain the precise values of the position and momentum of the microscopic particles at the same time, the 'Laplace monster' no longer exists. In this way, the analytical reduction method based on mechanical determinism has lost its last turf in the field of microphysics (Huang Xinrong 2007).

2.2.2 The rise of complexity science: Complexity becomes the object of scientific study

At the same time, research progress on biological organisms from a new perspective has also given a powerful blow to mechanical determinism. In his book 'General System Theory: Basics, Development and Application'. Bertalanffy (1987) the famous Austrian theoretical biologist Bertalanffy (Ludwig Von Bertalanffy 1901–1972) revealed several main characteristics of organisms that are different from simple systems:

1. *A biological organism is an indivisible whole with a complex structure.* Its holistic attributes are not equal to the sum of the attributes of its components. That is to say, the mechanistic method that attempts to reduce life and psychological phenomena to physical and chemical phenomena is theoretically untenable.

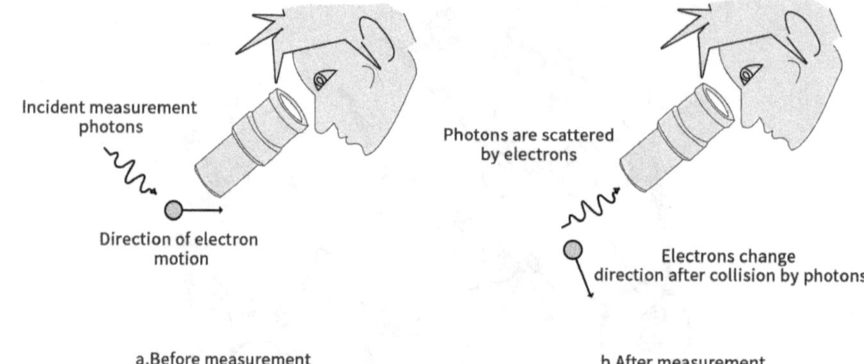

Figure 2.3. Heisenberg's 'Uncertainty Principle'.

2. *A biological organism is an open system.* It can maintain the homeostasis of its own temperature, metabolic rate, and water and electrolytes through the exchange of substances, energy and information through internal regulatory mechanisms and the environment. In this way, when a biological organism is subject to external disturbances or starting from different initial conditions, it can continue to grow according to its predetermined goal and achieve the same result. This is called equiconsequentiality.
3. *An organism is a system with hierarchical levels.* Each level as a whole has its own specific attributes. These different attributes distinguish one level of system from another. The low-level systems interact to form a high-level system, and the high-level system forms a higher-level system, finally forming a Chinese box-style hierarchical system.

In the 1970s, the Belgian physicist Prigogine's dissipative structure theory, Hacken's synergetics, Eigen's (Manfred Eigen 1927–2019) hyper-cycle theory, Thom's (René Thom 1923–2002) catastrophe theory, etc., were successively established, and gradually converged into self-organizing system theories that revealed the evolutionary phenomena and internal laws of complex giant systems. In the research at this stage, the method used by scientists was no longer a reduction and decomposition method, but a holistic research method, that is, they began to use physical experiments, mathematical models, computer simulations and other means to study complex systems while maintaining the overall integrity of the system.

As different aspects of the theory of self-organization, the theory of dissipative structure focuses on exploring the conditions and environment under which self-organization appears. It believes that the open system realizes self-organization through non-linear interactions and fluctuations under conditions far from equilibrium. The theory of synergy focuses on exploring the dynamics of self-organization, that is, the internal mechanism of self-organization. It explains how the subsystems in the system form an orderly structure through synergy. Catastrophe theory mathematically reveals the distinct forms of gradual and abrupt changes during the evolution of self-organizing systems. And the hyper-cyclic theory is a self-organizing theory about molecular evolution, which reveals the evolution process and internal law of the biological macromolecules to protocells, thus enabling mankind to take a solid step forward towards a comprehensive understanding of the origin of life. In addition, fractal theory and chaos theory study the complexity and picture of self-organization from the perspective of time series and spatial order (Huang Xinrong 2007).

In the 1980s, under the vigorous promotion of some physicists and economists, including three Nobel Prize winners Gellman-Murray (Gell-Mann 1929–2019), Anderson (Philip W. Anderson 1923–2020), and Arrow (Kenneth J. Arrow 1921–2017), the Santa Fe Institute, a research institute specializing in complexity science, was established in New Mexico, United States. This institute gained renown as the center of world complexity research. Some of the concepts and methods they put forward are considered to "represent a new attitude, a new perspective of looking at problems, and a brand-new worldview". At present, complexity research has become a world-scale scientific trend of thought, a cultural movement, which is driving the science based on reductionism to undergo earth-shaking changes (Huang Xinrong 2007).

With the establishment of the Santa Fe Institute, the terms complexity and complexity science gradually became recognized by the scientific community and spread rapidly. Since its establishment, the Santa Fe Institute has established the purpose of promoting the unification of knowledge, eliminating the opposition between science and humanities, and comprehensively carrying out the exploration of complexity. Although the researchers of the Santa Fe Institute have different research aspects, their research objects, research methods and tools are the same. Their research object is complex systems, and their research methods are not reduction and decomposition, but transcend reductionism, introduce simulation, model, metaphor and analogy methods into complexity research, and use computers as the main research tool. Just like the scientific revolution was triggered by the use of microscopes in biology and the use of telescopes in astronomy, computers have made complex systems the objects of scientific research for the first time. The Chinese scientist Qian Xuesen, in the 1980s when complexity research was just emerging, keenly proposed exploring the methodology of complexity science. He believed that the study of open complex giant systems must adopt a new method, that is, a comprehensive integration method from qualitative to quantitative (Qian Xuesen 1988).

Science in the 20th century, on the one hand, is deepening people's understanding of the origin of the world from the microcosmic and cosmoscopic levels in the field of traditional science; on the other hand, with the birth of new theories and methods such as system theory, cybernetics, information theory, dissipative structure theory, synergetics, and catastrophe theory, the boundaries of science have also been expanded in breadth.

As the 'science of the 21st century', the rise of complexity science has triggered changes in scientific concepts and ways of thinking. It has broken the linear theory that had dominated the world since Newtonian mechanics, and has expanded science from the traditional field of linear, definite, and order to the field of non-linear, uncertain and disordered. Some fields that were not considered to be science or that were difficult for science to reach earlier are gradually being included in the scope of science, and the goal of science has now moved from the original pursuit of simplicity to understanding complexity. As the complexity of reality systems such as the universe and life has been revealed, scientists have gradually realized that relying solely on traditional scientific analysis and empirical evidence, human beings may never be able to understand such a profound and mysterious reality system as the universe and life. As a result, scientists have adopted some methods that were not considered scientific in the past, such as metaphorical analogy and philosophical speculation, when studying complex reality systems. Today, whether it is in the 'Self-Sufficient Solution of the Universe' by Stephen Hawking (Stephen William Hawking 1942–2018), the most famous contemporary theoretical physicist, or in the latest theories on complexity research launched by the well-known American Santa Fe Institute, we can all see the shadow of this ancient and modern human cognition and argumentation model from time to time. Hawking even predicted: "The 21st century will be the century of complexity science." As a result, it also brought a re-examination of the ancient holistic science characterized by observation and speculation. When scientists used the new thinking of complexity science to re-examine ancient sciences such as traditional Chinese medicine (TCM), divination, astrology, and physiognomy that have been flourishing in ancient China, they find that these human knowledge systems that have been considered pseudoscience or non-scientific in modern times, which contains more profound scientific truths that cannot be understood by the reductionist method characterized by analysis (Yuan Bing 2010).

2.3 Emergence: The most important concept in complexity science

The concept of 'emergence' is not unfamiliar, and has a long history. The ancient Chinese thinker Lao Tzu's (571 BC–471 BC) thesis of 'being born out of nothing' is an ancient and profound understanding and expression of emergence. Bertalanfi borrowed Aristotle's (Aristotle, 384–322 BC) famous proposition, 'The whole is greater than the sum of its parts', to express emergence. In complexity research, the most important concept introduced by scientists is 'emergence'. With the rise of complexity science, emergence has gradually evolved into a scientific concept that describes the formation of complex systems, which triggered extensive and in-depth research by scientists, and its nature is also being gradually recognized by people.

2.3.1 Characteristics of emergence and possibility of being recognized

In complexity science, the properties, characteristics, behaviors, functions, etc., that the system as a whole possesses, but any part or the sum of the parts do not possess, are called emergence. When the parts of the system follow simple rules and form a whole through partial interaction, some new attributes or laws will suddenly be born at the system level. When we restore the whole to its individual parts, these attributes, characteristics, behaviors, and functions of the whole are not reflected in a single part. An example is the ant kingdom. The nervous system of ants is very simple and can only perform simple thinking. However, when a large number of ants interact, a hierarchical ant kingdom will be formed. Studies have confirmed that the queen ant does not directly give orders to all the ants, and each ant does not know the blueprint of the entire ant kingdom to be built. As long as each ant can follow simple rules to interact, the behavior of a large number of ants will show a clever and orderly division of labor to complete tasks such as foraging and nesting. As shown in Figure 2.4, the ant kingdom is a phenomenon of 'emergence' that appears above the entire ant colony.

Emergences are widely present in nature, animals and social systems, such as snowflakes in various forms composed of simple water molecules (as shown in Figure 2.5), interference or diffraction patterns formed by a large number of photons, nectar gathering behaviors of bees, foraging and gathering behaviors of fish groups, the formation of public opinion on the internet, the formation of traffic network congestion, and the formation of different economic systems, etc. The Santa Fe Institute in the United States, known for its complexity research, clearly stated that the complexity science is essentially a science of emergence. The core issue of emergence theory is to reveal the origin and internal mechanism of emergence at the system level.

The generation of emergence requires the system to have at least the following characteristics (Huang Xinrong 2012):

1. *Non-linearity*: The ubiquitous non-linear interactions in complex systems give the system the possibility of small causes producing big results.
2. *Self-organization*: This refers to the creativity, self-growth, and adaptive behavior of a complex system.
3. *The system is in a state of imbalance or far from balance*: This state of imbalance or far from equilibrium increases the possibility of accidental events, which is also a key factor in the unpredictability of emergence.

The emergence theory can be said to be a middle way that has been explored and developed amidst the long-term controversy between mechanical theory and vitality theory, reduction theory and anti-reduction theory. It discards the vital matter of vitality theory, but, in a sense, retains the irreducible characteristics and processes of life. Based on emergence theory, a property or behavior of a system is said to emerge if and only if it is not only absent from lower-level components, but also cannot be explained by reduction. Emergence is an ontological and epistemologically irreducible and fundamental property possessed at the macro-level, that is to say, the interpretation

Figure 2.4. Termite tower.

Figure 2.5. The emergence phenomenon of snowflakes.

of emergence at the holistic level of the system cannot be achieved through downward reduction. Downward irreversibility means upward unpredictability, that is, the unpredictability of the functions and attributes displayed at the holistic level. This unpredictability is also reflected in the process of system evolution, which is the unpredictability of new things or properties.

Emergence is a mutation of the property and structure of a system during the transition from low level to high level. In this process, new properties can be produced from old properties. Once the system as a whole is decomposed into its components, these properties no longer exist. The emergence of new properties is the greatest feature of complex systems. Combining various elements to form an organic system will inevitably emerge new properties that a single element does not have. There is no need to introduce any mysterious factors to explain the emergence of this new nature, because it is the interaction between the various parts that leads to the new nature. The research of complexity science shows that no matter what type of emergence, they have the following common characteristics (Fan Dongping 2006):

1. Emergence is a new property that the system has but its components do not have, that is, the overall properties of the system at the macro level, which is the most basic feature of emergence. Some attributes of the higher level emerge from the lower level and have different 'novelty'

from the lower level. The new features of the higher level cannot be deduced or predicted from the description of the lower level. Just as moral phenomena and their laws cannot be deduced from psychological phenomena and their laws, psychological phenomenon and their unique laws cannot be deduced from physical phenomena and their laws.

2. The emergence of complex systems is expressed as the novelty of 'generating complexity from simplicity'. Novelty often refers to the emergence of new things in the course of evolution. The existing entities may form new combinations, produce new structures, and form new entities with new properties and behaviors. Holland (John Henry Holland 1929–2015) explained the basic characteristics of the emergence of the continuous generation of new things from simple to complex with 'simplicity breeds complexity'. "A few rules and laws have generated complex systems, and they have caused eternal novelty and new emergence in a constantly changing form." Therefore, emergence again and again promotes the evolution process of the system from a simple system to a complex new system.
3. Emergence is manifested as the non-derivability and unpredictability of non-iterative simulation. The emergence of complex systems is unpredictable, so researchers cannot deduce the macro structure or properties of complex systems from the composition and behavior rules of the micro level. This underivable or unpredictable novelty is a typical feature of complex systems. Mark Bedau (Mark A. Bedau 1950) believes that the un-derivability and unpredictability of the emergence stem from a large number of non-linear causal interactions and contextual correlations at the micro level. It is precise because of the aggregation and iteration of a large number of nonlinear causal interactions at the micro level that the emergence phenomenon at the macro level is caused and displayed. Complexity science currently demonstrates the aggregation and iteration of such micro-level causal interactions by establishing models and using computer simulations, to study how they emerge with macroscopic characteristics. Epistemologically, therefore, emergence is cognizable.
4. Emergence is manifested as irreducibility to a certain degree between levels. A system or its properties are considered to be emergent if and only if it not only possesses properties and behaviors not possessed by its lower-level components, but these properties and behaviors cannot be fully explained by understanding the lower-level components and their interrelationships. In addition, the emergent behavior and laws of complex systems require the autonomy of macro-level theoretical explanations, which cannot be replaced by reductive explanations. In other words, using low-level laws cannot fully explain and replace macro-level laws.

The revealing of the characteristics and laws of the emergence of complex systems proves from another angle that the reduction analysis method of modern medicine, which strives to finally grasp the human body as a whole through 'analysis-reconstruction' is not feasible. In the process of studying the emergence of complex systems, the computer simulation methods used by scientists also provide a new way for medical research to move from analytical methods to holistic methods. We will see in the following analysis that this new method will eventually lead to the birth of holistic medicine, a new medical system that integrates TCM and modern medicine.

2.3.2 Exploration of emergence mechanism

So, what is the internal mechanism for the emergence of new properties of complex systems? The mechanism of emergence is essentially to achieve a level leap, that is, a leap from a local, low-level object to a higher-level overall object. The emergence mechanism reveals the causal relationship between the levels of the complex system (Fan Dongping 2005).

First of all, the emergence of complex systems shows unique behaviors and laws at the macro level, which has a downward causal effect on the lower levels. It means that all processes at the lower levels of the hierarchy are constrained by the laws of the higher levels and act in accordance with these laws. Downward causality is a basic feature of emergence, a concrete expression and embodiment of the overall and high-level emergence.

The theory of complex adaptive system emphasizes and simulates on the computer the process that adaptability creates complexity and emergence. Once the interaction among low-level objects progresses to a higher-level global function, it should not only be regarded as the action among low-level objects, but as the attributes and functions of the high-level system. Adaptability is the basic characteristic of complex systems, including adaptability of the system to the environment, or the environment has constraints on the system, that is, downward causality. Various typical characteristics of 'emergence' such as irreducibility and unpredictability are also revealed from this.

Regarding the generation of 'emergence', the 'constrained generation mechanism' proposed by Holland reveals that the conversion of low-level system components from local to global actions and the mutual adaptability between components produce an overall structure, a new level, manifested as an emergence nature. These new levels can be used as building blocks, converging and constrained to generate new structures, that is, new systems and properties of higher levels. From this layer upon layer of emergence, not only is a hierarchical system produced, but also emergences of new attributes, new things, and new organizations, in an endless stream (Huang Xinrong 2012).

The high-level constraints on the low-level are embodied as the external environmental stimuli that the low-level components must adapt to. The changes made by the low-level components in form and function to adapt to this stimulus form a new steady-state structure, whereby the emerging new characteristics of the high-level are maintained and stabilized and become an inherent attribute of the system. With the continuous emergence of new features and the gradual formation of new constraints of the higher-level on the lower-level, the lower-level systems will have to continuously change their structure and function to adapt to these constraints. As a result, the system will gradually evolve from simple to complex.

Since the end of the 20th century, with the rapid development of interdisciplinary sciences such as complexity science, computer science, life science and cognitive science, the research methods of emergence have also undergone fundamental changes. Scientists have begun to regard emergence as a black box, an object that does not need to be explained and cannot be explained, using computer simulation methods to reveal its internal mechanism by establishing its behavior model. In recent years, in the research of chaotic phenomena, cellular automata, genetic algorithms, etc., the focus of scientists' research has changed from 'how' the emergence occurs to 'why' the emergence is possible, and from studying the static characteristics of emergence to studying its dynamics process.

With the help of advanced computers, complexity scientists have conducted in-depth exploration of the mechanism of emergence, and simulated the emergence of typical complex systems such as neural networks, artificial life, and ecosystems. On the basis of these studies, scientists have given common features and markers of emergence (Fan Dongping 2005):

1. Emergence is a holistic mode, behavior or dynamic structure of a complex system.
2. Emergence is a self-organized hierarchical transition process.
3. Emergence has the non-derivability of iterative simulation or the derivability of non-iterative simulation.
4. Emergence has the autonomy of macro-level interpretation.

With the deepening of research on the mechanism of emergence, scientists realized that although emergence is underivable and unpredictable in the strict sense, it is derivable and predictable in the sense of simulation. That is to say, emergence can be understood and explained, and in certain conditions and meanings, it has a certain degree of derivability and predictability.

As the Santa Fe Institute clearly put forward: "Complexity is essentially a science on emergence". Only by clarifying the laws of emergence phenomena in the system can we truly understand these complex systems. At present, through simulation research on the emergence of typical complex systems such as neural networks, artificial life, and ecosystems, some operable description methods have been formed, and emergence mechanism models and theories that conform to modern scientific norms have been established, which also makes emergence a scientific concept.

2.4 Self-organization: The inner power driving system evolution

Organization refers to the orderly structure in the system or the formation process of this orderly structure, that is, the process of 'organizing according to a certain purpose, task and form'. The organizational process and the resulting organizational structure have the following salient characteristics (Xu Guozhi 2000):

1. Compared with the state before the organization, the organized structure has increased order and reduced symmetry. After the messy bricks piled up together are organized to form a house, the arrangement and distribution of the bricks change: the relative position is determined, and the degree of order increases. After a group of people is organized into a team, the status and role of each person in the team is determined, the mutual relationship among individuals is changed, and the superior-subordinate relationship is determined.
2. Organizational process is the process of qualitative change of the system and the process of increasing the order of the system. System evolution is usually in two ways: one is quantitative change, that is, the state of the system changes continuously with time. When the time interval is infinitely small, the difference between the two states is also infinitely small. If a certain functional relationship can be used to describe the law of system state changes over time, the form of the functional relationship remains unchanged during the quantitative change. The other is mutation. When the state of the system is abruptly changed, the quantitative value of state variables and the relationship among them may change.

The German theoretical physicist Harken divided 'organizations' into two types: hetero-organization and self-organization, according to the evolutionary form of organization. He calls organizations that cannot organize, create, evolve, and move from disorder to order by themselves as hetero-organizations. Hetero-organizations can only rely on specific instructions from the outside to push the organization to evolve in an orderly direction, thereby passively moving from disorder to order. On the contrary, self-organization refers to the organizational process that can organize, create and evolve by itself without specific instructions from the outside world, and can autonomously move from disorder to order, forming a structure with certain functions.

Hetero-organization must have an organizer outside the system, usually with a definite goal. Therefore, the hetero-organization is actually a control process implemented by the organizer in accordance with the determined goals. The organizer either completes the organization process in an orderly manner in accordance with the organization plan formulated in advance; or adopts the negative feedback control mechanism we mentioned earlier, through multiple feedback control based on target difference, and finally achieves the organization's goal. In general, the organizational process where the organizer does not include people is called control, such as the control of man-made machines, equipment, and fully automated production lines; and the organizational process where the organizer includes people is called management, such as the organization of a school, a factory, or a company.

Self-organization is different. The direct reason for the emergence of the system organization structure lies within the system, which mainly depends on the interaction among the internal elements of the system. For example, the evolution of species is caused by genetic mutation and inheritance within organisms. Real systems inevitably have various connections with the environment, and the evolution of each system is also carried out in a certain environment. Therefore, the role of environmental factors in evolution cannot be ignored. Generally speaking, the external cause is the condition of change, the internal cause is the basis of the change, and the external cause plays a role through the internal cause. Without the proper temperature, an egg cannot become a chicken; but the proper temperature cannot make a rock become a chicken. Whether a chick or duckling is hatched is determined by the nature of the egg, that is, the internal factors of the system; while the change from an egg into a chicken or duckling requires the temperature of the environment, that is, the external factor of the system. Therefore, in the process of self-organization, internal and external factors are indispensable.

The hetero-organization theory is the traditional control theory, which is mainly applicable to artificial systems. Artificial systems are usually systems which be built based on the understanding of the relationship between control and response, and can achieve certain functions and response mechanisms. A successful artificial system can make appropriate responses and complete predetermined tasks under the action of input. In real systems, especially complex giant systems like life systems and social systems, it is difficult to figure out the mechanism of how external environmental conditions affect the internal structure of the system to lead to the organization process, which leads to the self-organization theory. Therefore, the self-organization theory mainly studies the formation and development mechanism of complex self-organization systems, that is, under certain conditions, how the system automatically changes from disorder to order, from low order to high order. It believes that the evolution from disorder to order must meet several basic conditions (Xu Guozhi 2000):

1. The system that can produce self-organization must be an open system. According to the second law of thermodynamics, the degree of disorder in an isolated system cannot be reduced. For an isolated system, the result of its evolution must be to reach a state of equilibrium with the greatest degree of disorder. Therefore, to make the system develop in an orderly direction, system openness is a necessary condition. Only through the exchange of matter, energy and information with the outside world can the system produce and maintain a stable and orderly structure.
2. A system developing from disorder to order must be in a state far from thermal equilibrium. According to the definition of thermodynamics, the equilibrium state is the most uniform and disordered state that an isolated system can exist in stably after an infinite time. According to Prigogine's dissipative structure theory: the isolated system in the non-equilibrium linear region, that is, the region close to the equilibrium state, the final result of the system evolution is to reach the equilibrium state, and it is impossible to form an ordered structure. Only in the non-linear region far away from the equilibrium state can it evolve into an ordered structure. Only in an open system, due to the input and influence of environmental energy, matter, and information, is it possible to push the state of the system away from the equilibrium state. In the non-linear region far from the equilibrium state, the system may form an ordered structure.
3. There are non-linear interactions among the various subsystems in the system, and this interaction does not satisfy the superposition principle. The non-linear interaction between the subsystems will make the system emerge with new properties that each subsystem does not have, so that the system changes from disorder to order. Obviously, non-linear interaction is the internal reason that the system forms the ordered structure.
4. The fluctuation of some state variables in the system is the inducement of the self-organization process. Fluctuation usually refers to the deviation of the state of certain variables in the system from the equilibrium state. Based on the stability principle mentioned above, when the system deviates from the equilibrium state for some reason, the effect of fluctuations is usually to quickly restore it to the original equilibrium state. But when the state of the system is in a region far from equilibrium, the non-linear interaction of fluctuations may cause the state of the system to cross the critical point of the stable state, undergo a qualitative change, and transition to a new stable ordered state. Therefore, fluctuations are the initial driving force that initiates the self-organizing process of the system. Fluctuations are mainly caused by the interference of some difficult-to-control complex factors inside or outside the system, with random

chance. However, they can cause the system to transition from disorder to order, from one ordered structure to another, thereby promoting the evolution of the system from low-level to high-level.

These conditions mentioned above are closely related. If the system is not open to the environment, it will not be able to exchange matter, energy and information with the environment, and its state cannot be far from the equilibrium state. In this case, any non-linear interaction among the subsystems cannot take the system far from the equilibrium state, and the fluctuations of the system can only stabilize the system in the equilibrium state, and cannot form a new ordered structure. Without staying away far from the equilibrium state, the interaction between the open system and the environment can only play a perturbative role, and it is difficult to make a qualitative change in the state of the system.

Throughout the history of scientific development, physics in the past was based on linear interaction, and systems which satisfied the principle of superposition. Under such a theoretical framework, it is impossible for the system to form a dissipative structure, and no calculation and analysis can draw the result of the evolution of the system to an ordered structure. Therefore, the existence of non-linear interaction is a necessary condition for the qualitative change of the system. Fluctuations are inevitable in an open system, but fluctuations do not have a tendency to change the direction of the system state. However, without fluctuations, it is difficult to maintain a steady-state equilibrium point for a system close to the equilibrium state, and it is also difficult for a system far away from the equilibrium state to cross the critical point of maintaining stability and reach another new steady-state equilibrium.

2.5 Methods and technical means for complexity science to deal with complex systems

The rise of complexity science has triggered tremendous changes in scientific concepts and research methods. It has not only had a significant impact on traditional philosophical ontology and epistemology, but also a huge impact on scientific methodology characterized by reductionism. So, how does complexity science study complex systems, and what role will traditional scientific methods play in complexity research?

2.5.1 The integrative approach of reductionism: Reconstruction based on analysis

In modern science, the method of analysis and reduction has basically occupied the status of mainstream science. Based on this method, no matter how complicated a scientific object is, it will eventually be reduced to its most basic constituent elements. As for the grasp of the whole, the reductionist method tries to achieve it through reconstruction on the basis of analysis. It is based on this approach that science in the past 400 years has created a whole set of operational methods, which have been applied to various fields of natural science and achieved great success.

The method of 'analysis-reconstruction' of reductionism is also a method of grasping the whole. In other words, reduction theory is not analysis for analysis, and decomposition for decomposition. Using this method, the system is first separated from the environment to study it in isolation; then the system is decomposed into parts, and the high level is reduced to the low level. On the basis of adequate analysis and decomposition of the system, parts are used to explain the whole, and the low level is used to explain the high level.

Systems science still uses this method to a large extent when studying less complex systems. The difference is that it performs reduction analysis under the guidance of a holistic view, and obtains an overall understanding by integrating the knowledge of relevant parts. This method is effective against simple systems and even simple giant systems, but it can do nothing for complex giant systems with many types of subsystems, complex interconnections and non-linear characteristics. The method of adding up the knowledge of the parts is inherently inappropriate to describe the overall emergence. The more complex the system, the less effective this method is for grasping the overall emergence.

Complexity science is developed by exposing and overcoming the one-sidedness and limitations of reductionism. The ancient naive holism did not and could not produce rigorous modern scientific methods, but it contains methodological ideas for understanding and handling problems as a whole that reductionism lacks. However, since complexity research is also scientific research, it is impossible to put aside the reductionist method that has been effective for hundreds of years, otherwise it will return to the ancient simple holism. Without reduction analysis, our understanding of complex systems can only be intuitive, speculative, and general, lacking scientific rigor. Without a holistic view, our understanding of things can only be fragmentary and not being able to grasp things as a whole, just like only seeing the trees but not the forest. To study open complex giant systems, the scientific attitude should be based on the holistic method as the main body, and organically combine the method of reduction analysis. As a result, the grasp of the whole will be established on the basis of rigorous empirical analysis, and the 'dialectical unity of reductionism and holism' is realized (Xu Guozhi 2000).

As the systemic philosopher Laszlo pointed out: Both thinking modes of traditional holism and reductionism have inevitable shortcomings: the former replaces detailed inquiry with faith and insight, while the latter sacrifices the thorough grasp based on comprehensive studies in exchange for analysis point by point (Laszlo 1978). In other words, reduction and integration are also dialectically linked, and simply emphasizing one aspect is one-sided. Modern science worships analysis and reduction, and almost equates the scientific method with analysis, which brings about its mechanical nature and makes it a hotbed of metaphysics. Traditional holism emphasizes the whole too much, so that its 'synthesis' often becomes an obstacle to in-depth research.

Due to the 'uncertainty' caused by human intervention in the decomposition and analysis process and the technical difficulties caused by the numerous variables in the reconstruction process, the 'analysis-reconstruction' method has encountered insurmountable obstacles in responding to the open and complex giant system of the biological world. To fully and fundamentally grasp the complex life system, scientists have to find another way.

2.5.2 Traditional metaphor and analogy methods enter the vision of science

Since the method of decomposition and analysis cannot grasp the internal laws of complex systems, it can only be studied as a whole and according to the original face of complex systems. However, in real life, it is very difficult to understand an unfamiliar complex object at once. The usual approach is to find a certain similarity between a complex thing that is known and a certain simple thing that is already known, and use metaphor and analogy to understand the complex thing. Metaphor and analogy are commonly used methods in ancient and modern Chinese and foreign novels and poems and are also commonly used methods in ancient science. Many theories in TCM were established by the method of 'metaphor and analogy'.

The tradition of modern science is to pursue logic and clarity, and it is best to be mathematicised and axiomatic. Logical positivism believes that science must be able to be falsified. Anything that cannot be clearly expressed, that is, cannot be logicalized or falsified, cannot be called science. Under this clear demarcation standard of science that was previously recognized by the scientific community, metaphor and analogy, although well-known ways of thinking, have always been excluded from science. TCM has always been considered not a science, which is also related to this concept.

Metaphor and analogy, whether in the East or the West, modern or ancient, are well-known ways of cognition. However, their use as scientific methods for the establishment of scientific theories has only come about in the last two decades. The Santa Fe Institute in the United States first introduced it as a method for studying complex systems from a holistic perspective, and used it as a basic scientific method for describing and dynamic simulation of real systems.

Holland pointed out in his work, 'Emergence': "Once we notice that a metaphor also involves a source and a target, we can relate this discussion of the use of models for innovation in science to the construction of metaphors. (I should note that I use the term 'metaphor' here in the broad sense, including similes and other tropes as special cases.) Even a simple metaphor such as 'iceberg, emerald of the sea', involves much more than a source object, the emerald, and a target object, the iceberg. Both the source and the target are surrounded by an aura of meaning and associations. The metaphor causes a kind of recombination of these auras, enlarging the perceptions associated with both the target and the source. When the metaphor is successful, the interaction produces connections and interpretations that are, variously, surprising, amusing, or exhilarating" (Holland 2001).

Metaphors work by extending certain characteristics of the source object to the target object by 'associated implication'. The basis for these hints is that there are some common elements or similarities between the source and the target. These common elements (or similarities) can be reflected in the similarity of constituent elements and related relationships, or even the similarity of certain imagery. It is because of this similarity that, with the help of the model of the target object established based on metaphors, we can perceive and understand the characteristics of the target object.

Metaphors and analogies are widely used in complexity science. Among the dozens of basic definitions in complexity science, more than ten are defined by metaphors and analogies, such as butterfly effect, fractal, artificial life, chaos edge, self-organized criticality, emergence, etc. The concept of complexity of metaphor types mainly uses metaphors and analogies to express complex things and complex phenomena that are difficult to express with precise language.

Complexity science often uses metaphors and analogies to construct theoretical models of complex systems. The difference between this modeling method and the theoretical model established by theoretical physics lies only in whether the introduction of the model structure is based on the ideas of scientists or on the metaphors and analogies of familiar systems. The subsequent process of perfecting and verifying the model is completely consistent. Today's science, whether it is Einstein's theory of relativity, Heisenberg's quantum model, Stephen Hawking's quantum gravitational model, or Holland's genetic algorithm and echo model, are all verified and evaluated in the same way. As we will see later, metaphors and analogies based on basic concepts such as yin and yang, five elements, and Eight Trigrams have played an important role in the establishment of the TCM human body model.

2.5.3 The holistic approach for modeling complex systems

For any system, if there is a way to transition from micro description to macro-overall description, it marks the establishment of the basic theory of the system. For a simple system, the description of the basic properties of its elements can be found in the basic theories of natural science. The descriptions of the system as a whole can be obtained by directly synthesizing the description of element characteristics. For simple giant systems, there is also a basic method to transition from micro description to macro description, that is, statistical description. For complex systems, due to the numerous types, huge quantities, and complex interconnections of microscopic components, research has shown that there is no description method for the transition from micro to macro.

Failure to transition from the micro to the macro does not mean that the overall synthesis cannot be achieved. Scientists often use another solution when studying complex reality systems: first jump to a certain holistic structure, and then look for empirical evidence in turn, which is the method of the macro to the micro, from the whole to the part. Today, complexity science also uses this method to establish the theoretical description of complex systems: construct a model of the complex system from the holistic level through metaphor and analogy, and then find evidence from the research on the input and output of the real system.

The purpose of constructing a model is to study the prototype, and objectivity and validity are the basic requirements of the model. All information reflecting the essential characteristics of the prototype should be shown in the model, so that the main characteristics of the prototype can be grasped through model research (Sun Xiaoli and Zhang Zengyi 2004). For complex giant systems with many types and quantities of subsystems and complex interrelationships, building their models from the whole often requires greatly simplifying the prototype and extracting its main features, thereby obtaining a simplified model of the real system.

Usually, a system as a whole is composed of parts or subsystems. The whole controls the part, the part supports the whole, and the behavior of the part is restricted and dominated by the whole. Therefore, to conceive a system model from the holistic level, the following steps must first be taken:

1. Comprehensively sort out the functions and attributes of the system to ensure that the established model can reflect these functions and attributes.
2. Determine which elements or components constitute the system, what functions and attributes these elements or components have, and how they are connected to form a unified whole.
3. Carry out environmental analysis to clarify the environment and functional connection of the system, how the system and the environment influence each other, and the characteristics and changing trends of the environment.

In complex systems, some of the functions and behaviors displayed by the system at the holistic level can be inferred and predicted based on the understanding of its low-level elements and components, and some are 'emergences' that cannot be predicted and inferred from the understanding of the low-level elements and components. The derivability and predictability of the emergence in the simulation sense profoundly show that for complex systems, although analysis and research for the lower-level of a system cannot fundamentally explain the functions and behaviors of the upper level, it is still very meaningful for establishing the model of the upper-level system from the perspective of functional simulation. Therefore, combining the functions of the whole system with the analysis and research of the elements and components at the lower level of the system is of great significance for rationally determining the setting of elements and the division of components in the holistic model. As we will see later, the internal structure of the human body in the theoretical model of TCM is set up based on the observation of human morphology and structure combined with the understanding of the physiological and pathological phenomena of the human body from a functional perspective.

The basic goal of science is to understand the reality system, and the purpose of understanding is to better control it. But in the process of scientific research, a situation may arise where we have a thorough understanding of the system, but cannot effectively control it. Conversely, sometimes we can effectively control the real system and solve the problems we encounter, but we don't know anything about the reason and mechanism of its effectiveness. As people often lament when comparing the two systems of traditional Chinese and Western medicine: In many cases, Western medicine can understand the causes and pathological changes of diseases, but there are few effective treatment methods; while TCM can cure the disease, it is not clear how the disease was cured.

Unlike traditional science, which pays more attention to understanding the object, the methodology of complexity science puts the control of the system in the first place, and it does not require that the system be fully understood before the method of adjustment and control is studied. It does not even pursue understanding at the micro level of reduction analysis. The model it builds is often not directly related to elements, the structure, and their inter-relationships that people can observe at the micro level of the real system, but applying it can effectively grasp the system and effectively regulate/control the state of the system. As we will discuss later, as a medical system with thousands of years of history and a great contribution to the reproduction and prosperity of the Chinese nation, the 'theory' of TCM on which the diagnosis and treatment of diseases is based is essentially such a model.

At present, the research of complexity science usually builds the model of complex systems on the basis of metaphorical simulation. Holland's model for studying complex adaptive systems was established in this way. By studying the different methods of selecting building blocks and reorganizing these building blocks, he established some rules to create an easy-to-understand system model governed by certain rules, i.e., echo model. The echo model clearly shows the complexity and emergence phenomena of the simulated system. It is through the echo model that Holland clearly explained how complex adaptive systems evolve, adapt, gather, compete, and cooperate, and at the same time how to create rich diversity and novelty. The echo model uses very few principles to construct an extremely beautiful model. It simply and clearly reproduces the process of how complexity emerges and adapts with a vivid and detailed picture. On this basis, Holland expertly used metaphorical methods to continuously construct various models, thus revealing the laws of complexity hidden behind the emergence phenomena.

Per Bak (1948–2002) established the famous 'sand pile model' on the basis of the metaphorical concept of self-organized criticality (as shown in Figure 2.6). Sand piles are something we take for granted, but Bak used them to build a model that reflects the evolution of complexity. The data simulation of the sand pile model shows that an open dynamic system with multiple degrees of freedom and far from equilibrium operates in a critical state, evolves in an intermittent, chaotic, avalanche-like form, and can evolve to a stable self-organized critical state. This model has been widely used in phenomena such as solar flares, volcanic eruptions, economics, biological evolution, turbulence, and the spread of infectious diseases. It is through this simple sand pile model that Bak revealed the secret of the evolution of complex systems.

Researchers also use the metaphorical modeling method when studying the complex phenomenon of artificial life. For example, on the basis of the metaphorical concept—the edge of chaos, Langton (Christopher Langton 1949) and other scholars explored various models of artificial life generation and evolution, such as self-reproducing cells, automata, bird colony models, ant colony models, Tierra models, Avida model, amoeba world, etc. It is through these models that scientists represented by Langton discovered that the essence of life lies in the organization form of matter rather than the specific matter itself. If we create conditions in a certain medium to produce the edge of chaos, then we may create life in this medium.

Today, when we take metaphor and analogy as a scientific method, and reexamine the ancient TCM that has a history of thousands of years in China with new scientific concepts brought about by complexity science, we noticed that since the Warring States Period 2000 years ago, the methods

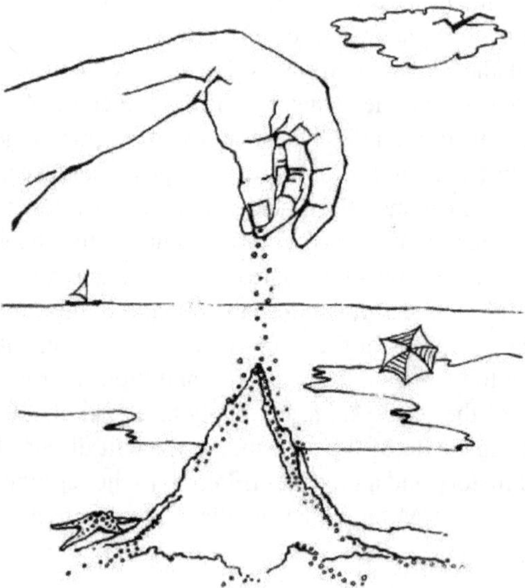

Figure 2.6. Sand pile model.

of metaphor and analogy have been widely used in the process of establishing the theoretical system in TCM. The model systems involved in China's earliest masterpiece of medical theory 'Huangdi Neijing, The Yellow Emperor's Inner Canon', are constructed based on the analogy of yin and yang, the Five Elements, and the Eight Diagrams, through observation of the body's entity structure and basic physiological and pathological phenomena. This theoretical system has been effectively guiding the clinical practice of TCM for thousands of years. In this sense, it is hard to say that it is less scientific than the theoretical model established by the same method of complexity science today.

2.5.4 The application of quantitative analysis methods and computer simulation techniques in complex system modeling

The model established by the method of metaphor and analogy needs to be attributed to the expectation of the real system through rigorous logical reasoning, and verified through observation and experiment. The acceptance of models as scientific theories by the scientific community requires the support of empirical evidence. Secondly, the deepening and precision of the model's revealing of various phenomena and laws of the real system requires the establishment of precise qualitative and quantitative relationships between the state of the model and the observable variables of the real system. Sometimes it is also necessary to establish this relationship with the aid of quantitative analysis and statistical analysis, which are the effective mathematical methods developed by reductionism, so as to accurately describe the real system and grasp the state of the real system using model.

All systems have both qualitative and quantitative characteristics. Qualitative characteristics determine quantitative characteristics, and quantitative characteristics express qualitative characteristics. With only qualitative description, it is difficult to grasp the system behavior characteristics deeply and accurately. However, qualitative description is the basis of quantitative description. If the qualitative knowledge is incorrect, no matter how precise and beautiful the quantitative description may be, it is useless, and it may even lead the understanding of the system astray. Quantitative description serves qualitative description. With the help of quantitative description, qualitative description can be deepened and refined. The combination of qualitative description and quantitative description is one of the basic methodological principles of complexity scientific research.

The system theory that successfully applied quantitative methods shows that only with a basic understanding of the qualitative characteristics of the system can we correctly determine how to express them with quantitative characteristics. In the natural sciences, even if it is recognized as a discipline with the most complete application of quantitative description, at least its basic assumptions are the result of qualitative thinking. The key to establishing a quantitative description system is to appropriately select state variables on the basis of correct qualitative knowledge (Xu Guozhi 2000).

Since Newton successfully used mathematical formulas to describe the laws of motion of objects, quantitative methods have received more and more attention and have been greatly developed. For hundreds of years, the natural sciences regarded qualitative methods as less scientific, temporary and expedient methods which had to be used before quantitative methods were found. However, with the development of complexity science, research objects are becoming more and more complex, and the difficulty of quantitative description is increasing. As a result, complexity science requires a re-evaluation of qualitative methods, and there is an increasingly strong voice against the one-sided pursuit of precision and quantification in complexity research.

Based on the profound understanding of the role and limitations of qualitative and quantitative methods in the study of complex systems, Qian Xuesen summarized the research and practice of four complex giant systems, including social systems, geographic systems, human body systems, and military systems. In the 1980s, he proposed a comprehensive integration method from qualitative to quantitative, confirming the principle that qualitative judgment and quantitative description should be organically combined in the study of complex giant systems.

For a model that needs to comprehensively describe a complex system, especially an open complex giant system, even if compared with the real system, the level of the model and the number of variables extracted are greatly simplified, it will still be quite complicated. There is not only qualitative analysis, but also quantitative description. To establish the model according to scientific principles, the correctness of the model requires that the expectations based on the model are consistent with the observations of the real system. Therefore, every adjustment to the model requires a lot of observation and verification. For complex systems, this means a huge amount of work, and it is almost unimaginable to rely entirely on manual methods. Modern computer technology and advances in computer-based knowledge engineering, expert systems, and artificial intelligence have opened up new ways for us. We can make full use of modern mathematical tools and various modeling methods to build models that include a large number of parameters based on systematic observation data (including various statistical data).

These models must also be based on the actual understanding and experience of the system to ensure the rationality and accuracy of the model, and to ensure that accurate qualitative and quantitative results can be obtained through computer simulation, experiment and calculation. At the same time, it should also make full use of information technology such as knowledge engineering, expert systems, and artificial intelligence to realize comprehensive knowledge integration based on expert experience and man-machine integration. The knowledge here includes empirical knowledge and professional knowledge of scientists, qualitative and quantitative knowledge, rational and perceptual knowledge. This empirical model of man-machine combination needs to be put into practice for testing, and constantly revised, based on the test results. Repeated this many times, the fit between the model and the real system will be continuously improved (Qian Xuesen 2001).

Computer simulation is to simulate and imitate the behavior of the actual system through computer programs. It can experiment and test the system in a virtual environment to predict the performance and behavior of the system. Today, the rapid development of electronic information and artificial intelligence technology provides powerful analysis tools for complexity research. The discovery of chaos is itself a miracle of computer applications. In the research of fractal theory, complex fractal graphs are also produced by some simple algorithms through repeated iterations of computers. Without a computer, the completion of these tasks is unimaginable. In addition to traditional differential equations and formal logic, some unique methods are also used in complexity research, such as decision-making technology under uncertain conditions, comprehensive integration technology, overall optimization technology, computational intelligence, non-linear technology, system simulation technology, etc. These methods are also inseparable from computers, especially the simulation technology is a new method that only appeared due to the application of computers. This method is different from the general experimental method. It performs system simulation completely through the model, thus becoming an important method for scientific discovery and research, and is widely used in the research of complexity science. The more famous ones include artificial life, meta cellular automata, competition-cooperation model, and swarm simulation tool SWARM, among others.

Complex systems often contain a large number of elements and interactions, making them difficult to analyze directly. Model-based simulation simplifies complexity by abstracting key aspects of a system into a mathematical or computer model, making it easier to understand and analyze. Model-based simulation can test the behavior of the system under different conditions and scenarios. This allows researchers to explore the impact of different decisions, policies or parameter settings on the system to predict possible outcomes. Simulation models can be used to predict the future behavior of a system to support decision making. It can help make smarter decisions, reduce risk, and optimize resource allocation. Model-based simulation can be used to verify whether the expected behavior of the system is consistent with the actual situation, to test the validity of the theoretical model, and to find potential problems or errors. The behavior of many systems is dynamic, involving interactions and feedbacks that vary over time. Model-based simulations allow the analogy of these dynamic processes, revealing the interplay within the system. Simulations can be used to find the optimal settings or parameters of a system to achieve a specific goal. This allows testing multiple scenarios under different conditions to find the best solution (Sterman 2000, Law and Kelton 2020).

Model-based system simulation is a powerful method for studying and analyzing complex systems in the real world. It involves constructing an abstract model of a system and predicting how the system will behave under different conditions by analogizing the behavior of the model. Explain how people can better understand the behavior of complex systems, predict future trends, and provide support for decision-making and optimization. Among the outstanding contributions of the Santa Fe Institute to complexity science, Holland's genetic algorithm, complex adaptive system theory, echo model, Langton's meta-cell automata, artificial life research, Per Bak's research on self-organized criticality through the sand pile model, and Arthur's research on economic complexity have all been done with the support of computers. It can be said that without computers, there would be no current achievements of the Santa Fe Institute, and there would be no current situation of complexity research. Therefore, we can say that computers are the most important technical means for complexity science or complexity research (Huang Xinrong 2007).

Chapter 3

The Human Body under the Vision of Natural Science and Complexity Science

Human beings are the most advanced form of life evolution on earth. While science pushes mankind's understanding of the natural world towards deeper levels, it also pays constant attention to human beings themselves. However, through the ages, human beings' knowledge of themselves has been branded by the mainstream worldview of the times without exception. What are we, humans? What is the relationship between mankind and the world in which it lives? Whether it is ancient philosophers or today's science, these are all issues that must be faced.

3.1 The human body revealed by biology and physiology

From a biological point of view, humans have already a basic understanding of the human body itself. The most basic elements of the human body are cells and intercellular substance. Cells and intercellular substance with similar functions and structures are organically combined to form tissues with specific functions; various tissues are combined into organs with certain morphological structure and physiological functions, such as skin, muscle, heart, liver, brain, etc. The structural characteristics of organs and tissues are compatible with their functions. The sum of multiple organs and tissues to accomplish certain physiological functions is called a system. For example, the mouth, pharynx, esophagus, stomach, intestine, and digestive glands make up the digestive system, and the nose, pharynx, larynx, trachea, bronchi, and lungs make up the respiratory system. The entire human body can be divided into more than a dozen systems, including the epidermal system, muscular system, skeletal system, respiratory system, digestive system, circulatory system, urinary system, reproductive system, nervous system, endocrine system and lymphatic system (Wikipedia contributors 2020). Table 3.1 lists the main human systems and the organs and tissues included in them. All systems are interdependent, coordinated, interact with each other, and work together to maintain the life process of the human body. The human body is such a complete unity composed of many organs and systems, and no one organ can survive independently from the whole. The reason why the various systems of the human body can closely cooperate and coordinate activities is due to the regulation of nerves and body fluids.

Human beings live in nature and constantly exchange materials and energy with the outside world. Cells, as the most basic life unit constituting the human body, also exist in certain environments, and are constantly exchanging material and energy with their environment. However, most of the cells that make up our body have no direct contact with the external environment, and cannot directly exchange material and energy with the external environment. What is the environment in which these cells live?

Table 3.1. Human body systems and their functions.

No.	System name	The main structure and function
1	Integumentary system	Skin, hair, nails, sweat and other exocrine glands.
2	Muscular system	Enables the body to move using muscles.
3	Skeletal system	Bones maintain the structure of the body and its organs.
4	Respiratory system	Brings air into and out of the lungs to absorb oxygen and remove carbon dioxide.
5	Digestive system and excretory system	System to absorb nutrients and remove waste via the gastrointestinal tract including tract, including the mouth, esophagus, stomach and intestines.
6	Circulatory system	Circulates blood around the body via the heart, arteries and veins, delivering oxygen and nutrients to organs and cells and carrying their waste products away.
7	Urinary system	The system where the kidneys filter blood to produce urine, and get rid of waste.
8	Reproductive system	The reproductive organs required for the production of offspring.
9	Nervous system	Collects and processes information from the senses via nerves and the brain and tells the muscles to contract to cause physical actions.
10	Endocrine system	Influences the function of the body using hormones.
11	Immune system and lymphatic system	Defends the body against pathogens that may harm the body.

Body fluids are liquids in the human body with water as the main component, which contains a large number of ions and compounds. Two-thirds of the body fluid of the human body exists in the cell, which becomes the main component of the cell, intracellular fluid; the remaining one-third exists outside the cell, and is named extracellular fluid. This is mainly composed of plasma, tissue fluid and lymph fluid. In order to distinguish it from the external environment in which humans live, the environment composed of extracellular fluids is usually called internal environment. When we take out a cell or a piece of tissue that normally lives in the body, it will die soon if no special environment is provided. But in the internal environment, why can cells live normally? (Figure 3.1).

Studies have shown that extracellular fluid is a salt solution. About 90% of the plasma is water, and the remaining 10% are inorganic salts, proteins, and nutrients and metabolites carried by the blood, such as glucose, hormones, oxygen and carbon dioxide. The composition of tissue fluid and lymph fluid is similar to that of plasma, but with less protein in them. In a healthy human body, the content of the main components in the extracellular fluid and the osmotic pressure, pH and temperature are always kept within a certain range. As an open system, cells can directly exchange materials with the internal environment, continuously obtain materials needed for life activities, and at the same time discharge waste generated by metabolism, thereby maintaining their normal life activities (Figure 3.2). The internal environment of the human body completes the material

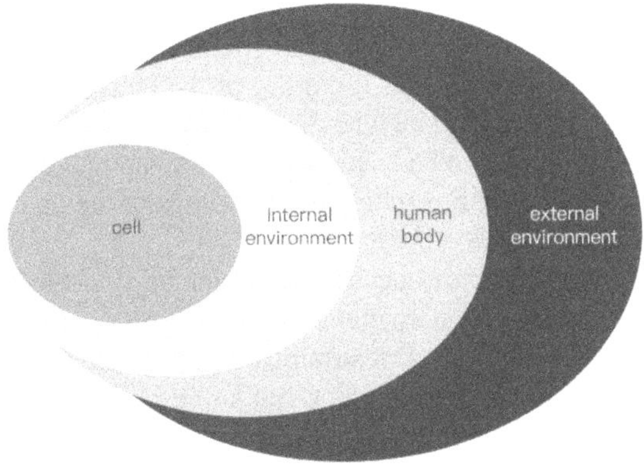

Figure 3.1. The structural relationship between cells and the environment.

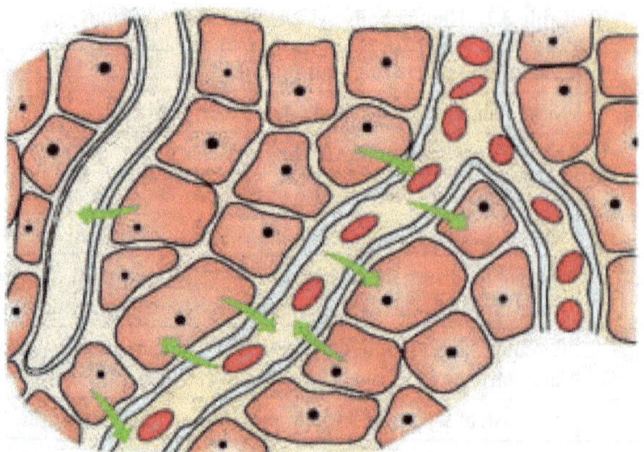

Figure 3.2. Material exchange between cells and internal environment.

exchange with the external environment through the normal functional activities of digestion, breathing, circulation, and the urinary system.

This is the human body that biology shows us. It is the basic understanding of the human body that biology and medicine have obtained through more than 300 years of research in anatomy, histology, physiology, cytology and related disciplines.

3.2 The mystery of the human body that science is gradually revealing

However, the advancement of natural sciences, especially theoretical physics, is approaching such a reality step by step: what biology and medicine have revealed is only the 'shell' of human beings, which are parts of human beings that appear as material existence. But in fact, in the process of human life, it is far more than such visible and tangible material that plays a role. There are many things that we can't see and touch, we feel their existence, feel that they are working, but today's science has not been able to clearly reveal what they are.

3.2.1 Mysterious meridians and acupuncture points

Today, the acupuncture therapy originating from traditional Chinese medicine (TCM) has spread all over the world and has been accepted as an important treatment method by modern medicine. In TCM, the acupuncture therapy is based on meridians and acupoints. There are 14 main meridians/channels, and more than 300 acupuncture points with different names along the routes of these meridians. There are also a large number of acupuncture points that are not on the route of these meridians, which are called 'extraordinary points' in TCM. From ancient times to the present, TCM has clearly marked and illustrated the routes of these meridians and the locations of these acupoints. Figure 3.3 shows a bronze figure of acupuncture and moxibustion dating back to the Song Dynasty (1026 AD) in China. It was cast by the National Medical Administration, the highest medical management institution at that time, under the order of the emperor of the time, Song Ren Zong (1010–1063). The bronze figure is marked with 354 acupoint names, and all acupuncture points are drilled through small holes. Inside the body cavity, there are internal organs and bones carved with wood. At that time, it was not only a teaching aid for acupuncture teaching, but also a model for assessing acupuncture practitioners.

In TCM, the therapeutic effect of acupuncture is far beyond the boundary of pain relief of 'trigger point therapy' accepted by modern medicine. TCM textbooks clearly record the therapeutic indications of each acupuncture point, and these indications are often not limited to the local locations where the acupuncture points are. For example, Guangming Point (GB37) for treating eye diseases is located on the outside of the calf; Neiguan (PC6) for treating gastric, heart and chest diseases is located on the forearm. TCM practitioners and acupuncturists have been relying on the guidance

Figure 3.3. Acupuncture bronze man from the Song Dynasty.

of ancient medical texts and these textbooks to effectively treat all kinds of human diseases since ancient times. After acupuncture needles are pierced into specific acupoints of the human body, the patient will feel sore, numb, distended in the local area, or conduction sensation along a certain route. This phenomenon is called 'got qi' in TCM, which means that the effect of qi is triggered. Tests on meridian sensitive people show that the sense of conduction is usually consistent with the route of the meridian recorded in TCM (People's Health Publishing House 1979). In a certain kind of Qigong that has been circulating in China for thousands of years, the Xiao Zhou Tian of Taoist 'inner alchemy', is formed by connecting the Ren Meridian (CV) and Du Meridian (GV) of the 14 meridians. After the trainees undergoes '100 days of foundation building' training, and once their 'Xiao Zhoutian' is opened, there will be a phenomenon of warm air flow running along the human body's Du Meridian (GV) and Ren Meridian (CV) cycle, as shown in Figure 3.4.

So, what is the meridian? What are acupoints? Do they exist objectively in the human body, and can their existence be proven in a scientifically recognized way?

Since the 1950s, China has begun the research of 'meridian essence' aimed at finding the meridian in human entities. However, the results of anatomical studies show that the route of

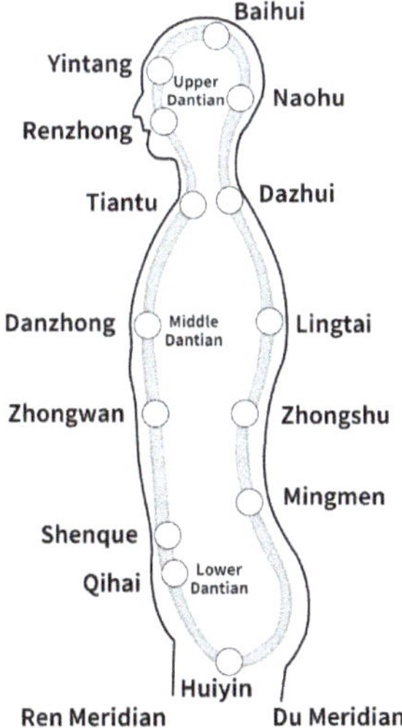

Figure 3.4. The 'Xiao Zhoutian' cycle of the human body.

meridians is not consistent with the distribution of human channels such as nerves, blood vessels, and lymphatics that have been revealed by anatomy, that is, meridians are neither nerves, blood vessels, nor lymphatic vessels.

In the 1970s, electro-physiological research on meridians found that the resistance of the meridian route is usually lower than that of the surroundings, which seems to provide a certain objective basis for the existence of meridians (Zhu Zongxiang 1989). But low resistance may only be a feature of the meridian, not the meridian itself.

In 2015, Dr. Ding Zhiwei from Weifang City People's Hospital in Shandong Province, after systematically studying the distribution of vascular skin perforators in the human body, was surprised to find that the distribution of acu-points in TCM highly coincides with the position of vascular skin perforators. This discovery has been confirmed with the help of B-ultrasound and other modern perspective imaging techniques (Meng et al. 2023, Ding Zhi Wei et al. 2016). Professor Wang Zengtao, an expert in microsurgery at the Provincial Hospital Affiliated to Shandong Medical University, further found that most of the places on the human body where acupuncture has an obvious sense of qi are the places where nerves and blood vessels enter muscles or bones, also known as muscle gates or bone gates (Wang Zengtao 2023). These research progresses seem to suggest that although meridians and acupoints cannot be equated with anatomical organs such as blood vessels, nerves, and lymphatic vessels, they also have an anatomical basis. The revelation of the anatomical characteristics of acupuncture points has undoubtedly laid the foundation for its objective and precise positioning.

In 2020, the research team of Fudan University published a research paper titled 'Infrared imageries of human body activated by teas indicate the existence of meridian system', which for the first time published a systematic image that conforms to the traditional description of human body meridians, allowing people to 'see' meridians (Jin et al. 2020).

Although these studies are not enough to reveal the mysteries of meridians and acupoints as a whole, the revelation of the anatomy and physical characteristics of meridians and acupoints undoubtedly provides strong evidence for their objective existence and the scientific nature of the treatment methods based on them. In fact, in seemingly unrelated fields, scientists began to explore and study the phenomenon of human energy fields as early as the beginning of the last century.

3.2.2 Research on human body aura and energy field

The existence of the bio-energy field has been confirmed by modern science decades ago. Any energy body with a temperature higher than absolute zero, including animals and plants, will radiate energy outward and 'glow', just like the sun emits light. The light emitted by the energy body includes visible light, infrared rays, ultraviolet rays, X-rays, and gamma rays. Usually, the human eye can only see 2% of light, which is the visible light part of the electromagnetic spectrum, and light in other spectral ranges is invisible to humans.

In 1911, when the British medical practitioner Wald·Kilner (Kilner, Walter John 1847–1920) painted a glass screen with anthocyanidin dye, he accidentally found a halo about 15 mm thick around the human body. This halo is colorful and illusory, sometimes bright and sometimes dark, giving people a sense of mystery. This was the first time that a human glow had been discovered.

The research of modern bio-photonics shows that the human body can spontaneously emit electrons and photons, producing a glow invisible to the naked eye. Scientists regard the human glow as a manifestation of human energy, which is usually difficult to measure. However, when the human body is in an electromagnetic field, the emission of such electro-photons will be excited and can be photographed. This is the principle of Kirlian photography. In 1939, S. V. Kirlian (Semyon Davidovich Kirlian 1898–1978), an engineer from the former Soviet Union, placed the subject to be photographed in a high-frequency electric field generated by a high-frequency high-voltage generator, and found that a living human body emits colorful halos and light spots at a certain rhythm. When a person dies, after a period of time, the halos and spots will disappear. They found that different parts of the human body have different glows, for instance, arms are blue, buttocks are green, and the heart is dark blue. The brightness of each part of the body is different: the hands and feet are brighter, and the arms, legs and torso are darker. Kirlian's discovery caused a sensation in the world. Around 1976, Kirlian invented a camera based on electromagnetic technology. The human body glow captured by this camera clearly shows a flame-like halo around the human head, and each person's halo has its own uniqueness. This halo is similar to the halo we see behind the heads of statues of Buddha, Taoism and God in temples around the world today. It's just that the halo on the statues is usually golden, and brighter and bigger than that of ordinary people. Just imagine, if the ancestors of human beings had not seen it, how could they draw it so consistently? After years of tracking and research, scientists have discovered that the color and intensity of a human body's glow are the inner quality of a person, that is, a comprehensive manifestation of morality, cultivation, health, mood, luck and other aspects. When a person's internal quality changes, the glow will change, and the life track will change accordingly.

Research by scientists has found that the color and shape of the human body's glow will change according to the body's location, age, health status, physiological and psychological conditions, etc. Generally, the halo of the human head is light blue, but the arms are cyan, and the halo of the hands and feet is brighter than the halo of the arms, legs and torso. The glow of young adults is brighter than that of the elderly and babies, those who are healthy have a brighter glow than those who are weak and sick, and most athletes have brighter halos than ordinary people. When a person is sick, the glow will become faint and dim, and the gray color indicates a serious disease condition. Therefore, according to the glow information emitted by the human body, not only can a person's health condition can be judged, but diseases can be diagnosed too. The human body glow also changes with brain activities such as emotions: it is light blue when calm, orange yellow when angry, orange red when afraid, and with alternate colors flashing when the person is lying (as shown in Figure 3.5). The study also found that the human body glow has a variety of cycles that change over time, including 7-day, 14-day, 32-day, 80-day and 270-day physiological rhythms as well as day, night, weekly, and monthly physiological rhythms. In a day, the lowest luminous intensity occurs at 10 am, the highest luminous intensity occurs at 4 pm, and then the luminous intensity gradually decreases. From birth to growing up, a person's glow gradually increases with age, and then gradually weakens after middle age.

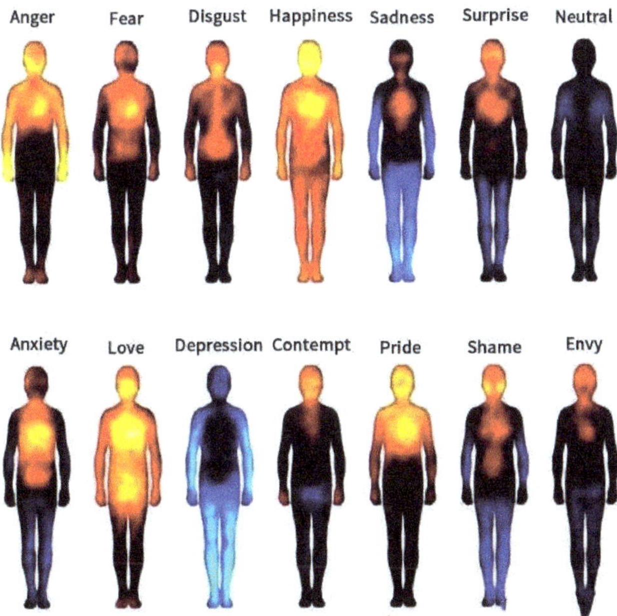

Figure 3.5. The relationship between human glow and emotion.

Scientists further discovered that the bright and shining halo in the human body's glow coincided with the acupuncture points marked on the TCM Meridian Acupoint Map. This seems to indicate that the meridians and acupoints have an alternative and objective existence different from the nerves, blood vessels, and lymph. Scientists have also conducted studies on some animals and plants and found that the glow phenomenon is not unique to humans. As long as animals and plants are alive, their bodies can emit such ultra-faint light. After the death of people, animals, and plants, this glow disappears. This biological energy field light appears to have a certain natural close relationship with the qi in ancient Chinese philosophy and traditional medicine (including Qigong). The glow emitted by ordinary people is only about 20 mm, which is not easy to be seen by human eyes under normal environment and conditions. But the glow of people who practice Qigong is very strong, and it may even appear as visible light in the dark. Qigong masters often emit light invisible to ordinary people from acupoints such as Baihui (GV20), Laogong (PC8), and Yintang (GV29), and auras and beams of light even appear on the top of their heads or around their bodies. The higher the skill, the brighter the glow. Whether in the East or the West, there are reports that many functional people have the ability to 'see the light' and predict illness, disaster, and death through the glow of the human body. Many ordinary people have achieved the ability to see the light, see through or even observe the meridians through Qigong and meditation (Ke Xuan 2017).

The original Kirlian photography was a photographic technique that used high voltage to directly expose the discharge image of an object to photographic paper. But because the effect is unstable, it was not accepted by the scientific community. In 1995, combining the most advanced optics, digital TV matrix and computer technology available at the time, the Russian quantum physicist Korotkov's (Konstantin Korotkov 1952) team devised digital Kirlian photography, using gas discharge imaging (GDV), to realize the stable imaging of human energy field light.

In the electronic image of GDV, the energy field of a healthy or calm person is very strong, and the periphery of the field is very round; for an emotional person, spark-like spikes will appear around the energy field; and when a person has health problems or is sick, there will be abnormalities such as holes and gaps in his energy field. Abnormal energy fields in different parts and shapes correspond to different organ systems lesions, so they can reflect the source of the lesion. Figure 3.6 is a comparison image of the human energy field under normal and negative emotions. Korotkov believed that by observing the energy field of the human body, physical problems and even potential problems can be found as early as possible, and treated as early as possible to maintain the health of the body (Li Daozhen 2020).

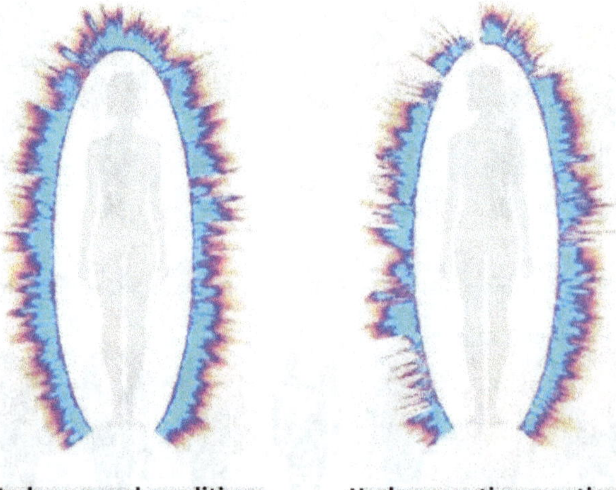

Figure 3.6. Human energy field.

3.2.3 Holography of living organisms

Acupuncture therapy in TCM is not limited to body acupuncture based on meridian points, but also auricular acupuncture based on ear acupoints to treat systemic diseases. The regional distribution of ear acupuncture points marked on the ear acupoint map clearly shows the shape of a baby lying upside down, as shown in Figure 3.7. The same is true for the corresponding scalp needles, nose needles, face needles, hand needles, and foot massage (as shown in Figure 3.8), as well as the distributions of their corresponding treatment areas. The effectiveness of these local acupuncture therapies for systemic diseases suggests that there is obviously a certain form of deterministic connection between the relatively independent parts of the human body such as ears, nose, hands, and feet, and the whole body (Guo Changqing 1998).

In TCM, the diagnosis of diseases through pulse (as shown in Figure 3.9), tongue, and face examination is a typical example of judging the state of the systemic disease based on the clues of a certain local part of the patient. In fact, as far back as the 5th to 4th centuries BC, Hippocrates (ποκράτης, 460 BC–370 BC), the originator of modern medicine, had a brilliant exposition of this phenomenon: "What exists in the largest part of the body also exists in the smallest part. This smallest part itself has all parts, and these parts are interrelated and can transmit all changes to other parts" (Baidu Encyclopedia 2022).

In 1957, Dr. Paul Nogier (1908–1996), a French neurologist and physician, published his paper 'Über die Akupunktur der Ohrmuschel'. The distribution map of acupuncture points in the shape of an inverted fetus depicted in this paper reveals the correspondence between specific areas of the ear and specific organs/parts of the body (Nogier 1957). Since then, auricular acupuncture therapy has been promoted in Germany and spread all over the world.

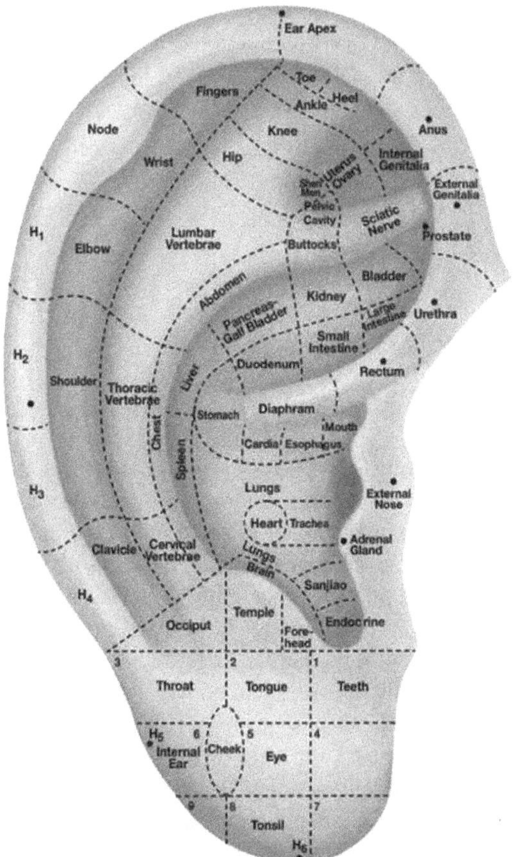

Figure 3.7. The body structure corresponding to the ear points.

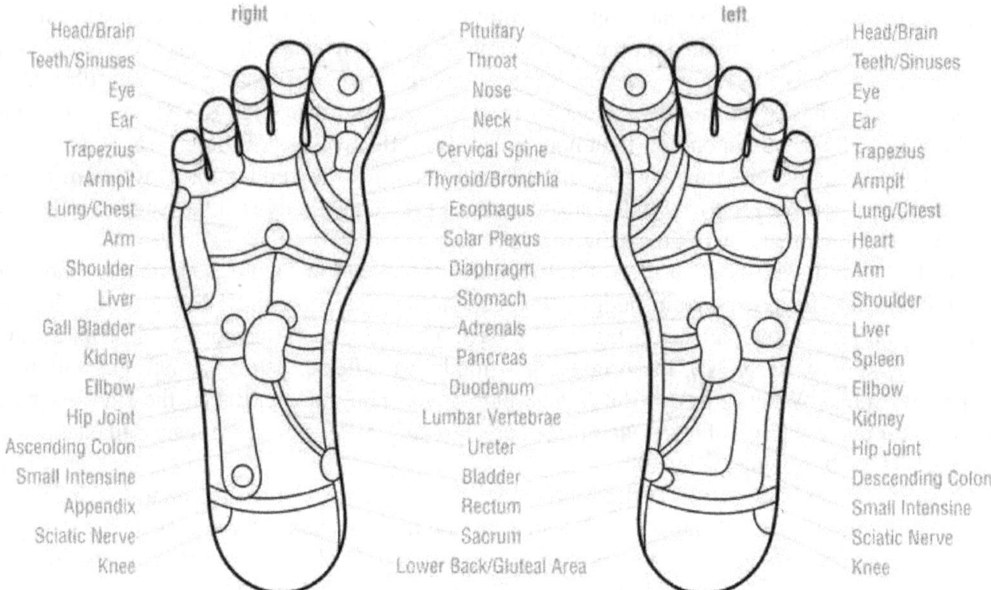

Figure 3.8. The body structure corresponding to the plantar reflex zone.

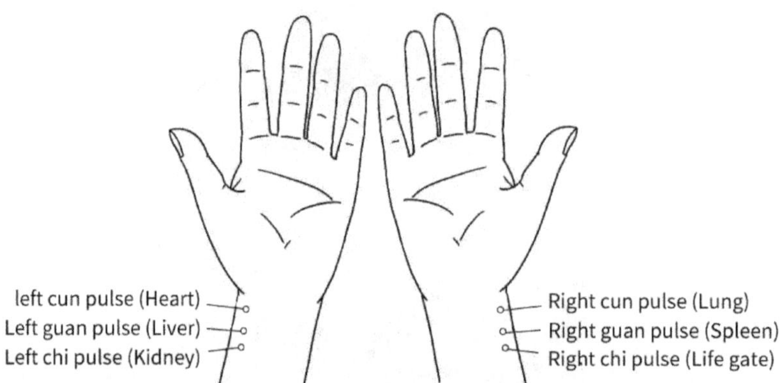

Figure 3.9. The distribution of viscera corresponding to pulse.

In 1981, Mr. Zhang Yingqing (1947–2004) published his famous work 'The Law of Bioholography' (Zhang Yingqing 1980). This theory believes that each part of each organism that has life functions and is relatively independent (also called holographic element) contains all the information of the whole. The holographic element can be said to be a microcosm of the whole to a certain extent. For example, the human upper limb humerus (upper arm bone), forearm bone, five metacarpal bones and lower limb femur, calf bone, ears, nose, palm, etc., are all holographic elements, which are all a microcosm of the human body.

This phenomenon exists not only in the human body, but also in other animals and even plants. The holographic phenomenon of plants has been proved in nature from many aspects such as morphology, biochemistry and genetics. Palm trees on the side of the road, for example, have leaves composed of thin fan-like leaves and long petioles. When you carefully observe the shape of the leaf and compare it with the shape of the whole plant on the ground, you will find how consistent their shape is, only the size ratio is different. It can also be seen that the shape of a pear is consistent with the outline of the entire pear tree. There are also obvious holographic phenomena in the biochemical composition of plants. For example, the distribution of cyanic acid on a leaf of sorghum is consistent with its distribution in the entire plant. In the whole plant, the upper leaf contains more cyanide, and the lower leaf contains less cyanide; on a single leaf, the upper part of the leaf contains more cyanide than the lower part.

The fractal theory created by Mandelbrot (Benoit Mandelbrot 1924–2010) extends the similarity between local and global forms to the entire natural world. When there are no buildings or other things as reference objects, a photo of a 100-kilometer-long coastline taken from the air and an enlarged 10-kilometer-long coastline look very similar (Mandelbrot 1967). Forms with self-similarity exist widely in nature, such as rolling mountains, floating clouds, rock fractures, Brownian motion of particles, tree crowns, cauliflower, cerebral cortex, etc. (Falconer 2003).

Obviously, from the revelation of the entity structure of the human body in modern medicine, we can't see any link between the part and the whole that is similar to the biological holographic correlation. However, the accuracy of diagnosis and the effectiveness of treatment based on this association in TCM have irrefutably confirmed the reality of the existence of these associations.

3.2.4 The scientific proof of the unity of body, Qi, and spirit and the immortality of the soul

In ancient Chinese philosophy, qi is the most basic element that composes all things in the world. For its greatness, nothing lies outside it; for its smallness, nothing lies within it. It is everywhere and everything. Qi gathers to take entity, which is the energy and power that composes the human body and all things and promotes life activities. Everything in the universe and its development and changes are the result of the movement and changes of qi. In TCM, qi is the basic substance that constitutes the human body and maintains the life activities of the human body. Qi is stored in the blood, can run outside the blood vessels, and can be gathered in acupoints. It maintains the functions of internal organs, nurtures the skin, muscles and bones, and resists external pathogenic factors. When the qi is insufficient, people will feel fatigued, have shortness of breath and weakness in speech and voice. After taking Chinese herbal medicine (CHM) with a qi-tonifying effect, their fatigue is usually improved, and the power in breathing and speaking is enhanced. Obviously, in reality, we can feel the presence of qi, but from the perspective of modern medicine, it is neither visible nor touchable.

In TCM, there is another important concept, spirit. In the category of ancient philosophy, spirit refers to the most fundamental power that governs the development and change of all things in the universe. The spirit in TCM is the master of human life activities and the most fundamental manifestation of life activities. It has an important commanding role in human life activities. The spirit is obviously related to people's spiritual consciousness activities, and seems to be related to the soul in Chinese Taoism and western Christianity. In TCM, there is a saying that 'the combination of the essence of the two sexes is called the spirit'. It is believed that it is produced by the combination of the reproductive essence of men and women, and its separation from the body means death.

The existence of the soul and its relationship with life have been confirmed in recent years from the records of many near-death experiences by scientists. According to a report in the British 'Independent' on October 7, 2014, a medical study conducted by British researchers showed that after a person died (the heart stopped beating), life did not actually stop, and the 'soul' was still active. In the four years before 2014, researchers worked to analyze the experience of patients with cardiac arrest. It was found that about 40% of survivors described that they had some form of 'consciousness' when they were declared clinically dead. The study also obtained convincing evidence that the patient still experienced a real event for two to three minutes after the heart stopped beating, and could clearly recall what happened after regaining consciousness. The study involved 2,060 patients from 15 hospitals in the United Kingdom, the United States and Australia. The results of the study were published in the medical journal Resuscitation (Withnall 2014).

A lot of evidence for the law of immortality of consciousness is now available. In 1990, Gallup conducted a questionnaire survey, and the result was that three-quarters of the Americans and half the Europeans surveyed believed that there were some life phenomena after death. Ian Stevenson (1918–2007), a world-recognized authority on the study of reincarnation, and a professor of psychiatry at the University of Virginia in the United States, spent 40 years investigating and collecting more than 3,000 cases of evidence about reincarnation from the perspective of a scientist. Stevenson's research institute also studies special mental phenomena, such as out-of-body experience (OBE), ghost phenomena (GS), extra-sensory perception (ESP), and near-death experiences (NDE). The purpose of the research is to use scientific methods to prove that consciousness can exist independently after being separated from the body. Professor Stevenson has written 14 monographs on reincarnation and published more than 300 papers, and his papers have been published in the journals of the American Medical Association (JAMA). The editor-in-chief of JAMA believes that Stevenson's research is very detailed, the research attitude is very objective, the methods used are impeccable, and the cases cited are difficult to attribute to other reasons besides reincarnation. Stevenson's collection of detailed cases in line with scientific norms fundamentally shakes the perception that 'life ends in death' that modern medicine has always adhered to, and provides convincing scientific proof for 'the existence of consciousness away from the flesh and the continuity of consciousness' (Wikipedia contributors 2021).

Besides Stephenson, a large number of scholars in the West are engaged in the study of reincarnation. For example, Dr. Brian Weiss (1944), chairman of the Mount Sinai Medical Center in Miami, USA, and a famous psychiatrist, who received his MD degree from the Yale University School of Medicine in 1970, and taught at the University of Pittsburg and the University of Miami, and has been engaged in clinical psychology medicine for more than 30 years. He published a large number of papers and works, and is an authority on modern spiritual psychology, especially reincarnation. In addition, there is Dr. Rick Brown, vice-chairman of the International Society of Psychological Regression Therapy and a famous American psychiatrist. Alastair (Bruce) Scott-Hill, a professional engineer dedicated to science, won the 2016 Ashton-Wylie Literary Prize in New Zealand for his book 'The Paranormal Is Normal: The Scientific Validation of Reincarnation, the Paranormal, and Your Immortality'. He followed rigorous empirical criteria that conform to scientific norms, and included only replicable scientific experiments, which serve as the basis for his investigations into reincarnation, the paranormal and the possibility of an afterlife and immortality for all. The book was a summary of the author's personal journey of exploration that began about 15 years ago and was not completed until recent research had a scientific breakthrough. The book provides welcome scientific support for a public that believes in a Creator, an afterlife, psychics, and the paranormal (Alastair 2013).

In TCM, the spirit is nourished by essence, qi, blood, body fluid and other substances, and can in turn act on these substances. The spirit has the role of commanding and regulating the normal metabolism of these substances in the body. The fullness of essence, qi, blood, body fluid and the exertion of normal nourishment, the metabolic balance of material transformation and energy transformation, the execution and coordination of visceral functions, the generation and regulation of emotional activities, and the maintenance of a peaceful and happy mental state, as well as anti-aging and longevity, all are inseparable from the command and adjustment of spirit. The spirit is the fundamental symbol of the existence of life, and if the biological body is separated from the spirit, it will die. The earliest TCM medical book 'Yellow Emperor's Internal Classic' states that those "who gain spirit will prosper, who lose spirit will die". This explains that the existence of spirit is of great significance for judging the rise and fall of positive qi, and the severity of illness and the prognosis. When a TCM practitioner examines a patient, if the patient has bright and attentive eyes, is alert and quick in thinking, and the speech is fluent, they are deemed as 'have spirit' or 'got spirit', which means that their positive qi is not harmed, the visceral function is not attenuated, and the prognosis is better, even if the condition is currently severe. If the patient's eyes are dull and the expression is sluggish, they are lethargic, unresponsive, weak in breathing, or even in a coma, it is called 'absence of spirit'. This means that the positive qi has been injured, the condition is serious, and the prognosis is poor. Obviously, spirit, as an invisible existence, has traces to follow in human life activities.

Today, with the progress of theoretical physics research, mankind's understanding of the origin of the world is getting closer and closer to the ideas of Eastern Taoism, Buddhism and ancient philosophy. The theory of relativity is revealed the unity of energy and matter: Matter is energy, energy is matter, and matter and energy can be converted into each other (Einstein and Infeld 1938). If we connect the concepts of 'qi' and 'entity' in ancient China with the 'energy' and 'matter' in Einstein's theory of relativity, we will find that the meaning of this sentence is surprisingly consistent with the ancient Chinese belief that 'qi gathers to form matter'. The phenomenon of 'quantum entanglement' discovered in quantum mechanics research shows the correlation of microscopic particles across time and space, and also provides a factual basis for the integrity of the universe (as shown in Figure 3.10). From the 'boot loop hypothesis' of quantum mechanics in the early twentieth century to the 'universe self-sufficient solution' of contemporary theoretical physics master Stephen Hawking, theoretical physicists have generally abandoned the idea that the universe is composed of a set of basic entities with basic properties, and replaced it with a new 'self-consistent' cosmology (Hawking and Mlodinow 2005). Based on this concept of the universe, every arbitrary part of the universe in size is composed of the whole of all other parts, and every arbitrarily tiny part includes the whole of the universe. In this sense, not only are living beings holographic, but the entire universe is also holographic; not only is the human body a closely-related whole, but the entire universe is also a closely-related whole of everything.

In ancient Eastern philosophy, matter and energy are a unity that can be transformed into each other, and they are all conscious, whether as large as the universe or as small as any tiny part of the universe. Consciousness, matter, and energy can all be transformed into each other, and they are all mutually inclusive along with the universe. Matter, energy, consciousness, macroscopic, microscopic, past, present, and future are all integrated into a whole and are indivisible (Xiong Shili 1985). It is precisely based on this kind of recognition that, unlike the outward-looking cognitive model based on observation and reasoning in Western science, there has always been an introspective cognitive model that seeks inward, or toward 'my heart' in the ancient East.

The theoretical physics research on the origin of the world is now approaching the last stronghold that has been dominated by Eastern philosophy in this field: the essence of consciousness, the relationship between consciousness and matter and energy. More and more physicists have gradually realized that microscopic particles and even all objective entities may have consciousness. Consciousness and energy, as well as consciousness, energy and matter, are mutually transformable, that is, the universe actually embodies the characteristics of matter, energy and consciousness at the

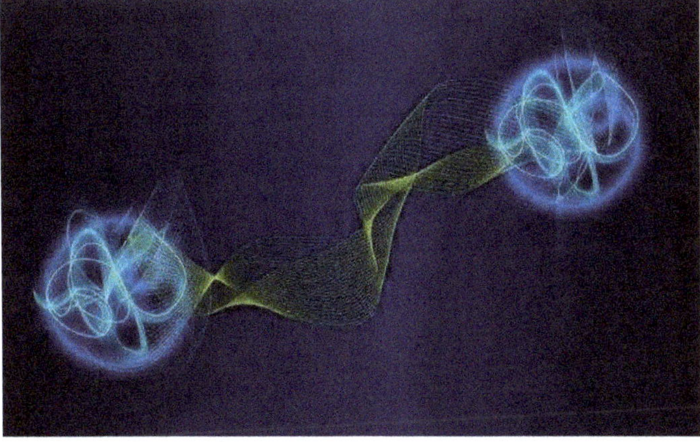

Figure 3.10. Diagram of quantum entanglement.

same time. The universe as a whole and any tiny parts of it are mutually contained, and the universe itself is completely self-sufficient.

As the most advanced form of life evolution on this planet, human beings are the most advanced form of the unity of matter, energy, and consciousness, which is the unity of 'material, spiritual and qi' from the perspective of ancient Chinese philosophy. The revelation of the mysteries of the human body by modern biology and medicine is still basically limited to the structure and relationship of the low-level 'entity'. The intangible structure and relationship beyond the tangible structure of the human body have yet to be revealed, and understanding the relationship similar to 'quantum entanglement' between the inside of the human body and between the human body and the natural environment is even further away. Obviously, current biology and medicine lag far behind the development of natural sciences, especially theoretical physics. And more and more scientists in the field of life sciences have realized that what we are facing is a much more complicated human body than what modern medicine and biology have revealed.

3.3 Negative feedback regulation and the maintenance of homeostasis in human body

Human beings live in nature, and the environment they live in is extremely complex and fickle. Physiological studies tell us that the normal human body temperature always fluctuates around 37°C, and the daily fluctuation range does not exceed 1°C; the plasma pH value can fluctuate between 7.35 and 7.45; the fluctuation range of various ion concentrations in plasma is also very small, such as blood potassium concentration is only between 3.5–5.5 mmol/L, and blood calcium concentration only fluctuates in a narrow range between 2.25–2.75 mmol/L. So how does the human body maintain the stability of the internal environment?

In 1926, American physiologist Cannon (Claude Elwood Shannon 1916–2001) put forward the concept of the steady state: it is not constant, but a dynamic balance. Any changes in the internal environment will cause the body to automatically adjust the activities of its tissues and organs, so that the changes will eventually be limited to a narrow range. At present, it is generally believed that the stability of the human body's environment is maintained by the interconnection, division and cooperation of the three major systems of nerve, endocrine, and immunity.

With the development of science, the concept of homeostasis has become a universal basic concept in biological sciences. Scientists have discovered that there are widespread steady-state phenomena at all levels of life activities. For example, at the molecular level, there is a steady state of gene expression; at the organ level, there is a steady state of blood pressure, heart rate, and cardiac activity; at the macro level, there is a steady state phenomenon in the growth and decline of biological populations, and the largest ecological system, the biosphere, also has steady states.

Ever since Norbert Wiener (1894–1964) published his famous book 'Cybernetics, The Science of Control and Communication in Animals and Machines' in 1948 (Wiener 1948), cybernetic thoughts and methods have penetrated into almost all fields of natural sciences and social sciences. Wiener regards cybernetics as a science that studies the general laws of control and communication in machines, life, and society.

The study of control systems in cybernetics ignores the physical, chemical, and biological characteristics of various systems, and conducts research purely from the basic elements, basic structure and control mechanisms of various control systems. Research on the human body using modern cybernetic methods has shown that there are thousands of control systems in the human body at the molecular, cell, tissue, organ, and system levels. Even in a cell, there are many sophisticated control systems that precisely regulate/control various functional activities of the cell.

In the framework of modern life sciences, homeostasis regulation at the cellular and molecular levels usually belongs to the research areas of disciplines such as cell biology, molecular biology, and biochemistry; while homeostasis regulation at the organ, system, and holistic level belongs to the research category of human physiology. Negative feedback regulation can be seen everywhere in the life activities of all levels and parts of the human body. Almost all the steady-state maintenance towards goals in the life process is based on negative feedback regulation. Physiologically, the regulation of human arterial blood pressure is an example (Zhu Danian 1979). When arterial blood pressure rises for some reason, the afferent impulse of blood pressure sensitive receptors increases. After the information is fed back to the nerve center by the afferent nerves, the cardiovascular center will issue corresponding control commands to reflexively inhibit the activity of the heart and blood vessels, weaken the activity of the heart and dilate the blood vessels, and consequently limit the rise in blood pressure. On the contrary, when the arterial blood pressure drops, it will in turn weaken the stimulation of the receptors, so that the afferent impulse is relatively reduced, and the cardiovascular center will enhance the activity of the heart and blood vessels through reflexes to increase the blood pressure. Due to the existence of this negative feedback adjustment mechanism, blood pressure is maintained at a relatively stable level, as shown in Figure 3.11.

In the human body, the hypothalamus acts as the regulatory center of the endocrine system. It acts on other endocrine glands to induce them to secrete hormones by causing the pituitary gland to secrete hormones. If there are too many hormones secreted by various endocrine glands, after the hormones enter the blood, they will in turn inhibit the functions of the hypothalamus and pituitary gland, thereby maintaining a relatively stable level of secretion of various hormones in the human body. For example, excess corticosteroids inhibit the secretion of corticotrophin-releasing hormone in the hypothalamus and corticotrophin in the anterior pituitary (as shown in Figure 3.12), too much sex hormone will inhibit the secretion of pituitary gonadotropin, and the rise and fall of blood sugar will feedback regulate the secretion of insulin and glucagon and so on. These are all done automatically by the body under the action of the negative feedback mechanism. Figure 3.13 shows the negative feedback regulation mechanism related to blood glucose regulation (Zhu Danian 1979).

It is precisely because of the extensive existence of negative feedback regulation mechanisms at all levels of the human body that humans living in complex and changeable natural and social environments can effectively deal with the interference of environmental changes on the human body, so that the relatively stability of internal environment and various vital indicators in human body can be maintained, and diseases are not prone to occur.

The stability of this balance is manifested in that if any external interference makes a certain (or some) state variable of the system deviate a little, then the action of other state variables on this variable will usually make it return to the normal state. This is why though we experience a lot of external disturbances every day, diseases do not occur frequently. However, the ability of the human body to maintain this steady—state balance is limited, in other words, the stability of a system is always relative to a certain range of interference. If the interference acting on the system is too strong or continuous, and exceeds the ability of the system to maintain a steady state balance, that is, the action of other variables on this variable is not enough to make it return to the normal state, then in turn, its action on other variables will make state deviations of other variables in different degrees. As a result, the state of the system will deviate from the original steady-state balance, which usually means disease.

The maintenance of the steady-state structure of the system usually depends on the state variables related to the specific steady-state structure. The state of these state variables is different, the equilibrium point of steady-state equilibrium will be different. As a result, there are qualitative and quantitative differences in the clinical manifestations of different patients in the same pathological homeostasis, while the physiological homeostasis of normal humans under certain conditions shows certain pathological characteristics in patients. The normal value range of various indicators detected by modern medical methods is usually the stable balance point (interval) of these indicators when the human body is in a normal physiological state. For example, the normal blood glucose value refers to the blood glucose value when a person is fasting, usually between 3.9 and 6.1 millimoles/L range. That is to say, although the body's blood sugar will change with diet, exercise and other life activities throughout the day, its fasting blood sugar will always remain within a certain range. The changes in these indicators when a disease occurs usually means that the patient's steady-state equilibrium

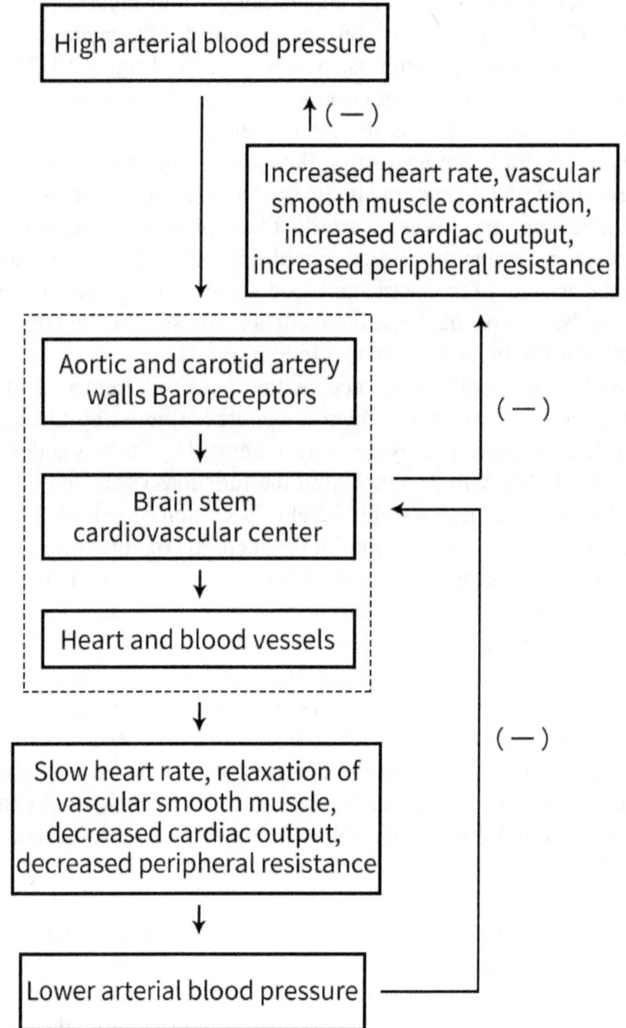

Figure 3.11. Negative feedback regulation of arterial blood pressure.

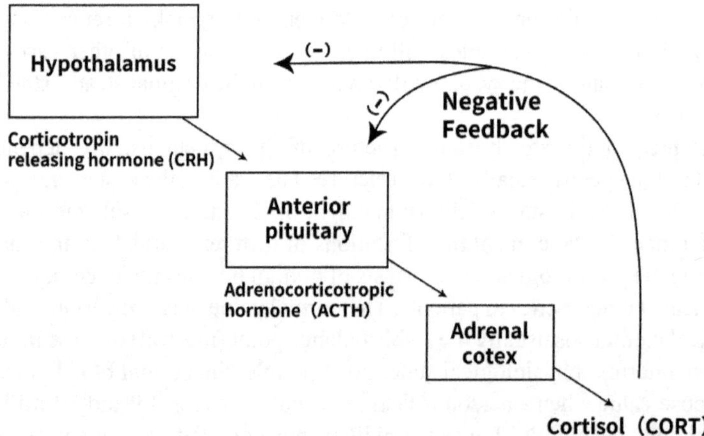

Figure 3.12. Negative feedback regulation of adrenal cortex hormones.

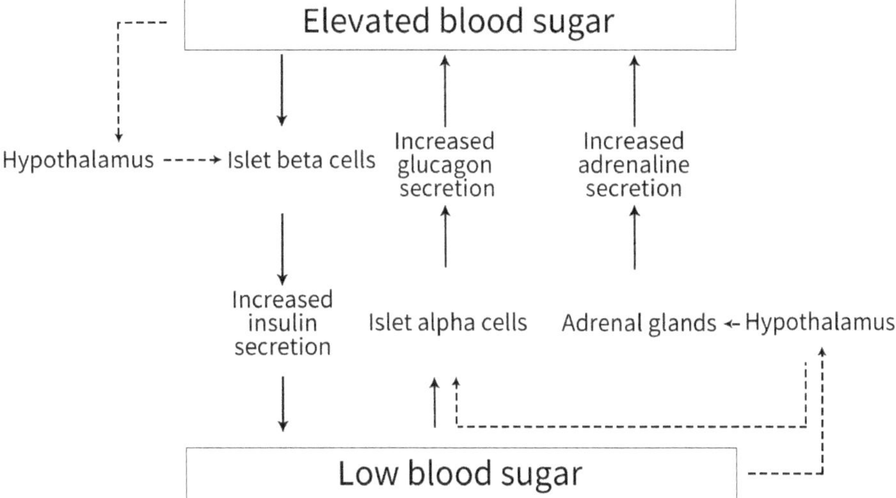

Figure 3.13. Negative feedback mechanism of blood sugar regulation.

point has changed. For example, when the fasting blood glucose exceeds 7.0 mmol/L, modern medicine can diagnose diabetes based on this indicator. At this time, the patient's blood sugar will still fluctuate within a certain range with daily diet, exercise and other life activities, but his fasting blood sugar will stabilize at a certain interval exceeding 7.0 mmol/L. Obviously, diseases in many cases mean changes in the balance point of certain physiological homeostasis.

The system always tends to a steady state structure automatically. In other words, as long as the system state has not reached a certain stable structure, it will be in constant motion and change. But the steady-state structure is not always good, and diseases that are difficult to cure are usually the steady-state structures that people try to avoid entering or try to escape from. It is precisely the purpose of treatment and rehabilitation to take appropriate intervention or treatment measures to promote the state of the body to break away from the pathological homeostasis, move to the direction of normal homeostasis or enter a stable state closer to the normal homeostasis.

3.4 The complex human body capable of self-organization, self-adaptation and self-regulation

Biological research tells us that life on earth has evolved from the most primitive cell-free organisms to cell-structured prokaryotes, from prokaryotes to eukaryotic single-celled organisms, and then developed in different directions to derive the fungi kingdom, plant kingdom and animal kingdom. The plant kingdom went from algae to naked ferns to ferns, gymnosperms, and finally angiosperms. The animal kingdom has evolved from primitive flagellates to multicellular animals, from primitive multicellular animals to chordates, and then evolved to higher chordates, vertebrates. The fish of vertebrates evolved to amphibians and then to reptiles, from which mammals and birds were differentiated. One of the mammals further developed into a higher intelligent creature, the human.

The development history of the biological world shows that biological evolution is a process from aquatic to terrestrial, from simple to complex, and from inferior to advanced, showing a trend of progressive development. Generally speaking, the progress of the evolutionary process has the following characteristics:

1. The biological morphological structures at different levels are gradually becoming more complex and perfect; correspondingly, the physiological functions are becoming more and more specialized, and their effectiveness is gradually improving.

2. On the whole, the amount of genetic information gradually increases with the evolution of organisms.
3. The continuous improvement of the regulation-control of internal environment and the development of response ways to environmental changes strengthen the body's autonomy in the external environment and improve its ability to adapt to environmental changes.

In 1859, C.R. Darwin (Charles R. Darwin 1809–1882) published the book 'Origin of Species', proposing the natural selection theory of 'survival of the fittest', and establishing the scientific theory of biological evolution.

In the long process of evolution, both new species are produced, and the original species get extinct. Various organisms have formed morphological structures and living habits adapted to their respective living environments during the evolution process. Similarly, their offspring inherited certain characteristics of their parents to adapt to the natural environment through inheritance, and passed on these characteristics to their offspring. Over several generations, the species gradually changed, taking on traits that were more adaptable to the environment, leading to the emergence of a new species.

Evolution is the process of the formation of new biological species, with characteristics that have accumulated through many small changes over a long period of time. Today, however, with the deepening of biology's understanding of living organisms at all levels, as the structure and function of various parts and aspects of living organisms are revealed, biologists with the evolutionary concept of 'natural selection' feel more and more confused: whether at the level of organs and systems or at the level of cells and molecules, the structure and function coordination of the various parts of the living body is so ingenious, just like a masterpiece of uncanny craftsmanship. Can't they survive if they were not built this way? For species that have become extinct in the past or are now on the verge of extinction due to industrial development and environmental destruction, are their structures and functions necessarily 'lagging behind' and 'inferior' to those that survived? If not, then besides the natural selection of 'natural selection', is there also an internal mechanism in the organism that promotes the individual organism to gradually change its structure and function in order to adapt to the environment?

As we mentioned earlier, the negative feedback control mechanism widely existing in the human body and even various organisms gives the whole and its relatively independent parts the ability to maintain a stable state in the constantly-changing external environment. However, the ability of the system as a whole, and each of its parts, to maintain a stable state is limited. When the interference caused by changes in the external environment is too strong, and the deviation of the system state from the stable state is large enough to reach or exceed a critical point, the original stable state is broken, and the system enters an unstable period.

A system leaving the previous steady-state balance, under the interaction of various elements and the constraints of the environment, will have to constantly adjust its structure and functions in a direction that can better adapt to the external environment. Eventually, a new homeostatic equilibrium is established with the environment. Under the action of the positive feedback mechanism, it gets farther and farther away from the previous steady-state balance, eventually leading to the disintegration of the system. Of course, the transition from the previous stable state to the new stable state can also be a gradual process. Under the constraints of the gradually changing external environment, the system continuously adjusts its structure and functions to adapt. In this process, the system slowly but continuously changes the steady-state equilibrium point. A system that has established a new steady-state equilibrium under environmental constraints will undoubtedly be better able to adapt to changes in the environment than the original system.

The 'evolution' of the living system is the process of constantly breaking away from the previous steady state and establishing a new steady state in the process of adapting to environmental changes. In the process of passively and actively adapting to environmental changes, the internal structure and function of the living system undergoes subtle changes due to the interaction between its parts and the constraints of the environment. These changes will simultaneously change the genes of living organisms and pass them on to future generations through genes, thereby promoting the slow evolution of biological species. In this sense, natural selection is only a screening mechanism for surviving species on earth, and is not a direct cause of the evolution of organisms from low to high.

In general system theory, a system that can form a new structure and function through its own evolution under the conditions of interaction with the environment is called a self-organizing system. In the real world, self-organizing systems are everywhere.

Our universe is a self-organizing system, and astronomical research shows that it originated from a big bang 15 billion years ago. At the beginning of the Big Bang, the temperature of the universe was very high, and its density was very high. With its rapid expansion, it gradually cooled down. After 300,000 years, the universe cooled to a certain temperature, and it had the conditions for the formation of atoms. After another one billion years, huge gas nebulae began to gather and gradually formed the relatively orderly structure composed of galaxies, nebulae, black holes, etc., that we can observe today.

The solar system where our planet is located is also a self-organizing system. It was a huge nebula made of gas and dust in the Milky Way. About five billion years ago, under the influence of its own gravity, it gradually gathered and organized into an orderly structure centered around the sun, with nine planets including the earth and their satellites constantly orbiting around it.

The biosphere composed of all living things on the earth and their living environment is also a self-organizing system. Over the past 3.5 billion years, from the emergence of the first organics, biological macromolecules and single-celled organisms on earth, after a long evolutionary process, it has been organized into an interconnected, interacting, and indivisible ecosystem that has multiple categories of plants and animals, and countless species.

Every living organism, including a human individual, is a self-organizing system. The process of embryonic development from fertilized eggs to birth to adulthood is the process of organism self-organization under the control of genes and the exchange of energy, matter and information with the outside world, as shown in Figure 3.14. In the same way, the knowledge and thoughts in our brain are also a self-organizing system. The learning and practical experience from elementary school, middle school, university and even throughout life is a self-organizing process in which each of our thinking system constantly evolves and perfects itself through interaction with external information.

The human body is a typical self-organizing system. The formation and development of embryos, the growth from infants to adults, and the adaptation of the human body in response to the changes of external environment are all self-organizing processes in different forms. The human body can use materials and energy ingested from the outside to form its own organism with complex functions, and, to a certain extent, can automatically repair damages and troubleshoot to restore normal structure and function. The human body can adjust its own functional state in response to changes in the external environment to maintain the stability of the internal environment. The human body can adjust its structure according to external environmental changes and stimuli, and change or strengthen certain functions to adapt to external environment.

The self-adjusting characteristic of the human body is its ability to continuously adjust the functional state of its various parts to changes in the external environment, and to maintain the steady-state balance of the system. As mentioned earlier, the basis of this function is the negative feedback adjustment mechanism established by the human body in the long-term adaptation process to respond to environmental changes. Body temperature regulation, blood sugar regulation, and the regulation of various hormones and electrolyte levels all reflect this self-regulating nature of the human body.

The self-adaptive process of the human body is a process in which it responds to the changes in the environment, adjusts the functions and structures of the corresponding internal organizations, and establishes a new steady-state balance. Generally speaking, changes in the environment and external stimuli start the body's self-regulation function first. The continuous stimulation of

Figure 3.14. Human growth process.

environment keeps the body in a new state that is different from the previous steady-state balance for a long time. Based on the principle of the law of 'use and disuse', relevant structures of the body and its functions make corresponding changes, making this state a new steady-state balance point. As a result, the human body has adapted to changes in the environment. The adaptation of people living in the plains for a long time to the plateau environment, the tolerance of people from a certain region to the food that is habitually eaten in that place, and human resistance to various drugs are all reflections of this adaptation.

Of course, there is another situation where the body has not yet established a strong enough negative feedback mechanism to maintain a steady-state balance for some new environmental factors. In this case, this kind of external interference can easily damage the structure and function of certain parts of the body, pull the body out of its normal steady-state balance, and cause disease. In the course of disease, the self-organization ability formed by the interaction of various parts of the human body is in constant struggle with external pathogenic factors and their damage to the human body. The outcome of the disease depends on the power balance between external pathogenic factors (evil) and the body's ability to repair damage and restore the previous steady-state balance (positive energy). If the 'evil' is strong and the positive qi is weaker, the body is likely to enter the 'vicious circle' of positive feedback, which will eventually lead to the disintegration of the system, that is, death. If the 'evil' and the positive qi are evenly matched and in a state of stalemate, the system may also enter a pathological steady state different from the normal steady state, which is a long-term chronic disease. If the positive qi is strong and the 'evil' is weak, the system will repair the damaged structure and function and restore the previous steady-state balance, that is, the disease is cured. The human body that has healed in the process of fighting against this pathogenic factor has established a corresponding feedback regulation mechanism in the body with the help of its self-organizing ability. Next time, when the same pathogenic factor invades, the body's negative feedback adjustment mechanism will be activated immediately to automatically maintain the steady-state balance of the system. This is the process by which the human immune system responds to various viruses and bacteria to establish corresponding immune mechanisms as revealed by modern medicine (Yuan Bing 2010).

Human beings are the most advanced form of evolution of life on earth. According to a rough estimate by scientists, the total number of cells in our body is about 40 trillion to 60 trillion, of which the number of brain cells is about 10 billion. Each cell is a complex subsystem, so the human body is a typical complex giant system. A wide range of cells are organized according to a certain structure, forming various tissues, organs and even systems. The human body is covered with many ubiquitous and meticulous networks such as nerves, blood vessels, and lymph, which closely connect all levels and parts of the human body. There are extensive and ever-present dynamic connections among cells, tissues, organs, and systems, through these networks or in the process of performing their respective functions. Such a system cannot be said to be not complex enough. The human body is constantly exchanging material, energy and information with the external environment through the five sense organs, seven orifices, skin, and other induction and contact methods that have not been recognized by science even today. In the process of adapting to the environment, by constantly adjusting and perfecting its own structure and function, the human body promotes its own evolution from low-level to high-level. The process of the continuation of individual human life is a vast and complicated network of causality formed by the intertwining of numerous causal relation chains formed by human life activities in the interaction among the various subsystems, the human body and the external environment (Yuan Bing 2010).

Obviously, the human body that biology, physiology, modern medicine and TCM face together is an out-and-out and open complex giant system, a living organism with self-organization, self-adaptation, and self-regulation capabilities.

Chapter 4

Traditional Chinese Medicine (TCM)

A Medical Tradition that only Today's Science can Understand

Acupuncture, which is gradually being recognized around the world, is just a branch of traditional Chinese medicine (TCM) treatment techniques. Today, the treatment system based on syndrome (also known as the syndrome differentiation and treatment system) with broader clinical value in TCM is not yet understood by mainstream medicine. The reason is that acupuncture is a treatment technique that relies less on TCM theories, and it is easy to graft onto the disease medicine system of modern medicine. However, the treatment system based on syndromes is based on a completely different methodology from modern medicine.

Today, natural science, which started from the study of simplicity, has begun to move towards the era of complexity science. On the basis of traditional biology based on observation and experiment, the emergence of systems biology has begun integrated research on living organisms; and the rise of precision medicine marks the beginning of the transformation of mainstream medicine from disease medicine into personalized medicine. In fact, both the cognitive approach by which complexity science establishes complex system models using metaphors and analogies, and precision medicine's medical model that formulates treatment plans based on the individualized state of the human body, have been widely used core concepts and methods of TCM, an ancient medicine with a long history. They are human cognitive models and medical traditions with a longer history than the experimental analysis methods of modern science and medicine.

It now appears that in the West hundreds of years ago, in the process of abandoning the 'unscientific' elements of traditional medicine to establish modern experimental medicine, due to the naive nature of science, discarded the more valuable things in traditional medicine. It's like accidentally throwing the baby out of the tub when pouring dirty water after the baby's bath.

4.1 Observing exterior to infer interior: A modeling method consistent with complexity science

Traditional medicine refers to various medical knowledge systems developed in different civilized societies before the rise of modern medicine. The World Health Organization (WHO) defines traditional medicine as "the sum total of the knowledge, skills, and practices based on the theories, beliefs, and experiences indigenous to different cultures, whether explicable or not, used in the maintenance of health as well as in the prevention, diagnosis, improvement or treatment of physical and mental illness" (World Health Organization 2013). There are many kinds of traditional medicine, including herbal medicine of many ethnic groups, such as Indian Ayurvedic medicine, Greek and

Arab Unani medicine, etc. The most famous of these is TCM including acupuncture and other medical methods. Traditional medicine popular in East Asian countries such as South Korea and Japan is in line with TCM or originated from a part of TCM. For example, the origin of Kampo medicine in Japan is the 'Shanghan Lun, Treatise on Cold Pathogenic Diseases' written by Zhang Zhongjing (About 150 AD–219 AD) a famous physician of TCM in the Han Dynasty.

In a broad sense, traditional Chinese medicine (TCM) is the collective name of medicine of various ethnic groups in China, including Han (Chinese) medicine, Tibetan medicine, Mongolian medicine, Uyghur medicine and other ethnic medicines. In China, the Han nationality has the largest population, the earliest written characters, and a long history and culture. Therefore, Han nationality medicine has the greatest influence in China and even in the world. What people usually call 'Traditional Chinese Medicine' refers to the medicine widely spread and applied among the Han people. It not only involves rich experience and knowledge of health, disease, health care, and disease treatment methods, but more importantly, it has formed a complete set of theoretical systems from the aspects of methodology, theory, treatment method, prescriptions, and medications. This is precisely the reason why TCM stands alone in the forest of traditional medicine in the world and has not declined for thousands of years. Even in today's highly developed modern science and technology, it still emits dazzling brilliance.

Today, when opening any textbook on TCM, the first thing that catches the eye is the introduction of the main characteristics of the theoretical system of TCM: the theoretical system of TCM has two most important characteristics, one is the holistic concept, and the other is the treatment based on syndrome (Beijing College of TCM 1974).

The holistic concept of TCM refers to the concept of the integrity of the human body and the integrity of the person and the external environment (including the natural environment and the social environment). TCM believes that the human body is an organic whole. The various parts that make up the human body are indivisible in structure, interdependent and coordinated in function, and influence each other in pathology. Therefore, in terms of treatment, a certain part of the disease cannot be treated in isolation. Prescribing medications must comprehensively consider the relationship between the various parts of the disease process, and formulate a treatment plan as a whole. On a larger scale, the human body, the natural environment, and the social environment are also an indivisible whole that influences each other. The human body adjusts its structure and functions all the time to adapt to changes in the external environment. The changes in the external environment will inevitably have an impact on the health of the human body, thus becoming the cause of many diseases and a factor that cannot be ignored in the process of formulating treatment plans.

The TCM theory on the internal structure of the human body was called the 'visceral manifestations' theory in ancient times, which means that 'the internal organs are hidden inside, and the signs of existence can be shown outside'. That is to say, although in ancient times, TCM scholars had a certain understanding of the anatomical organs located inside the human body, the theoretical model of TCM on the internal organs of the human body is more importantly based on the method of 'observing the exterior to deduce the interior' (si wai chuai nei).

4.1.1 Human body model of TCM

The 'outside' in 'observing exterior to infer interior' refers to the external manifestations of various physiological functions of the human body and the symptoms and signs that appear when diseases appear, 'inner' refers to the body structure with certain functions and their abnormal changes when the disease occurs. 'The internal changes will inevitably be reflected outside', so there is the method of 'observing exterior to infer interior' in the 'Lingshu•Lunjizenchi'. On the Disease Diagnosis Ruler'. And this method is precisely the method used by modern science to establish a system function model and identify the system's function status based on the inspection of the system's input and output.

Based on this method, TCM established a functional model of the internal structure of the human body with 'viscera-qi-blood-body Liquid-essence' as the core. In such a model, the internal organs of the human body are summarized into five central organs named Zang and six deputy organs called Fu. The five 'Zang' organs include the heart, liver, spleen, lung, and kidney (including life-gate), which are characterized by 'storage of essence and qi without wasting'; the six 'Fu' organs include gall bladder, stomach, small intestine, large intestine, bladder, and triple-jiao, which are characterized by 'conduct and excrete metabolites without storing' (Beijing College of TCM 1974). The functions of the five 'Zang' and six 'Fu' are shown in Table 4.1 and Table 4.2.

Based on such a structure, the maintenance of human life activities depends on some basic life substances. These substances mainly include qi, blood, body fluid, essence, and spirit (Beijing College of TCM 1974). Their main functions and their relations with modern medicine-related substances are shown in Table 4.3.

In TCM, the channels all over the organ, limbs and body are called blood vessels, which are channels for the transmission of matter, energy and information among the viscera and the body's limbs and the trunk. The blood vessel has a main trunk and branches, and the small branches are distributed throughout the body like a network, everywhere. Through the connection of blood vessels, all internal viscera, five sense organs, seven orifices, fur, muscles, bones and other tissues of the human body are closely connected to form an organic whole. The qi and blood running in the blood vessels play the role of nourishing the whole body, and at the same time transport metabolites for excretion. Obviously, the blood vessels in TCM cover part of the functions of blood vessels, lymphatic vessels and the neuro-humoral system of modern medicine.

We refer to the passages that connect viscera and the whole body and move qi and blood as blood vessels, instead of using the name commonly used in TCM—meridians. Because, as mentioned earlier, there is more and more evidence to support that there is a kind of channel in the

Table 4.1. 'Zang' organs and their functions.

Name	The main function	Corresponding structure and function of modern medicine
Heart	Promote the circulation of qi and blood in the blood vessels, disperse qi and blood and nutrients throughout the body through the blood vessels, and host the mental consciousness and thinking activities of the person	Heart pumping blood, cognition and conscious activities of the brain
Liver	Store blood; keep the body's qi running smoothly; it has attributes similar to 'wind' and is related to symptoms like 'wind' such as spasms, convulsions, and dizziness	Liver; blood; central nervous system related to mood regulation, sensation and movement
Spleen	In charge of the absorption and transportation of diet and water to produce qi, blood and body fluid, essence, and nourish the essential substances of human life	Absorption function of the small intestine
Lung	Governs qi; performs the breathing function; controls the dispersing of qi and liquid outward body surface and has the action of exorcising outward evil; lung qi is normal when it can move down smoothly	Respiratory function of lung
Kidney	In charge of storing the essence; promoting the growth and development of the bones; nourishing bone marrow and the brain; responsible for the filtration and production of urine	Kidney
Life-gate (Some scholars regard it as part of the kidney)	The gate of human life, the place where the birthplace of human yang qi, which is the fundamental driving force of human life. The life gate fire (also called kidney yang) resides in the life gate. It has the function of warming the spleen and stomach and helping the digestion of food; it is related to the sexual function of the human body and the reproductive system; it has the function of assisting in receiving inhaled air	No corresponding organ

Table 4.2. 'Fu' organs and their functions.

Name	The main function	Corresponding structure and function of modern medicine
Gall bladder	Store bile, help stomach digestion	Gall bladder
Stomach	Responsible for the temporary storage and digestion of food	Stomach
Small intestine	The place where food is absorbed	Small intestine
Large intestine	Transport and excrete the dross after the food is digested and absorbed	Colon, rectum
Bladder	Urine storage and urination	Bladder
Tri-jiao	Corresponds to the three parts of the body cavity-chest, abdomen and lower abdomen	Thoracic cavity, abdominal cavity

Table 4.3. Essential/vital substance for human life and their functions.

Name	Main function	Similar substances and relationships in modern medicine
Blood	It runs in the blood vessels and spreads throughout the body to nourish the internal organs and limbs.	Tangible; covering the concepts and functions of blood, lymph in modern medicine.
Qi	Permeates the whole body, big without outside, small without inside, it has the functions of motivation, nutrition, warmth, control, and consolidation.	Invisible; there is no similar concept in modern medicine. But its existence and its fullness or deficiency can be described in TCM.
Body fluid	It travels through the blood vessels and water channels, spreads throughout the body, and plays the role of moisturizing and nourishing.	Tangible, similar to body fluids in modern medicine.
Essence	It is the most fundamental substance in the life activities of the human body, and it is also the material basis for the functional activities of the various organs of the human body. It comes from parents and is constantly supplemented by nutrients absorbed from food.	Contains semen related to reproduction in modern medicine, and this part is visible.
Spirit	Dominating human life activities, the most fundamental manifestation of human life. Originates from innate, nurtured by acquired essence, qi, and blood.	Intangible, modern medicine has no corresponding concept. It is related to soul, consciousness, and thinking. The condition of 'with spirit' or 'lost spirit' can be described by clinic manifestation in TCM.

human body, and hundreds of acupuncture points are distributed along its route. When using certain methods (such as acupuncture or pressing) to stimulate certain acupuncture points, there will be the phenomenon of 'feeling conduction along the meridian route'. Although this channel has not a physical structure similar to blood vessels, lymphatic vessels, and nerves, it may be an objective reality that has certain anatomical properties (such as skin perforators, muscle gates, and bone gates) or physical characteristics and can be detected by certain methods/means. And the blood vessels here, as channels connecting various parts of the human body, are functional models similar to the internal organs, qi, blood and body fluid of TCM.

The integration of the viscera with various functions, the basic substances for maintaining human life, and the blood vessels connecting viscera and various parts of the human body form a simplified human functional model that can fully describe the basic physiological and pathological activities of the human body. Judging from the names of the internal organs in TCM, it is obvious that the model of the internal structure of the human body in TCM has a certain anatomical basis, but the functions of these organs are quite different from today's anatomical organs of the same name. For example, the heart of TCM includes the basic functions of the anatomical heart and part of the central nervous system of modern medicine; the spleen in TCM is completely irrelevant to the anatomical spleen. In other words, anatomical organs that are completely consistent with the functional description in TCM cannot be found in the actual human body. Therefore, based on the concept of modern science, people have doubted the scientific nature of TCM.

4.1.2 The scientific nature and practical significance of the human body model in TCM

As science deepens its understanding of nature, people are getting a clear understanding of the nature of science and the limitations of scientific capabilities. Today, the purpose of science is no longer to reveal the true face of the world, but to understand and recognize the laws of nature through the establishment of functional models. As a result, the models used by science to describe the laws of nature are often significantly different from the real existence in nature. For example, based on the atomic model that has ruled physics for nearly 100 years, nature world at the microscopic level is composed of an atomic nucleus and negatively charged electrons that revolve around it at high speed. The atomic nucleus is composed of positively charged protons and neutral neutrons. Regardless of whether they are protons, neutrons or electrons, they are all 'particles' that can be seen under a microscope with a sufficiently high magnification. With the collapse of the 'particle theory' triggered by the emergence of quantum mechanics, people realized that there are no solid particles at the micro level. They are all energy clusters with (or without) polarity. In other words, the structure revealed by the atomic model does not exist in the real world. But can we draw the conclusion that the atomic model is unscientific based on this? (Figure 4.1).

Almost all the scientific theories that are popular today are transitional models that will one day be overthrown by new theories that can describe the laws of nature more accurately, comprehensively, and reasonably. If a theoretical model is overthrown, it is no longer scientific, then no theory can be called science. However, many scientific discoveries and development applications based on this atomic model that is inconsistent with the real world still have practical significance and practical value.

Therefore, based on our understanding of the nature of science today, if the description of the internal structure and functions of the human body in TCM is regarded as a functional model like the atomic model, we will not demand that it be consistent with the anatomical entity structure, nor will we doubt its scientificity based on this. The syndrome differentiation and treatment system based on such a theoretical model has been effectively guiding the diagnosis and treatment of diseases in TCM for thousands of years, and has created countless medical miracles that amaze today's highly developed modern medicine. Based on the consistency of the modeling methods of complexity science and TCM, the valuable work we can do is: Introduce mature modeling methods and technical means of complexity science, and verify the accuracy of the model fitting the real system based on the comparison between the prediction made by the model and the actual situation presented by the

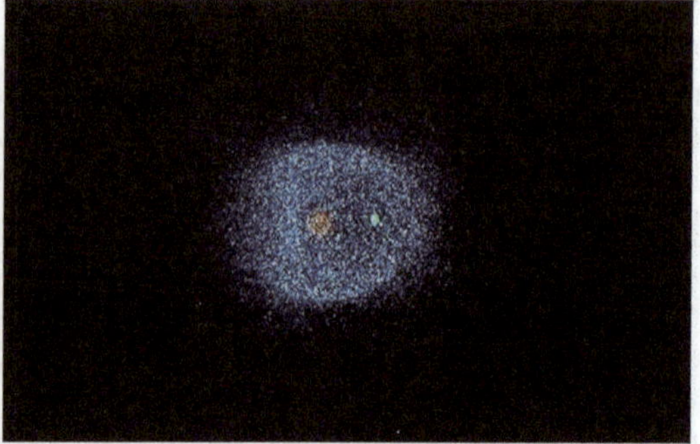

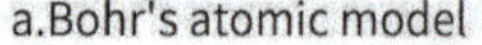

a.Bohr's atomic model b.The atomic model of quantum mechanics

Figure 4.1. The historical changes of the atomic model.

real system. On this basis, the model is modified and perfected where it does not conform to the real system or does not conform to scientific norms.

4.2 The theory of yin and yang: A model analysis method based on the concept of the unity of opposites (Beijing College of TCM 1974)

The viscera and basic life substances of the human body each have different functions. As a whole, there are various relationships among the various internal organs of the human body, and among the organs and various parts of the whole body, and between the human body and the external environment. How to understand the functions of these viscera and the relationship among them? How to effectively control the disease process caused by abnormal viscera function based on theoretical models? To this end, TCM introduced the yin-yang and five-element theory, a popular philosophy in ancient China, and established a unique model analysis method.

4.2.1 The concept and analogy of yin and yang

In ancient Chinese philosophy, yin and yang represent two aspects where things are mutually opposed and interdependent. It can represent two opposing things, or two opposing aspects that exist within the same thing. The theory of yin and yang holds that everything in the universe contains two opposite aspects of yin and yang, such as day and night, sunny and cloudy, hot and cold, activity and stillness, and so on. The movement and changes of yin and yang constitute everything and promote the development and change of things. So 'The manifestation of yin and yang from the macrocosm to the microcosm' (Suwen•Yin Yang Yingxiang Dalun): states that "yin and yang, the way of heaven and earth, the outline of all things, the parents of change, the beginning of life and death, the house of spirit". The Tai Chi diagram in Figure 4.2 vividly shows the concept of yin and yang.

Generally speaking, everything that is active, external, rising, warm, bright, functional, and hyperactive, belongs to yang; calm, internal, descending, cold, dark, material, and function attenuated things, belong to yin. For example, the sky is above, so it is yang, and the earth is below, so it is yin; water is cold and descends, it is yin, and fire is hot and flaming upward, it is yang; the tangible is yin, and the intangible is yang; rising and outward are yang, and descending and inward are yin; the static is yin, and the moving is yang; when it comes to the life activities of the human body, everything that has the effect of pushing, warming, and excitement belongs to yang, and everything that has the effect of cohesion, nourishment, and inhibition belongs to yin.

The yin and yang attributes of things are not absolute, but relative. This relativity is manifested in that under certain conditions, yin and yang can be transformed into each other, yin can be transformed into yang, and yang can be transformed into yin. This relativity is also reflected in the infinite separability of things. For example, the day is yang, and the night is yin; while in the morning, it is the yang in the yang, the afternoon is the yin in the yang, the first half of the night is the yin in the yin, and the second half of the night is the yang in the yin. That is, there is still yin and yang in the yin and yang, which are endless.

According to the theory of yin and yang, the two aspects of yin and yang are mutually opposed and restrict each other. 'Treating cold syndrome with hot-natured drugs, treating heat syndrome with cold-natured drugs' clearly expresses the mutually opposed and restrictive relationship between cold and heat. On the other hand, yin and yang are interdependent, either cannot exist independently of the other. The upper is yang, and the lower is yin. If there is no upper, there is no lower; if there is no lower, there is also no upper. Similarly, if there is no heat, there is no cold; if there is no cold, there is also no heat. Yang depends on yin, and yin depends on yang. Each regards the other's existence as the condition of its own existence. The external yang is the external manifestation of the inner material movement, and the internal yin is the material basis of the function. If the yin and yang lose the opposing side they rely on, they will no longer be able to survive and grow. That is, 'lone yin cannot survive, lone yang cannot grow'.

Figure 4.2. yin and yang in Tai Chi Diagram.

In addition, the opposing and interdependent two sides of yin and yang are not in a static state, but in the movement of 'yang diminishes and yin grows' or 'yin diminishes and yang grows'. For example, in the changing climate of the four seasons, from winter to spring and summer, the climate gradually shifts from cold to hot, which is a process of 'yin diminishes and yang grows'; from summer to autumn and winter, the climate gradually shifts from hot to cold, which is a process of 'yang diminishes and yin grows'. As far as the human body is concerned, the production of various functional activities (yang) must consume a certain amount of nutrient substances (yin), this is the process of 'yin diminishes and yang grows'. The metabolism of various nutrient substances (yin) must consume a certain amount of energy (yang qi), this is the process of 'yang diminishes and yin grows'. Under normal circumstances, this yin and yang ebb and flow is in a relative balance with a certain degree of stability. If this growth and decline exceeds a certain limit, and the balance and stability cannot be maintained, a certain aspect of yin and yang will be prosperous or declining, which is a disease.

Because of the ebb and flow of yin and yang, the yin and yang attributes of things will be transformed in the opposite direction, yin can be transformed into yang, and yang can be transformed into yin. If the 'growth and decrease of yin and yang' is a quantitative change process, then the transformation of yin and yang is a qualitative change process. In 'Suwen·Yin and Yang Yingxiang Dalun', 'heavy yin must be yang, heavy yang must be yin'. 'Cold extremes generate heat, and hot extremes produce cold' refers to this phenomenon of qualitative changes in the attributes of things. This kind of yin and yang transformation is often seen in the development of diseases. Certain acute infectious diseases, such as toxic pneumonia, toxic dysentery, etc., due to serious heat toxicity, consume a lot of positive qi in the body. In the case of persistent high fever, patients with these diseases may suddenly experience a sudden drop in body temperature, pale complexion, cold limbs, and weak pulse, which belong to the yin and cold crises, which is a change from yang to yin. If the rescue is in time, the complexion and pulse condition will become calm, the yang qi will be restored, and the condition will turn for the better.

4.2.2 Practical application of yin-yang theory

In TCM, the theory of yin and yang, as an important model analysis method, is widely used in the description of the human body's structure, physiological functions, and the occurrence and development of diseases, and is used to guide clinical diagnosis and treatment.

(1) Explain the structure of the human body

In TCM, the human body is an organic whole. All the organizational structures of the human body are not only organically connected, but also can be divided into two opposite parts of yin and yang.

As far as the human body is concerned, the upper part is yang and the lower part is yin; the body surface is yang, and the inside is yin; the back is yang, the abdomen is yin; the outer side of the limbs is yang, and the inner side is yin. In terms of viscera, the six Fu organs belong to yang, and the five Zang organs belong to yin. The five Zang organs can be also divided into yin and yang, the heart and lungs belong to yang, and the liver, spleen and kidney belong to yin. Specific to each viscus, there are also yin and yang aspects, such as heart yin and heart yang, and kidney yin and kidney yang. In short, although the relationship between the upper and lower, internal and external organizational structures of the human body, and each organizational structure itself, is complicated, it can be summed up with yin and yang.

(2) Explain the physiological functions of the human body

TCM believes that the normal life activities of the human body are the result of maintaining a harmonious balance between the two aspects of yin and yang. The physiological activities of the human body are based on the substances belonging to yin. Without yin essence, yang qi cannot be produced. As a result of the physiological activities promoted by yang qi, yin essence is continuously produced and supplemented. Therefore, only when yin and yang are in a stable dynamic balance that restricts and promotes each other, can there be a continuous material basis to sustain life, and life activities will be vigorous and orderly. If yin and yang lose this mutual restraint and mutual promotion relationship and separate from each other, the essence and qi that sustain life activities will lose the source of metaplasia, and life will cease.

(3) Explain the pathological changes of the human body

The theory of yin and yang believes that the occurrence of disease is the result of the loss of relative balance between yin and yang, and the occurrence of partial prosperity and partial decline. The occurrence and development of diseases are related to the body's disease resistance and pathogenic factors. The interaction and mutual influence between them can be summed up with yin and yang. The human body's disease resistance has two parts: yang qi and Yin essence. The pathogenic factors are also divided into yang heat and yin cold. Excessive yang-heat will damage the yin-essence and result in heat syndrome; and excessive yin cold will damage the yang qi, the syndrome of cold will appear. If yang deficiency cannot control yin, the deficiency-cold syndrome of yang deficiency and excess yin will appear; if yin deficiency cannot suppress yang, the deficiency-heat syndrome of yin deficiency and yang hyperactivity will appear. It can be seen that although the occurrence and development of diseases are complex and diverse, many of them can be explained by 'yin and yang imbalance'.

In addition, if any one of the yin and yang of the body is depleted to a certain extent, it can often lead to the deficiency of the other. It is the so-called 'yang damage and yin', 'yin damage and yang', so that 'yin and yang are both deficient'. For example, some chronic diseases, in the course of their development, due to the weakness of yang qi affecting the insufficient metaplasia of yin essence, or due to the loss of yin essence affecting the production of yang qi, this situation will occur.

(4) Used for disease diagnosis

In the process of identifying syndromes in TCM, the first step is to distinguish the yin, yang, surface, interior, cold, heat, deficiency, and excess of the disease, which is the so-called 'eight cardinal syndrome differentiation'. Among the eight outlines, yin and yang are the general outlines, which govern the surface and interior, cold and heat, and deficiency and excess, namely, the surface, heat, and excess belong to yang, and the interior, cold and deficiency belong to yin. To make a correct diagnosis, yin and yang must first be distinguished, so that the essence of the disease can be grasped, and the problem can be simplified. As 'Jingyue Quanshu · Chuanzhong Lu' says: "Syndrome differentiation must first distinguish between yin and yang, which is the principle of medicine. If the identification of yin and yang is accurate, then how can the treatment go wrong? Although medicine is complicated, it can be summarized by yin and yang. The syndrome has yin and yang, the pulse condition has yin and yang, and the medication has yin and yang... . Although medical theories are very mysterious and complicated, as long as you truly understand yin and yang, you can grasp its core."

(5) Used for diseases treatment

Since the excess and decline of yin and yang are the root causes of the occurrence and development of diseases, the basic principles of treatment are to adjust yin and yang, supplement the deviation and remedy the disadvantages, and restore the relative balance of yin and yang. If the yang heat is too excess and damages the yin fluid, the surplus of yang can be suppressed, using the method of 'treat heat with cool'; if the yang energy is damaged due to the overwhelming yin and cold, the excess yin can be suppressed, using the method of 'treat cold with heat'. Conversely, if yin fluid is insufficient and cannot restrict yang, resulting in hyperactivity of yang, or due to insufficient yang qi, which cannot restrict yin and causes yin excess, the deficiency of yin or yang must be supplemented. This is the treatment principle of 'for yang diseases, treat yin; for yin diseases, treat yang; supplement the source of fire to eliminate yin haze', so that yin and yang can restore a new relative balance.

The concepts of yin and yang are not only used to establish principles of treatment, but also to categorize the property, taste, and function of medications (such as Chinese herbal medicine, CHM) as a basis for guiding medication. For example, cool and moisturizing CHMs belong to yin, warm and dry CHMs belong to yang; CHMs that taste sour, bitter, and salty belong to yin, those that taste pungent, sweet, and light are yang. CHMs with astringent and depressing actions belong to yin, and those with lifting and divergent actions belong to yang. Treating diseases is to determine the principle of treatment according to the nature of the yin and yang of the disease, and then choose the corresponding CHMs in combination with the yin and yang properties of the CHM to achieve the purpose of treatment.

4.3 The Five Elements Theory: A functional model and related analysis methods based on the five elements metaphor and analogy (Beijing College of TCM 1974)

There are extremely complex interconnections and mutual influences in all things in nature, and the same is true of the internal tissues and organs of the human body, the human body and the external environment. How to understand these intricate inter-relationships and mutual influences, so as to effectively guide the diagnosis and treatment of diseases? In ancient China, the ancients divided all things in the universe into five categories: wood, fire, earth, metal, and water, according to their characteristics, collectively called the five elements. The relationship among the five elements of metal, wood, water, fire, and earth, which mutually produce and restrict each other, deduces and explains the complex connection among everything in the universe, as shown in Figure 4.3.

Today, building simplified models of complex systems through metaphors and analogies, and understanding and explaining the complex relationships among complex things through the analogy of simple things in the simplified model has become the main method for complexity science to understand and grasp complex systems. The human model system in TCM theory is constructed based on the principles revealed by the five-element model and through the observation of the basic physiological and pathological phenomena of the human body. The significance of the theory of the five elements in TCM is to understand and explain mutually dependent and restrictive relationships between the various parts of the human body, the human body and the external environment through the analogy of the five-element model.

4.3.1 Five-element classification of object attributes

Ancient TCM scholars used metaphors and analogies to attribute the organs, physiology, and pathological phenomena of the human body, as well as the natural things related to human life,

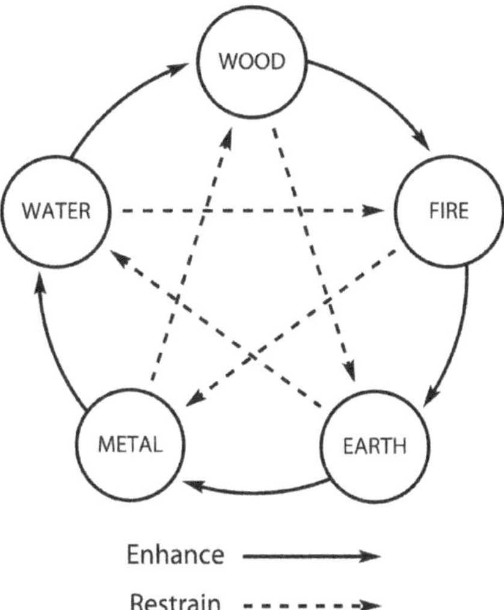

Figure 4.3. The interconnection of the five elements.

to the five elements of metal, wood, water, fire, and earth, according to their nature, function and form. Involved in this method of summarizing things with the five elements are no longer the five substances of wood, fire, earth, metal, and water themselves, but the attributes of different things abstractly summarized based on their characteristics. For example, woodiness is characterized by growth, extension and softness, and anything with these characteristic is classified as 'wood'; the characteristic of fire is yang heat and inflammation, and anything with these characteristics is called 'fire'; the characteristics of earth are nurturing and change, and anything with these characteristics is called 'earth'; the characteristic of metal is pure and strong, and anything with this characteristic is called 'metal'; water is characterized by coldness and downward movement and anything with this characteristic is called 'water'. The five elements used in medicine are also a summary of the attributes of these five different things. See Table 4.4 and Table 4.5.

Based on the five-element classification and analogy of the attributes of things based on 'imagery similarity', the principles revealed by the five-element model are extended to various fields of nature and the human body. This is the most convincing basis for the rationality of this expansion in ancient China, where the holistic philosophy including Yin-Yang and the Five Elements was dominant. However, everything in nature is very different, and its attributes and relationships have their own particularities. Forcibly including the specific attributes and correlations of all things in nature into the framework of the five substances of metal, wood, water, fire, and earth and their imagery extensions is inevitably far-fetched and leads to some conclusions that are inconsistent with reality.

4.3.2 Mutual generation and restriction, over-restriction and reverse restriction among the five elements

The theory of the five elements is mainly based on the mutual generation and restriction of the five elements to explain the relationship among things. Mutual generation means mutual production and promotion; mutual restriction means mutual opposition and restraint. The order mutual generation of the five elements is: wood produces fire, fire produces earth, earth produces metal, metal produces water, water produces wood, which generates in sequence, and the cycle is endless. The order of mutual restriction of the five elements is: wood restrains earth, earth restrains water, water restrains fire, fire restrains metal, and metal restrains wood, which restraint relationship is also cyclical. The movement and changes of all things in the natural world have this kind of mutual generation and mutual restraint relationship. Only when there is restriction in generation and there is generation in restriction, and the opposite is mutually compatible, can it survive and develop continuously.

Table 4.4. Attribution of the five elements of things in nature.

Five elements	Wood	Fire	Earth	Metal	Water
Five flavors	sour	bitter	sweet	pungent	salty
Five colors	green	red	Yellow	white	black
Five climatic agents	wind	heat	dampness	dryness	Cold
Five transformations	sprouting	growth	transformation	reaping	storing
Five directions	east	south	central	west	north
Five seasons	spring	summer	late-summer	autumn	winter

Table 4.5. Attribution of the five elements of human body elements.

Five elements	Wood	Fire	Earth	Metal	Water
Five Zang-viscera	liver	heart	spleen	lungs	kidneys
Five Fu-viscera	gall bladder	small intestine	stomach	large intestine	Urinary bladder
Five sense organs	eyes	tongue	mouth	nose	ears
Five constituents	tendon	vessel	muscle	skin	bone
Five emotions	anger	joy	anxiety	sadness	fear
Five fluids	tear	sweat	Thin saliva	snivel	Thick saliva

Without generation, there will be no occurrence and growth of things; without restriction, changes and development under a state of normal coordination and balance cannot be maintained.

In the mutual generation relationship of the five elements, any element has the relationships to be generated and generate others. The one which generates me is the mother, and the one which is generated by me is the child. In the relationship of mutual restraint among the five elements, any element has the relationship to restrain others and be restrained. As for over-restriction and reverse restriction, they are abnormal phenomena in the development and change of things. The over-restriction, that is, the restraint is too strong, exceeding the degree of normal restriction. For example, when the wood's qi is too strong and metal cannot exercise normal restraint on the wood, the wood that is too strong will over-restrict the earth to make the earth more deficient. The reverse restriction is the restriction in the opposite direction. For example, 'metal restricts wood' belongs to the normal relationship of mutual restraint. If the metal qi is insufficient, or the wood qi is too strong, wood will reverse restrict metal in turn.

The relationship of the five elements of generation, restriction, over-restriction, and reverse restriction deeply reveals the realistic possibility that each element has a relationship with the other four elements, and this relationship includes two directions of promoting or inhibiting. The balance of yin and yang in the human body and the maintenance of normal physiological homeostasis are achieved through continuous two-way regulation under the action of the negative feedback regulation mechanism that exists widely in the human body. The treatment of diseases also needs to comprehensively consider this relationship, so as to find the best solution to restore the pathological state to a normal homeostasis through the intricate appearance of the disease.

4.3.3 The practical significance and scientific limitations of the five elements theory

Applying the theory of the five elements to TCM is to use the five elements classification method of the attributes of things and the relationship of mutual generation, mutual restriction, over-restriction and reverse restriction, through metaphors and analogies, to explain human physiology and pathological phenomena, and guide clinical diagnosis and treatment.

(1) Explain the physiological functions and relationships of the viscera

The Five Elements Theory assigns the five internal organs of the human body to the five elements respectively, and uses the characteristics of the five elements to explain the physiological characteristics of the five internal organs. For example, the liver has the function of relieving venting, and the wood has the properties of growth, so the liver belongs to 'wood'; the heart yang has the warming effect, and fire has the characteristics of yang-heat, so the heart is 'fire'; the spleen is the source of growth and nourishment, and the earth has the characteristics of nourishing all things, so the spleen belongs to the 'earth'; the lungs should be clear and their qi should descend, and metal has the characteristics of clarity and contraction, so the lung belongs to 'metal'; the kidney has the function of controlling water and storing essence, and water has the characteristics of moisturizing and downward-moving, so the kidney belongs to 'water'.

The Five Elements Theory is also used to explain the internal connection of the physiological functions among the human viscera. For example, the kidney essence (water) can nourish the liver, the liver (wood) can store blood to help the heart, the heat of the heart (fire) can warm the spleen, the nutrients produced by the spleen (earth) can nourish the lungs, and the lungs (metal) purify downward can help kidney water. This is the relationship of mutual generation and nourishment among the five viscera. Lung (metal) qi is descending, which can inhibit the hyperactivity of liver-yang; the liver (wood) is free development, which can relieve the stagnation of the spleen (earth); the function of transformation and transport of the spleen (earth) can prevent the flood of kidney water; the nourishment of kidney (water) can prevent the exuberance of heart fire; the yang heat of the heart (fire) can restrict too much downward convergence the lungs (metal). This is the mutual restriction of the five viscera.

In addition, the five-season, five-qi, five-direction, five-color, and five-flavor relationships between the human body and the external environment can also be explained by the theory of five elements. As a result, through the metaphors and analogies of the five elements, it deeply reveals the interrelationship among various human viscera, and among human viscera and the natural world. Through the mutually complementary theory of the five elements and the theory of yin and yang, TCM's holistic concept, which regards the human body as an organic whole, and the human and nature as an organic whole, has been vividly and concretely embodied.

(2) Explain the pathological influence of viscera

The theory of the five elements can also be used to explain the mutual influence of the viscera in the case of illness. For example, liver disease can be passed on to the spleen, which is the wood over-restraining the earth; spleen disease can also affect the liver, which is the counter-restriction to wood by earth. Both ways can cause the disease of liver and spleen in same time. In addition, liver disease can also affect the heart, which belongs to 'the disease of a mother-organ affects a child-organ'; it affects the lungs, which belongs to 'wood reverse restricts metal'; it affects the kidneys, which belongs to 'the disease of a child-organ affects a mother-organ'. The same is true for the pathological changes of other organs, and the relationship between the five elements can be used to illustrate their mutual influence in the disease process.

(3) Used in the diagnosis and treatment of diseases

Abnormal changes in the functional activities of the human internal organs and their inter-relationships can be reflected in a person's complexion, voice, taste, and pulse. In other words, the disease can be diagnosed from changes in the patient's complexion, voice, taste, and pulse condition. Based on the classification and attribution of the five elements, there is a certain correlation between the five internal organs and the changes of five colors, five sounds, five flavors and related pulse conditions. Therefore, when clinically diagnosing a disease, the information obtained from the four methods of looking, smelling, asking, and touching can be integrated, and the condition of the disease can be inferred based on the belonging of the five elements and the law of their generation, restriction, over-restriction and reverse restriction. If the patient's face is blue, he likes to eat sour things and his pulse is like the strings of a piano, liver disease can be diagnosed; if the patient's face is red, the mouth is bitter, and the pulse is huge, exuberance of heart fire can be diagnosed.

The occurrence and development of diseases are often associated with abnormalities in the relationship of mutual generation and restriction among the internal organs. Therefore, during treatment, in addition to the direct treatment of the diseased internal organs, it should also be considered the indirect treatment of the diseased internal organs with the help of the relationship among the internal organs or the preventive treatment of other internal organs that may be affected by the diseased internal organs. As stated in 'Nan Jing; Classic of Questioning', 'seeing liver diseases, knowing that it will affect the spleen, and taking the lead in strengthening the spleen' is a concrete manifestation of using the theory of five elements to guide clinical treatment. Later generations of physicians used the law of generation, restriction, over-restriction and reverse restriction of the Five Elements formulated many more specific treatment methods, such as cultivating the earth to produce metal, nourishing water to moisturize wood, strengthening water to restrain fire, and so on.

As a simplified model established by the use of metaphors and analogies, the Five Elements Theory deeply reveals the intricate interrelationships and interactions between the human body's viscera, viscera and various parts of the human body, and between the human body and the external environment. The integrity of the human body contained therein, and the integrity of the natural world, including humans, undoubtedly have positive significance. But this simplified model is too simplified, far from being able to objectively reflect the complexity of the human body, the natural world, and the inter-related complexity inside the human body, between the human body and the natural world. Based on the five-element analogy, the inferences made on the attributes of various parts of the human body, the relationship between various parts in the human body and the relationship with the outside world will inevitably lead to some far-fetched and unrealistic conclusions. In this regard, ancient TCM scholars had a clear understanding. In the TCM we see today, the number of Fu-organs is not five, but six. The relationship among the viscera is not entirely based on the inference of the theory of five elements, and only those parts that are consistent with clinical observations are spread to the present. For example, the lack of appetite of the spleen deficiency which caused by the stagnation of liver qi, is considered to be 'the wood over-restrains the earth'; the edema of the dysfunction of the water metabolism of the kidney caused by the spleen deficiency is considered to be 'the earth over-restrains water'. The treatment principles derived based on the association of the five elements only retain the clinically valuable parts, such as cultivating the earth to produce metal, nourishing water to moisturize wood, and so on. There are also some explanations that be justified through 'conceptual disguised displacement', which have been retained in TCM to this day. For example, the 'water' in 'strengthen water to balance fire' usually refers to the kidneys, and the 'fire' here is not necessarily the heart, but internal heat caused by yin deficiency.

Today, complexity science builds models, evaluates models, and improves and perfects the structure and function of the models, all based on whether they conform to the actual situation of the real system. The most important significance of introducing the Five Elements Theory in TCM is to establish the feasibility and practicality of the model based on metaphor and analogy in guiding effective treatment of diseases. However, it is not widely recognized in the clinical practice of TCM to limit the specific attributes and correlations of each element in the human body model of TCM to the range that can be explained by the five-element model based on the five substances of metal, wood, water, fire, and earth. Based on the investigation of the physiological and pathological activities of the human body, the continuous revision and perfection of this model provide a broad space for the development of TCM.

4.4 The treatment based on syndrome: Model-based state description and regulation-control system

In TCM, a functional model of the human body was established with Zang-organs, Fu-organs, qi, blood, body fluid, essence as the core. By introducing the theory of yin and yang, the eight-dimensional model analysis method based on yin and yang, external and internal, cold and heat, and deficiency and excess is established (Beijing College of TCM 1974). The five-element model based on metaphor and analogy, establishes the extensive connections and influences among the internal organs of the human body, among the internal organs and various parts of the human body, and between the human body and the external environment. The next step is how to describe the various physiological and pathological states of the human body based on the model, and find ways to adjust and control these states. For this reason, TCM has introduced a unique concept of 'syndrome'. So, how does syndrome describe the physiological and pathological state of the human body?

4.4.1 The essence of syndromes: state variables describing the state of the human body

In modern science, to describe the state of a system, it is necessary to introduce a set of state variables. The essence of TCM syndrome is the state variable introduced by TCM to describe the human health state. The description of the human body state by the human body model mainly has two levels: the holistic level and the viscera level. Therefore, the syndromes describing the state of the human body are also divided into syndromes at the holistic level and syndromes at the organ level. At present, in TCM, syndromes at the holistic level include qi-deficiency, blood-deficiency, yin-deficiency, yang-deficiency, qi-stagnation, blood-stasis, damp-heat, cold-dampness, phlegm-dampness, fiery-heat, etc., while syndromes at the organ level include heart-qi deficiency, deficiency of heart and blood, stagnation of liver qi, deficiency of kidney yang, etc. The combination of all the abnormal syndromes that appear in the patient at a certain time completely describes the patient's health status at that time.

Today, syndromes of the holistic level, such as deficiency of qi, deficiency of blood, deficiency of yin, deficiency of yang, stagnation of qi, and blood stasis, have been used to describe the body's constitution. Using this as a frame of reference, TCM can effectively identify and regulation-control the individual's constitution state and improve the morbid constitution. As a result, effective treatment methods have been found for many diseases caused by constitution differences that modern medicine can do nothing for. The more detailed syndromes of the viscera level have been widely used in TCM treatment based on syndrome from ancient times to the present, forming the unique state regulation-control method of TCM. This is the research field and regulation-control method that modern medicine has not been involved in for hundreds of years before the emergence of precision medicine.

Take for example, the most important organ of the five internal viscera is the heart. In order to characterize its functional state, TCM introduces a set of state variables: heart qi, heart blood, heart yin, heart yang, heart blood flow status (with or without stasis), heart fire status (with or without heart fire), and thus constitutes the state space of the heart. The value of each state variable usually has a normal state and several abnormal states (syndrome). For example, the state variable heart qi has several state values, normal, heart-qi deficiency, and serious heart-qi deficiency. Each state value is defined by a set of symptoms and signs that can be felt or observed. For example, the symptom groups corresponding to the heart-qi deficiency are palpitations, fearful throbbing, shortness of breath, fatigue, aggravated after activities, pale complexion, etc. Serious heart-qi deficiency corresponds to palpitations, sweating, shortness of breath, aggravated after activities, listlessness, fatigue, weak pulse or intermittent, etc., as shown in Table 4.6 (Yuan Bing 2010).

The collection of these state variables constitutes the state space of the heart. In this way, a set of appropriate state variables that characterize its functional state are introduced to each viscus and

Table 4.6. Symptom groups corresponding to cardiac state variables.

State variables	Value range	Syndrome name	Corresponding symptom/sign collection
Heart-yang (H1)	H10	Normal	No abnormal clinical manifestations
	H11	Heart-yang Deficiency-1	Palpitations, shortness of breath, chills, cold limbs, chest tightness and pain, etc.
	H12	Heart-yang Deficiency-2	Suddenly dripping cold sweat, cold limbs, pale complexion, weak breathing, heart palpitations, severe heart and chest pain, confusion or coma, blue lips and tongue, and weak pulse
Heart-qi (H2)	H20	Normal	No abnormal clinical manifestations
	H21	Heart-qi Deficiency-1	Heart palpitations, shortness of breath, fatigue, especially after activity, pale complexion
	H22	Heart-qi Deficiency-2	Palpitations, sweating, shortness of breath, especially after exercise, listlessness, fatigue, weak or intermittent pulse
Heart-yin (H3)	H30	Normal	No abnormal clinical manifestations
	H31	Heart-yin Deficiency-1	Palpitations, insomnia, dreaminess, hot flashes, night sweats, upset, fever on the center of palms and feet bottom, red cheeks, dry throat, red tongue, little moss, thin pulse
Heart-blood (H4)	H40	Normal	No abnormal clinical manifestations
	H41	Heart-blood Deficiency-1	Heart palpitations, insomnia, dizziness, forgetfulness, pale or chlorosis, pale lips and nails, weak pulse
Heart blood flow status (H5)	H50	Normal	No abnormal clinical manifestations
	H51	Heart-blood stasis	Pain in the heart and chest, which can radiate to the shoulders, back and left upper limb, lips and nails are blue-purple, petechiae and ecchymosis on the tongue
Heart fire status (H6)	H60	Normal	No abnormal clinical manifestations
	H61	Heart fire	Upset, insomnia, dreaminess, red face, thirst, sores in the mouth and tongue, short urine
	H62	Heart fire Excess	Irritability, insomnia, red face, thirst, coma, delirium, etc.

basic living matter, and the set of all these state variables constitutes the state space of the human body. Based on such a batch of state variables, TCM has established the corresponding relationship between the state interval of each state variable and the observable symptoms and signs of the human body. As a result, clinically based on the patient's symptoms and signs, it can be determined whether each state variable of each viscus and basic vital matter has deviated from the normal state, the direction and extent of deviation, so as to determine the state of the body. The state variables at the holistic level describe the functional state of the person at the holistic level, that is, the constitution.

4.4.2 The Inevitability of Science: A syndrome differentiation and treatment system from 'Separatist Regime' to 'Great Unification'

TCM introduces state variables to describe the state of the body, and influences these state variables through appropriate intervention measures to achieve the regulation-control of the state of the body. Obviously, the description and regulation-control of human disease in TCM is consistent with the methods and ideas of modern science on the description and regulation-control of various systems. In contrast, the description and regulation-control of diseases in modern medicine has little to do with modern scientific methods. In this sense, the claim that 'medicine is not science' held by many scientists is not unreasonable.

With the help of the state description based on the functional model, although we do not know the exact morphological and functional abnormalities of the internal organs and tissues of the human body during the disease process, we can still grasp the state of the body through the understanding of the patient's symptoms and signs during the disease process, and adopt corresponding state regulation-control methods to treat the disease.

In this state description system (SDS), state variables are introduced based on the functional provisions of the human body model. The completeness of the human body model's description of human functions determines the integrity of the state variable system; the structured and rigorous logic of the human body model ensure the independence of the introduced state variables. In fact, TCM still has some problems in the integrity and independence of the state variable system. For example, the SDS based on the Zang-organs, Fu-organs, qi, blood, body fluid, essence as the core cannot express the law of the occurrence and evolution of infectious diseases in the human body. Therefore, people have to expand and supplement the model structure and state variable system.

The viscera-qi-blood-body fluid-essence system established in the 'Huangdi Neijing, The Yellow Emperor's Inner Canon' mainly involves the basic physiological functions of the human body. It can describe the pathological changes when these functions are abnormal, but it is far from enough to describe the development, transmission and treatment of epidemic and infectious diseases (hereinafter referred to as exogenous diseases) caused by external pathogenic factors. Zhang Zhongjing, a famous doctor in the Han Dynasty (approximately 150–219 AD), in order to effectively deal with the large-scale outbreak of infectious diseases at that time, on the basis of expanding the connotation of the concept of six meridians of lesser yang, sunlight yang, greater yang, lesser yin, reverting yin, greater yin in the 'Huangdi Neijing, The Yellow Emperor's Inner Canon', created a six-meridians syndrome differentiation model which can describe the occurrence, development, transmission, and treatment of diseases caused by exogenous cold pathogens. Since the characteristics of epidemic infectious diseases in the late Ming and early Qing and Zhang Zhongjing eras were significantly different, the six-meridians syndrome model established in 'Shanghan Lun, Treatise on Cold Pathogenic Diseases' could not cover its development, transmission, change and regulation-control laws. The famous doctor Ye Tianshi (1667–1746 AD) in the Qing Dynasty, introduced the concept of wei-qi-ying-blood - the human body level from the surface to the inside of the human body, established a wei-qi-ying-blood syndrome model, which can describe the laws of occurrence, development, transmission, and treatment of diseases caused by exogenous hot pathogens. After Ye Tianshi, Wu Jutong (1758–1836), a famous doctor in the Qing Dynasty, introduced the concept of tri-jiao in the 'Huangdi Neijing, The Yellow Emperor's Inner Canon' into the stage division of epidemic and infectious diseases, established the tri-jiao syndrome model, which expresses the laws of occurrence, development and treatment of diseases caused by exogenous warm-heat and damp-heat pathogens. As a result, these three syndrome differentiation systems suitable for exogenous febrile diseases have been produced. In each system, each syndrome corresponds to a different symptom/sign group. In TCM, the segregated situation of 'all heroes standing together' in the syndrome differentiation and treatment system has continued to this day.

The diseases caused by different types of pathogenic factors sometimes enter the same state in the process of the development and transformation in the human body. For example, the yangming meridian syndrome (belonging to the six meridian syndrome system) caused by wind-cold changed to heat, and the syndrome of exuberant heat in the qi (belonging to wei-qi-ying-xue syndrome system) caused by wind-heat entering the inside, the clinical manifestations and treatment methods of the two are completely consistent. Although the disease process caused by the same type of pathogenic factor is given different names under different models and SDS, the law of development and transmission of the disease reflected by changes in symptoms/sign groups, and treatment methods are consistent. For example, the diseases caused by wind-heat can be described by the tri-jiao syndromes model or wei-qi-ying-xue syndromes model. Although the names of the Upper-jiao or Middle-jiao diseases described by the tri-jiao model and diseases in wei or qi described by the wei-qi-ying-xue model are different, their clinical manifestations are basically the same.

Due to the incompatibility of the several syndrome differentiation systems, especially the difference in the scope of application between the syndrome differentiation system for exogenous diseases and the viscera-qi, blood, body fluid essence syndrome differentiation system, certain diseases can usually only be treated with the corresponding syndrome differentiation system. For example, when a patient with diseases of internal injuries develops a new exogenous disease, since there is no unified syndrome differentiation system to guide comprehensive syndrome differentiation and treatment, a TCM practitioner usually cures the exogenous disease first and then treats the inherent internal injuries. This order of treatment seems to have become an established routine in the field of TCM so that no one doubts its rationality. In fact, there are often various relationships between exogenous diseases and internal damage diseases that occur in the same person. New-onset exogenous diseases often worsen the original internal damage diseases, and the method of sweating and relieving the surface usually used in the treatment of exogenous diseases can sometimes adversely affect the original internal damage diseases. As a supplement to the insufficient coverage of the syndrome differentiation system, in the treatment of exogenous diseases, TCM usually had some treatment methods that deal with the patient's abnormal constitution at the same time, such as specific herbal formulas for cold with qi deficiency, blood deficiency, yin deficiency, yang deficiency, or qi stagnation, etc. However, the condition of patients in the course of the disease varies greatly, and this 'patching' method is obviously not the best solution based on the comprehensive analysis of the unified model. We often see that in patients with severe internal damage diseases, when complicated by exogenous diseases, it is difficult to achieve the expected results by simply using conventional treatments for exogenous diseases. However, if the treatment for internal damage is interrupted due to infection, the patient's original internal damage diseases could worsen after the exogenous disease is cured. From the perspective of TCM, some interventions for the treatment of internal damage diseases will have certain adverse effects on the initial treatment of external infections. For example, when the initial external infections are on the body surface, high-dose CHMs with lowering qi or laxative actions are not suitable, to avoid the risk of chest binding syndrome due to external evil invagination. Although TCM also has the treatment principle of 'symptomatic treatment in acute condition, radical treatment in chronic case', it will be the best choice to make a comprehensive treatment plan by using a unified human model and SDS to grasp the relationship between exogenous syndromes and internal damage syndromes. For a treatment plan based on this kind of thinking, as long as the ratio of medications eliminating exogenous evil and medications of lowering-qi and laxatives is appropriate, the effects of eliminating exogenous evil and lowering qi and laxatives can be achieved simultaneously, without the risk of external evil invagination.

Because the viscera-qi-blood-body fluid-essence syndrome system is not enough to deal with the state description and treatment analysis of exogenous diseases, these three systems of syndrome differentiation and treatment have emerged. Similar phenomena are not uncommon in natural sciences. In physics, there was a period when two incompatible theories coexisted: the general relativity describing the large-scale universe, and quantum mechanics describing the structure of microscopic matter. Regarding the theory of light, there also used to be disputes between the 'particle theory' and the 'wave theory'. However, judging from the trend of scientific development, this phenomenon of coexistence of multiple incomplete theories has been or will eventually be replaced by a unified structured theory with the development of science. If we regard TCM as a scientific system, then replacing the four incompatible systems with a unified structured human body model and a unified system of syndrome differentiation and treatment compatible with it will also be an inevitable development.

4.5 Differentiation of symptoms and signs to identify etiology: The etiology model compatible with the human body model

People live in nature and society, forming a closely related and indivisible whole with their environment. If the changes in the external environment exceed the body's self-adjustment and self-adaptation capabilities, the body's state can deviate from the daily steady-state balance. The pathogenic factors that can cause disease include physical, chemical, biological, and psychological factors; and the mode of action includes continuous, random, and cumulative ways. The discovery of some pathogenic factors is limited by the development level of science and technology and the depth of the understanding of the human body; some diseases are the result of the combined action of multiple factors in different time sequences.

Research on these pathogenic factors themselves and their pathogenic actions is undoubtedly necessary to eliminate the cause, prevent and treat the disease caused by these pathogenic factors, which is the main method of modern medicine to deal with the cause. But even entity pathogenic factors, such as bacteria, viruses and other microorganisms, were invisible to people at the level of ancient science and technology. Moreover, some pathogenic factors are non-physical, and will not remain in the human body after the disease is caused. What is left in the body is the deviation of the state caused by it. Therefore, it is impossible in many cases to base the understanding of the pathogenic factors on the analysis of these factors themself. But the problem is that these pathogenic factors must be effectively dealt with. How to do it?

4.5.1 Etiological model based on the body's response state

In TCM, the method called 'differentiation of symptoms and signs to identify etiology' is adopted to establish its unique etiology theory. This method is to investigate the changes in human body structure and function caused by pathogenic factors acting on the human body, and, in turn, establish the model about the characteristics of pathogenic factors and their pathogenic actions. Specifically, the body's states before and after the pathogenic factors act on the human body are used as the input and output of the black box of the pathogenic factor respectively, and based on the analysis of the corresponding relationship between the input and output, the behavioral model of pathogenic factors is established through metaphors and analogies.

As mentioned earlier, in order to make up for the insufficiency of the viscera-qi-blood-body fluid-essence syndrome system in dealing with epidemic and infectious diseases, TCM has developed the 'six meridians', 'wei-qi-ying-xue' and 'tri-jiao' syndrome systems. Compatible with the syndrome differentiation system for exogenous diseases, TCM has perfected the 'six evils' etiology models that mainly cause exogenous diseases: wind, cold, summer heat, dampness, dryness, and fire. And according to the pathogenic characteristics of the 'six evils' described by the model, its application is extended to the treatment and analysis of internal damage diseases. Based on similar methods, TCM has established a 'seven emotions' model that can cause internal damage diseases: happiness, anger, worry, thought, sadness, fear, and shock. It has also established models of 'phlegm' and 'blood stasis', which are both metabolite of human life activity and also pathogenic factors. As a result, the etiology model system based on the difference of the body's response state caused by the etiology has been perfected (Yuan Bing 2010).

The 'six evils' pathogens have common characteristics, such as entering from the mouth and nose or fur and spreading from the outside to the inside. However, some of these pathogenic factors such as wind, cold, dampness, dryness, and fire can also be produced in the body, due to deficiency of yin, yang, qi, blood and body fluid in the body or their abnormal delivery and distribution. In order to distinguish the pathogenic factors which come from the external environment, TCM calls those pathogenic factors that arise from the inside internal wind, internal cold, internal dampness, internal dryness, and internal fire.

Taking 'wind' as an example, let's look at the process of establishing etiological models in TCM. In clinical manifestations, the symptoms of exogenous wind syndrome are fear of wind and cold, fever, pulse floating, etc., internal wind manifests as dizziness, convulsions, tremors, spasms, itching, etc. Both external wind and internal wind have the characteristics of rapid onset, erratic location of symptoms and more changes in the condition. In the treatment of external wind, TCM usually uses the method of dispelling wind to expel external evils from the body surface; the treatment of internal wind includes calming the liver to quench the wind, nourishing yin or nourishing blood to quench the wind. Based on the clinical features of this type of syndrome and the metaphors and analogies on this basis, TCM calls the pathogenic factor that causes this type of pathological state as 'wind', thus establishing a 'wind' behavior model:

1. The wind is a yang evil, and its characteristics are open and venting. Wind is the main qi in spring, with the characteristics of rising, upward and outward. Wind evil can easily invade the upper part of the human body (such as the head and face) and body surface, and cause the skin to vent, causing symptoms such as sweating and fear of wind.
2. The attributes of wind are good at moving and changeable, so the disease caused by wind evil has the characteristics of moving, and no fixed disease location. For example, in arthralgia caused by the mixed wind, cold and dampness, if the pain location is not fixed, it is caused by wind evil, which is called 'migratory arthralgia'. The variability of wind evil is vividly manifested in rubella (urticaria): skin itching occurs from time to time, and there is no fixed area for the appearance of rashes. The appearance in one place is often accompanied by dissipation elsewhere, and it sometimes appears and sometimes disappears. At the same time, diseases mainly caused by wind evil generally have rapid onset and rapid transmission. If the wind evil attacks the head and face, the mouth and eyes can be slanted suddenly (facial nerve palsy). Another example is wind-water syndrome, when the onset only has symptoms similar to a cold, but within a short period of time, symptoms such as swelling of the head and face and short urination can occur (this situation may occur in some patients with acute nephritis).
3. The disease caused by wind evil has the characteristic of constantly swinging. If wind evil invades the human body, facial muscle twitching, or dizziness, tremors, convulsions, stiff neck, opisthotonos, and binocular upward vision often appear.
4. Wind evil is the forerunner of external evil, and it often harms the human body by combining with other evils. Because the nature of wind is open and diffuse, the evils of cold, dampness, dryness, and heat are often attached to the wind and invade the human body together, thus forming exogenous wind-cold, wind-wetness, wind-heat, and wind-dryness. The second is that wind evil is always present all year round, so there are many opportunities to cause diseases. The wind evil that invades the human body is pervasive, can spread throughout the internal organs and peripheral tissues of the human body, and causes a variety of diseases.

The above description is full of metaphors and analogies between the clinical manifestations of a specific disease and the attributes of 'wind'. In TCM, the etiological models of the 'six evils' of wind, cold, summer heat, dampness, dryness and fire are all established by this method, as shown in Table 4.7.

In the etiology theory of TCM, the 'six evils' are etiological models established by analogy among six climate-related pathogenic factors and corresponding disease characteristics. The 'seven emotions' refer to the seven emotions of people: joy, anger, melancholy, worry, grief, fear, and fright. The 'Seven Emotions' model describes the pathogenic actions of the seven emotions, which are mainly reflected in the excessive or continuous seven emotions, which can lead to the dysfunction of the internal organs and the imbalance of qi and blood. For example, overjoy will hurt the heart, anger will hurt the liver, overthinking will hurt the spleen, and excessive fear will hurt the kidneys.

As for phlegm and blood stasis, it is a pathological product formed by the body's body fluid metabolism or abnormal circulation of qi and blood. After they are formed, they accumulate in the body and can act as secondary pathogens to cause new diseases. Both sputum and rheum are the pathological products of water-liquid metabolism disorders: the thick one is sputum and the clear one is rheum. The phlegm is divided into the tangible phlegm and the intangible phlegm. The tangible phlegm refers to the spitted sputum which is visible and audible. Intangible phlegm means that there is no visible thing, but its pathological manifestations can be eliminated by using phlegm-resolving methods, such as nodules formed between the skin and muscles, and stubborn dizziness. The main manifestations of diseases caused by phlegm-rheum include stuffy chest, cough,

Table 4.7. 'Six evils' etiological models.

Pathogenic factors	**Clinical manifestations**	**Pathogenic characteristics**
Wind	External wind is manifested as fear of wind and cold, fever, pulse floating, etc. For internal wind, the symptoms are dizziness, convulsions, tremors, spasms, and itching.	The wind is open and divergent, and the wind evil tends to invade the upper part and body surface of the human body; the attributes of wind are good at moving and changeable, and the pain is characterized by wandering and is unfixed; the diseases caused by wind evil have features of rapid onset, rapid transmission, and with many changes; the diseases caused by wind evil show signs of constantly swinging, such as dizziness, tremors, convulsions; wind evil often combines with other evils and can cause many diseases.
Cold	External cold is manifested as fear of cold, fever, no sweat, headache, body pain, floating and tight pulse, etc. Internal cold is manifested as cold limbs, fear of cold and liking for warmth, cold pain in the abdomen which reduces with warmth, clear urine and long urination, loose stools, not thirsty, or like hot drinks, etc.	Cold is yin evil, easy to damage the yang qi, and yang deficiency leads to cold; cold causes coagulation and block, so that it is easy to cause blood stasis and pain; cold shrinks so that tends to cause cramps, and the sweat holes are closed without sweat.
Summer heat	Manifested as high fever, thirst, upset, yellow urine, hyperhidrosis, and strong pulse.	The summer heat is a yang evil, and its nature is hot; the summer heat rises and spreads, disturbing the spirit, depleting the body fluid, and hurting the qi; summer heat often causes diseases with wetness together.
Wetness	External dampness is characterized by heaviness of the head and body, joint and muscle pain, full stomach stuffiness, etc. Internal dampness causes abdominal distension and discomfort, heavy pain in the limbs and joints, fixed pain, tasteless in mouth, not thirsty, nausea, vomiting, stool sticky and not smooth, short urination, etc.	The nature of wetness is heavy, turbid, and sticky, and diseases caused by it appear easily clinical manifestations characterized by a heavy feeling, and show the phenomenon of muddy and sticky secretions and excrements. Disease caused by wetness evil has a slow onset, longer course, recurring or lingering attacks.
Dry	External dryness is characterized by dry mouth, dry nose, dry lips, dry throat, itchy throat, thirst, dry cough without sputum, or hypochondriac pain, etc. Internal dryness is manifested as dry mouth, dry throat, dry lips, dry skin, and chapped skin, constipation, short urine, yellow hair is not prosperous, upset and thirsty, thin body, red tongue, dry tongue, less coating, thin pulse, etc.	Dryness evil readily damages body fluid and lungs.
Fire	Fire is similar to heat, but hotter. Exogenous fiery heat is characterized by fever, thirst and preference for cold drinks, and even high fever, irritability, redness of the face, red eyes, small urine volume and yellow color, or delirium, bleeding, spitting blood, erythema, red or crimson tongue, and thin yellow or black-yellow coating, fast pulse. Internal fire (internal heat) is characterized by fever, thirst and preference for cold drinks, constipation, short urine, sore throat, ulcer in mouth and nose, or abdominal distension and pain, increased pain when exposed to heat, red tongue, yellow coating, fast pulse.	Fiery heat is the yang evil, and its nature is upward and hot. Therefore, most of the diseases caused by it are heat syndrome, and the symptoms of the upper part of the human body often appear; clinical manifestations such as dizziness, convulsions, and bleeding to appear; it is easy to disturb the spirit, and causes sores and carbuncle.

panting, spitting sputum, nausea, vomiting, palpitation, dizziness, mental disorders, numbness of limbs and body, pain or swelling of joints, subcutaneous tumor or ulcer with pus, edema, ascites, diarrhea, etc. There are many reasons for blood stasis. For example, the invasion of cold evil can lead to blood stasis; anxiety and qi stagnation can cause blood stasis; blood stasis can also be caused due to weakening of the driving force of qi deficiency or insufficient warmth of yang. The main clinical manifestations of diseases caused by blood stasis are black complexion, dark purple lips and claws/ nails, blue skin, local fixed tingling, refusal to press, purple hematoma, continuous bleeding, dark purple with blood clots, black stools, amenorrhea, dark purple tongue or petechiae, etc. (Beijing College of TCM 1974).

As mentioned above, whether it is the six evils model based on metaphors and analogies, the seven emotions model based on emotional factors, or the models of phlegm and blood stasis as metabolite, the descriptions of their pathogenic actions are based on the human body model of TCM. In other words, TCM's understanding of the cause of disease is based on the body's response state caused by the pathogenic factors. Compared with the etiology theory of modern medicine, which is established through the analysis and research of pathogenic factors, what are the advantages of TCM's method of establishing an etiology theory? We can see clearly by comparing the methods to deal with infectious diseases in TCM and in Western medicine.

4.5.2 The superiority of establishing etiological models based on the body's response state in disease regulation-control

Most of the current epidemics are caused by viruses. As in the past diseases caused by pathogens, modern medicine puts the hope of overcoming them on the study of the pathogen (virus). It seems that as long as a method can be found to kill the virus, everything will be fine. Research on methods to deal with viruses usually separates the virus from the diseased body first, and then studies medications that can treat it, or makes vaccines that can kill or inactivate the virus. By vaccinating the human body, it can produce corresponding antibodies to achieve the purpose of preventing virus infection. However, this research concept is currently clearly challenged in several aspects:

1. The virus mutates too quickly during transmission. Vaccines against previous virus types have appeared in multiple different virus variants before they were clinically available. Since 2021, the vaccines against COVID-19 that have been inoculated around the world are all developed based on virus strains isolated a year ago. After several rounds of large mutations, the virus that causes infection will mainly be Delta and Omicron variants by 2022, and the existing vaccines have basically been unable to prevent them.
2. Medications that can kill viruses in *in vitro* tests always have side effects of one kind or another when used in humans. Just like SARS in 2003, after the epidemic period passed, a large number of victims were left with severe sequelae due to medications such as interferon, Tamiflu, and high-dose hormones. At present, the side effects of various vaccines for COVID-19 have also been widely reported, and cases of serious side effects and even death due to vaccination have also been reported.
3. At present, a large number of vaccinated people have clinical presentations similar to viral infection, such as chills, fever, headache, muscle, joint pain, and fatigue, are no less severe than those of patients infected with COVID-19.

On August 26, 2014, Michaeleen Doucleff, a reporter on the National Public Radio (NPR) website, published the article 'How Ebola kills you: It's not the virus.' Combined with the latest research by scientists, the article described the pathogenic mechanism of Ebola virus, and made a surprising discovery: It was not Ebola virus itself that killed the patient in the end, but the patient's own immune system. It is the immune cell's response to the virus damage and the resulting secondary damage to the body that kills the patient. When studying the role of viruses in the onset, development and evolution of diseases caused by them, the following facts cannot be ignored:

1. The existence of 'healthy virus carriers' shows that even if the virus invades the human body, it does not necessarily cause disease.
2. Although the disease-causing virus can be found in the blood of some patients who have recovered from a viral infection, it no longer causes damage to them. The reason is that the patient's immune systems have developed immunity to this virus.
3. In the process of the development and evolution of virus-induced diseases in the human body, not all damages are directly caused by the viruses. Many damages are the result caused by the virus's damage passing along the inherent chain of causality in the human body.

Obviously, during the disease process, even if we can kill the virus, it does not mean that all the problems of viral diseases have been solved. Without killing the virus that has invaded the human body, the infected human body also does not necessarily get sick or definitely not recover. From this, it can be concluded that the role of the virus itself in the development of viral diseases is not always decisive. The research on the virus itself is not the only way to solve the disease caused by the virus (Yuan Bing 2014).

In recent years, clinical studies on infectious diseases caused by bacteria and viruses, have shown that in many cases, although the viruses that invade the human body are different, the body's response state is the same or similar. For example, SARS, H3N2 Covid-19 and other infectious diseases caused by different viruses may all have clinical manifestations of fever, headache, muscle and joint pain, sore throat. Even the early clinical manifestations of many bacterial infectious diseases (such as infections caused by streptococcus and pneumococcus) are similar. The same clinical manifestations usually mean the same response state of the body to pathogens. This just reflects the isofinality of life organisms revealed by Bertalanffy in his works.

Using the same TCM treatment method, such as relieving wind-heat, these diseases caused by different pathogens can be improved. This provides a possibility: as long as the deviation of the body state caused by the pathogenic factors is corrected, the purpose of improving diseases caused by pathogens can be achieved. In other words, if we focus the treatment of infectious diseases on the body's response state caused by pathogens, not only do we ignore the mutation and type of the virus, but we can even ignore whether it is virus or bacteria that caused the disease. As long as we find a way to deal with this response state and correct it in time, prevent the secondary damage that this pathological response state may cause, and stop the process of disease transmission along the chain of fixed causality, the purpose of treatment can also be achieved (Yuan Bing 2020).

It is like a big rock appearing in the middle of the road, blocking the passage of vehicles. Of course, we can first explore where this stone came from: is it a meteorite falling from the sky? Is it a rock falling from a landslide? Did it slip off a passing transportation vehicle? Or was it carried on the road by somebody intentionally to obstruct traffic? Then determine the solution according to different situations. One can also ignore all these, get a crane truck, remove the stone, or move it to the side of the road, and the problem will be solved. The latter method seems much simpler, and is equally effective. The focus of dealing with infectious diseases is on the body's response state. This is exactly the secret of how ancient China, where science and technology lag far behind modern times, could still find effective ways to deal with influenza and even large-scale plagues.

On the other hand, the human body has the ability of self-organization, self-adaptation and self-regulation. Under the action of the same pathogenic factors, it is uncertain whether a person will get sick, and the nature and extent of the disease. It not only depends on the difference in the quality and quantity of the pathogenic properties of the pathogenic factors, but also has an extremely important relationship with the body's self-regulatory ability and the state of the regulatory mechanism. Therefore, it is not objective from a control point of view to grasp the pathogenic factors only by analyzing the pathogen itself.

Pathogens that invade the human body form an interconnected whole with the human body, and cause harm through interaction with the human body. TCM studies the changes in the human condition caused by pathogenic factors based on the symptoms and signs obtained from the outside of the human body, which can accurately grasp the results of the interaction between pathogens and the human body. Using this result as a controlled amount, the methods and medications found to improve and eliminate them are precisely aimed at the damage caused by pathogens to the human body.

In TCM, the pathogenic factors of infectious diseases are classified according to the response state of the human body. The 'six evils' of wind, cold, heat, dampness, dryness, and fire, and the types of pathogenic factors such as wind-heat, wind-cold, and damp-heat compounded on this basis, are the division of the body's response state caused by pathogens. Methods and medications that can improve or eliminate a certain type of pathological response state can obviously treat diseases caused by corresponding pathogens. In other words, the treatment methods and medications developed by using these response states as controlled amounts can undoubtedly improve or eliminate the damage caused by the corresponding pathogenic factors. In this process, these pathogenic factors would surely be killed or inactivated. If these pathogenic factors still play a harmful role, it would be difficult to improve the abnormal response state of the body, and repair the damage caused by the pathogenic factors.

The clinical practice of TCM has fully proved that the method of establishing the knowledge of pathogenic factors based on the body's response state is quite effective. When we are unable to determine the causative factors or the causative factors are too complicated and difficult to grasp, through this method, we can still accurately understand the nature of the causative factors, find appropriate counter-measures, and implement timely and effective treatment. In recent years, in response to various infectious diseases caused by frequently mutated viruses, TCM's method of 'respond to all changes with the same' can always find appropriate counter-measures in time, which is a strong proof of the effectiveness and scientific nature of this etiological research method.

4.6 Medication properties and efficacy: The functional models of herbal medicines compatible with the human body model

Treatment usually requires some kind of intervention to influence the body in order to change its pathological state. Therefore, before using certain interventions (such as medications), it is necessary to understand their properties and actions on the human body. In the treatment methods developed in TCM in the process of fighting against diseases for thousands of years, the most important ones are CHMs, and the herbal formulas system based on CHMs. And in the process of long-term application of them to treat diseases, their functional models have been established and improved.

4.6.1 Medication models

In ancient TCM, limited to the level of scientific and technological development, it was impossible to use analytical methods to study the role of CHMs in the human body and the mechanism of action. Therefore, the only way to understand the properties and actions of CHMs can be through constant 'taste' and 'try'. The so-called 'taste' is to understand the nature, breath, and taste of the CHM through tasting, and then to deduce the properties of the CHM; the so-called 'try' is to summarize the reaction of the CHM after it acts on the human body, and then deduce and verify its performance and effect. The ancient Chinese legend 'Shen Nong tastes a hundred herbs and encounters seventy poisons in a day' is a vivid description of this process. TCM describes the properties and actions of CHM by establishing a CHM function model, and then establishes the theoretical system of CHM. The properties and tastes of CHMs (four properties and five tastes) describe the basic properties of CHMs. The knowledge accumulated in the long-term clinical application about the various aspects of the role of the CHM relative to the TCM human model, combined with the basic properties of the CHM, constitutes a complete function model of the medication. The medication model's description of CHM properties and actions mainly includes the following aspects (Ling Yikui and Yan Zhenghua 1984):

(1) Properties and tastes

Properties and tastes, also known as four properties and five tastes, is a description of the basic properties of CHM. The property and taste of CHMs is usually strongly related to their efficacy and therapeutic actions. The four properties include four medicinal properties: cold, heat, warm and cool. Generally, cold-cool CHMs have the actions of clearing away heat, purging fire and detoxification, and are mostly used to treat heat or febrile diseases; warm-heat CHMs have the actions of warming, helping yang and dispelling cold, and are mainly used to treat cold or cool diseases. In addition, there is a kind of medicine that is gentle in nature and mild in action, called mild medicine.

Five tastes include pungent, sweet, sour, bitter, and salty:

1. *Pungent*: It has the effects of sweating, promoting qi, and promoting blood. Medicines that induce sweat and dispel exogenous evils, such as ephedra (*Ephedrae herba*), ginger (*Zingiberis rhizoma recens*), peppermint (*Menthae haplocalycis herba*); medicines for promoting qi and blood, such as costusroot (*Aucklandiae radix*), safflower (*Carthami flos*).
2. *Sweet*: It has the effects of tonifying, relieving muscle tension, and relieving pain, and is also used to tone medicinal properties. Nourishing and strengthening medicines such as Codonopsis pilosula and Prepared rehmannia root; medicines for relieving spasms, relieving pain, and toning medicinal properties, such as caramel and licorice (*Glycyrrhizae radix et rhizoma*).
3. *Acid*: It has astringent actions and is used to stop sweating and diarrhea. For example, cornus officinalis (*Corni fructus*), schisandra (*Schisandrae chinensis fructus*) to strengthen essence and stop sweating, and Galla chinensisl to relieve diarrhea.
4. *Bitterness*: it has the action of purging fire, clearing away heat, drying dampness, and promoting excretion promotion. For clearing heat and purging fire, such as gardenia (Gardeniae fructus); for laxative, such as rhubarb (*Rhei radix et rhizoma*); for drying dampness, such as Rhizoma atractylodis, Scutellariae radix and Coptidis rhizoma.
5. *Salty*: It has the actions of purging and laxative, softening and resolving lumps. For example, Ark shell (*Arcae concha*) softens and resolves lumps, and Glauber's salt (*Natrii sulfas*) is used as a laxative, both are salty medicines.
6. There is another kind of light taste, which has the effect of penetrating water and dampness and facilitating urination, such as Poria and Rice paper pith (*Tetrapanacis medulla*).

In the process of establishing the theoretical system of CHM, the taste of CHM is first derived from the taste in the mouth, but after the corresponding relationship between taste and efficacy is formed, the taste of some CHMs is determined based on their clinical efficacy. Therefore, in the CHM system, the properties of some CHMs are not consistent with their taste, such as the sour taste of red stone fat (*Halloysitum rubrum*), the salty taste of Oyster shell (*Ostreae concha*), and the spicy taste of ephedra (*Ephedrae herba*).

(2) Lifting, lowering and floating, sinking

The lifting, lowering and floating, sinking refer to the trend of CHM action. In the process of various diseases, the pathological states exist in different trends including upward (such as vomiting, asthma and cough), downward (such as diarrhea, metrorrhagia, prolapse of the anus), outward (spontaneous sweating, night sweats), and inward (for example, disease evil invades inward), etc. To improve these trends, it is necessary to understand the properties of lifting, lowering, floating, and sinking of CHMs. The lifting trend is upward, the lowering trend is downward, the floating trend is outward, and the sinking trend is inward. Lifting and floating CHMs have the actions of lifting yang, expressing, and dispelling cold, such as Ephedra (*Ephedrae herba*), Cassiabarktree twig (*Cinnamomi ramulus*), and

Astragali (*Astragali radix*). Most CHMs that action lowering or sinking have cold/cool property and bitter/sour tastes, such as Rhubarb (*Rhei radix et rhizoma*), Glauber's Salt (*Natrii sulfas*), and Amur corktree bark (*Phellodendri chinensis cortex*). Most of CHMs with lifting and floating tendency are mosaic leaves and light-weight CHMs, such as magnolia (*Magnoliae flos*), lotus leaf (*Nelumbinis folium*), cohosh (*Cimicifugae rhizoma*) and so on. Most CHMs with lowering and sinking tendency are seeds, fruits and heavy-mass medicines, such as Perilla seed (*Perillae fructus*), Immature bitter orange (*Aurantii fructus immaturus*), Gypsum (*Calcitum*), etc.

For clinical prescription, it is necessary to select CHMs with corresponding action trends according to the disease site on the body surface or inside the body, as well as the upward or downward trend of the disease, in order to effectively treat the disease. Specifically, if the lesion is on the surface of the body, it is appropriate to use diaphoretic CHMs to eliminate external pathogens rather than sedimentation CHMs. For example, Wild Mint Herb (*Menthae haplocalycis herba*) and chrysanthemum (*Chrysanthemi flos*) should be used for exogenous wind-heat. If the disease is on the rise trend, the CHMs with lowering rather than lifting trend should be used. For example, if the liver-yang is hyperactive and dizzy, the lowering trend CHMs such as red ocher (*Haematitum*) and Sea-ear Shell (*Haliotidis concha*) should be used. If the disease is on lowering trend, the CHMs with lifting rather than lowering trend should be used. For example, if chronic diarrhea causes qi deficiency to sink, in turn leading to prolapse of the anus, astragalus (*Astragali radix*), cohosh (*Cimicifugae rhizoma*), bupleurum (*Bupleuri radix*) should be used to lift the prolapsed anus.

The description of efficacies and therapeutic actions of CHM is based on the theoretical model of TCM. For example, the effects of invigorating qi and nourishing blood are aimed at the syndromes of qi deficiency and blood deficiency; the effects of activating qi and promoting blood circulation are aimed at the syndromes of stagnation of qi and blood stasis. To aim at cough and asthma, there is the effect of relieving cough and asthma, and to aim at bleeding and sweating, it has the effect of stopping bleeding and stopping sweating. Through the correspondence between the CHM efficacy and syndromes/diseases, and the correspondence between syndromes and symptoms/signs, TCM practitioners can flexibly apply these CHMs in clinical practice according to the actual condition of the patient.

For example, Angelica (*Angelicae sinensis radix*), which ranks first among blood tonics, is described in Chinese Materia Medica as: its property and taste are sweet, pungent, and warm. It has the actions of nourishing, warming, and dispelling cold, which is consistent with the description of its efficacy: nourishing blood, promoting blood circulation, relieving pain, and moistening the intestines. Its clinical applications:

1. Used for various syndromes of blood deficiency.
2. Used for irregular menstruation, amenorrhea and dysmenorrhea.
3. Used for abdominal pain due to deficiency and cold, pain caused by blood stasis, injury caused by falling and being hit, as well as arthralgia and numbness.
4. Used for carbuncle, gangrene, sore and ulcer, it can reduce swelling, relieve pain, drain pus and promote muscle regeneration.
5. Used for constipation due to blood-deficiency and intestinal dryness.

With the aid of the functional model of the medication, although we do not know the composition of a CHM and the actual mechanism of its action in the human body, we can still grasp the property of the CHM, and apply it appropriately in clinical treatment. Currently, there are about 400–600 types of CHMs commonly used by TCM practitioners. In the sixth year of Wanli in the Ming Dynasty (1578), the Compendium of Materia Medica by Li Shizhen (1518–1593), the most famous medical scientist, pharmacist and naturalist in Chinese history, included 1892 kinds of CHMs in various categories. The 'Big Dictionary of Chinese Herbal Medicine' published since the founding of the PRC contains 5767 kinds of CHM. In these books about Chinese Materia Medica, the source, place of origin, nature, taste, efficacy, and indications of each CHM are stated.

4.6.2 The herbal formulas models

The action of each CHM is limited. It is usually not enough to use one kind of CHM to deal with complicated diseases, especially when the patients have multiple syndromes and diseases at the same time. From ancient times to the present, when TCM practitioners use CHM to treat patients, they usually need to formulate a variety of different CHMs into prescriptions according to the patient's specific condition. Targeted selection of prescriptions composed of a variety of single CHMs can not only greatly enhance the therapeutic actions of specific aspects, but also reduce the side effects of certain CHMs to produce a better overall effect. Even for patients with only one syndrome or disease, it is sometimes necessary to organize multiple CHMs with different mechanisms of action to be applied together to produce a synergistic effect.

In the long history, ancient TCM scholars created a large number of effective herbal formulas on the basis of summarizing clinical practice. The 'Formulas of Fifty-two Diseases' found in the Mawangdui Han Tomb in Changsha is the earliest extant herbal formulas book, and it was written about the third century BC. The book contains 283 herbal formulas of various CHMs, and also describes the dosage forms of soups, pills, and powders. Although the 'Huangdi Neijing, The Yellow Emperor's Inner Canon' in the Warring States Period contained only 13 herbal formulas, it had detailed discussions on the principles of TCM treatment, the composition and structure of prescriptions, the coordination among CHMs, and the appropriate way and contraindications of taking CHMs, which laid a foundation for the formulas theory of TCM. The 'Shen Nong's Herbal' of the Han Dynasty, the earliest monograph on Chinese Materia Medica in China, contains theories on how to choose dosage forms. Zhang Zhongjing's 'Shanghan Lun, Treatise on Cold Pathogenic Diseases' contains 113 herbal formulas, and 'The Synopsis of the Golden Chamber' contains 262 herbal formulas. These herbal formulas were regarded as classic formulas by later generations due to their reasonable composition, precise selection of CHMs, accurate dosage, clever changes, and outstanding curative effects. The dosage forms made of these herbal formulas include decoctions, pills, powders, suppositories, ointments, wines, vinegars, enemas, lotions, baths, fumigants, ear drops, nasal lavages, nasal sprays, etc., including almost all traditional dosage forms except injections. Ge Hong's (283–343) 'Formulas Behind the Elbow', the only surviving book from the Jin Dynasty, contained a large number of convenient, cheap and effective prescriptions. The book proposed the concept of finished medicine for the first time, advocating that the medication be processed into a certain dosage form and stored for emergency use. The 'Essential Recipes for Emergent Use Worth Thousands Gold' written by Sun Simiao (541/581–682) in the Tang Dynasty contained 5,300 herbal formulas. Wang Tao's (670–755) 'Waitai Miyao Fang; Arcane Essentials from the Imperial Library' contains more than 6000 herbal formulas. In the Song Dynasty, the 'Taiping Shenghui Fang; Taiping Holy Formulas for Universal Relief' compiled by the government organization contained 16,834 herbal formulas, and the 'Sheng Ji Zonglu; General Records of Holy Universal Relief' contained more than 20,000 herbal formulas. 'Heji Bureau Formula' contains 297 herbal formulas, which is the first written pharmacopoeia of proprietary Chinese medicines issued by the government. 'Puji formulas' (edited in the 23rd year of Hongwu, 1390 AD), compiled under the organization of Zhu Su (1361–1425), the king of Zhou Ding in the Ming Dynasty, contains 61,739 herbal formulas, and it is the book with the most herbal formulas.

In the history of thousands of years with written records, the number of herbal formulas contained in ancient TCM books is huge, and usually described as 'the vast sea of smoke'. More than one CHM is combined into an herbal formula for the patient to take together, and the CHMs organization in it is not messy and irregular. A good herbal formula for treating a certain disease or a certain pathological state is usually not a simple accumulation of similar CHMs with the same or similar therapeutic actions, but is often composed of CHMs with different properties and actions. Due to the mutual synergy or antagonism among CHMs of the prescription, the prescription usually forms therapeutic actions that cannot be achieved by any one of the CHM ingredients. For example, Four Gentlemen Decoction combines ginseng which has the action of invigorating qi and invigorating the spleen, atractylodes macrocephala which can invigorate the spleen and dry dampness, poria cocos which can deal with diuresis and dampness, and licorice which harmonizes various CHMs. These CHMs cooperate with each other and have a good effect on patients with spleen and stomach deficiency, which is manifested by pale complexion, weak voice, loss of appetite, loose stools, and weakness of the limbs.

So far, the composition of the prescription still follows the principles established in the 'Huangdi Neijing, Yellow Emperor's Inner Canon' and is divided into four parts: monarch, minister, assistant and guide. The monarch medication is the CHMs that plays the main therapeutic action on the main syndrome in a prescription. It embodies the main direction of the prescription, its power ranks first in the prescription, and it is an indispensable part of a prescription. The minister medication assists the monarch medication to enhance the action of treating the main syndrome. The significance of assistant medication has three aspects: one is as an adjuvant medication, used to treat secondary syndromes and symptoms; the other is to eliminate the toxicity or slow-down the potency of monarch and ministerial medications; the third is according to the needs of the disease to use CHMs that are contrary to the properties of the monarch medications and can play a synergy and cooperation role in treatment. The role of the guide medication is to guide the medications in the prescription to the place where the disease is, or to harmonize the CHMs in the prescription to make it work better (Guangzhou College of TCM 1979). For example, the Ephedra Decoction in 'Shanghan Lun, Treatise on Cold Pathogenic Diseases' consists of four CHMs: ephedra (*Ephedrae herba*), cassia twig (*Cinnamomi ramulus*), bitter almond (*Armeniacae semen amarum*) and licorice (*Glycyrrhizae radix et rhizoma*). Indications: fear of cold, fever, headache, body pain, no sweating, wheezing, thin white tongue coating, floating and tight pulse, etc., which manifestation belongs to the wind-cold exterior repletion syndrome. In the prescription, ephedra is pungent and warm in nature, which has the action of expelling exterior evil, dispersing lung qi, and calming wheezing. It is the monarch medication aimed to the main syndrome. Cassia twig is pungent and warm in nature, which has the action of expelling exterior evil, dredging the ying and wei. It helps ephedra to increase sweating power and thus is a minister medication. Bitter almonds purify the lung and descend lung qi, which helps ephedra to calm wheezing and thus is an assistant medication. Licorice can reconcile the strong sweating of ephedra and cassia twig, so it is a guide medication.

TCM practitioners determine the principle of treatment on the basis of syndrome differentiation, and then organically combine CHMs with different or similar actions into prescriptions according to the needs of treatment. Prescriptions can exert a synergistic action that is difficult to achieve with a single CHM, and effectively restrict the side effects of a single CHM, thereby achieving better overall efficacy. However, after the prescription is composed, can it play the expected therapeutic effect in patients? The formation of a handed down formula usually requires the continuous improvement and adjustment of its composition and ingredient ratio by TCM practitioners in repeated clinical practice. The process of verification and improvement of a formula in clinical practice is actually the process of establishing a formula model through the investigation of the input and output of the formula black box. In this process, the patient's body state before and after taking the formula is regarded as the input and output of the formula black box. Since the description of the human body state by TCM practitioners is based on the theoretical model of TCM, the conclusions on the properties and actions of the formula obtained in practice are naturally relative to the theoretical model of TCM. Therefore, the formula models thus established are convenient to learn and understand by later TCM practitioners who are familiar with the theoretical model of the human body of TCM, so that these handed down formulae can be widely spread and used.

4.6.3 Models for therapeutic interventions must be fitted to human models

To use a certain medical theory to direct the application of medications, the controlled amount in medication action research must come from this medical theoretical model and SDS. This is an unquestionable way from a methodological point of view and is accustomed to modern medicine and TCM. However, after modern medicine was introduced into China and began to collide with TCM, this point became blurred, and sometimes even caused great controversy, which made people confused.

For example, aspirin is one of the three classic medications in the history of medicine. It has been used for a hundred years and is still the most widely used antipyretic, analgesic and anti-inflammatory medication in the world. It is also a standard preparation for comparing and evaluating other medications. The description of its role in modern medicine mainly includes the following aspects:

1. *Antipyretic and analgesic.* It can relieve mild or moderate pain, such as headache, toothache, neuralgia, muscle pain and menstrual pain. It can also be used to reduce fever in colds, flu and other diseases. It can only relieve symptoms and cannot treat the causes of pain and fever, so it is often used in combination with other medications in clinical practice.
2. *Anti-inflammatory and anti-rheumatic.* Aspirin is the first choice for the treatment of rheumatic fever. It can relieve fever, reduce inflammation, improve joint symptoms and decrease erythrocyte sedimentation rate, but it cannot remove the basic pathological changes of rheumatism, nor can it prevent heart damage and other complications.
3. *Treat arthritis.* In addition to rheumatic arthritis, it is also used to treat rheumatoid arthritis, which can improve symptoms and create conditions for further treatment. In addition, it can also be used for osteoarthritis, ankylosing spondylitis, juvenile arthritis and other non-rheumatic inflammation of skeletal muscle pain to relieve symptoms.
4. *Anti-thrombotic.* It has an inhibitory action on platelet aggregation and prevents thrombosis. It can be clinically used to prevent transient ischemic attack, myocardial infarction, atrial fibrillation, artificial heart valve, arteriovenous fistula or other thrombosis after surgery. It can also be used to treat unstable angina pectoris.

Salvia miltiorrhiza (*Salviae miltiorrhizae radix et rhizoma*) is a commonly used CHM that has the action of promoting blood circulation and removing blood stasis and has been used in TCM for thousands of years. The description on it in Chinese Materia Medica is as follows:

Property and taste: bitter, slightly cold.

Actions: activating blood and regulate menstruation, dispelling blood stasis and relieving pain, cooling blood to eliminate carbuncle, clearing heart and removing annoyance, nourishing blood and calming nerves.

Indications: Irregular menstruation, amenorrhea, dysmenorrhea, tumors and masses, chest and abdomen tingling, heat arthralgia pain, sore, swelling and pain, upset and insomnia.

According to this description, aspirin can only be used in modern medicine as a Western medication, while salvia can only be used in TCM as a CHM. In the Republic of China era, Mr. Zhang Xichun (1860–1933), a well-known TCM doctor in Hebei Province, created a gypsum aspirin decoction to treat wind-cold exterior syndrome, and used potassium bromide with calcined keel, calcined oyster, dogwood to treat dream emission (Zhang Xichun 2009). In fact, this is the use of aspirin and potassium bromide as CHM. However, modern pharmacological studies of CHM have found that salvia miltiorrhiza extract has the action of enhancing myocardial contractility, dilating coronary arteries, and is anti-thrombosis. When modern medicine clinically uses the processed products of salvia miltiorrhiza extract for cardiovascular patients, it is actually used as a Western medication.

In this sense, the existing medications and treatment means belong to the two systems of modern medicine and TCM respectively, and are not determined by the characteristics of these medications and treatment means themselves. In principle, which system is used as the reference system for the description of the action of a medication, and this medication can be used for clinical treatment under the guidance of this system. In other words, if through clinical research, the actions of a medication can be respectively described with the TCM system or modern medicine system as the reference frame, it could be freely used in clinic treatment of both TCM and modern medicine. For example, according to the characteristics of aspirin and with reference to the theoretical model of TCM, the role of aspirin relative to the human body model of TCM can be summarized as:

Property and taste: pungent, warm.

Efficacy: sweat to expelling exterior evil, dredge blood vessels, dispel rheumatism, promote blood circulation and remove blood stasis.

Indications: exogenous wind-cold, body pain, headache, rheumatic arthralgia, chest pain and heartache.

Due to the sweating effect of aspirin, continuous use can easily lead to spontaneous sweating, and excessive sweating can lead to deficiency of body fluid and qi. Its pungent and warm property tends to fuel the fire, which may cause coughing, chest tightness, and skin rash due to lung heat, and may also cause irritability and mental excitement due to inflammation of the heart fire. Long-term use of pungent and warm products may also injure the yin and move the blood, leading to bleeding tendencies.

Here, the generalization of the performance of aspirin based on the human body model of TCM is only a preliminary summary of the understanding of its pharmacological effects and adverse reactions based on modern medicine, without clinical observation and statistical analysis using the TCM human body model as a frame of reference. But it can be seen from this that using the human body model of TCM as the frame of reference and through the long-term observation of various clinical reactions of aspirin, we can establish the functional model of aspirin relative to the human model of TCM. Based on such a model, we can use aspirin as a CHM in TCM's clinical practice. In fact, the gypsum aspirin decoction of Mr. Zhang Xichun is a successful example of using aspirin as a CHM in this way.

From this perspective, it is not difficult to understand the current embarrassing situation of CHM research using chemical analysis and pharmacological experiments. In recent years, the research on the chemical components and pharmacological actions of commonly used CHMs has been carried out quite in-depth using analytical and experimental methods in China. But why can't these achievements touch the theoretical system of traditional Chinese medicine, reveal its mysteries more deeply in essence, and can't push the application of CHMs to TCM one step further? The reason is that these studies are based on the human body theory of modern medicine as the frame of reference and the actions and action mechanisms of the CHM studied are relative to modern medical human models (such as alleviating smooth muscle, expanding coronary arteries, suppressing immune responses, etc.), not to the human body model of TCM.

4.6.4 Discussion on the methodology of research on treatment means aiming at etiology, symptoms, and syndrome

TCM's treatment of diseases mainly focuses on three aspects: eliminating the pathogenic factors, improving symptoms, and correcting the deviated body state (syndrome). Therefore, the research on treatment methods and medications is divided into three aspects including to target the pathogenic factors, disease, and syndrome (Yuan Bing 2010).

(1) Research on treatment methods for pathogenic factors

Modern medicine's understanding of pathogenic factors is based on the direct analysis of pathogenic factors. Therefore, the controlled quantity of research on the treatment methods (especially medications) target to causes is naturally the actual pathogenic factors, such as bacteria, viruses, parasites, etc. As mentioned earlier, TCM's understanding of pathogenic factors is established on the body's response state caused by pathogenic factors, so the controlled amount is actually the initial state deviation caused by pathogenic factors. Due to the different understanding of pathogenic factors and the different ways of stipulating, the two major systems of modern medicine and TCM have different characteristics in the treatment method (or medications) of pathogenic factors.

The specificity of the treatment methods developed by using the pathogenic factors in entity form such as bacteria, viruses, parasites, and cancer cells as controlled quantities is quite high and very sensitive to these pathogenic organisms. However, the drawbacks of medications developed by this research method cannot be ignored. Due to the structural similarity between these pathogenic factors and some tissues of the human body, and the medications usually spread to the whole body through the blood to exert their effects, they often also have effects on the tissues with similar structures where they go. However, these effects cannot always be beneficial to human health, that is to say, side effects are often unavoidable.

TCM bases the understanding of pathogenic factors on the body's response states to them. Therefore, the research on treatment methods (or medications) for pathogenic factors also takes the body response state caused by pathogenic factors as a controlled quantity. The effective methods (or medications) developed by this method can obviously eliminate the deviation of the state caused by the pathogenic factors, and will inevitably directly or indirectly affect the entity pathogenic factors. Because in most cases, if these pathogenic factors that cause the body state to deviate are not eliminated, the deviated state variables cannot be restored to normal. Research of TCM on the treatment methods (or medications) target to pathogenic factors are carried out without taking the entity's pathogenic factors as the controlled amount. Obviously, this method is easier to find some treatment methods (or medications) that eliminate the entity's pathogenic factors by mobilizing one's own disease resistance ability compared with the research method that directly regards the entity's pathogenic factors as a controlled quantity.

Different from the research method that takes the pathogenic factors as the controlled amount, the effective treatment method that uses the response state of the body as the controlled amount is usually the method of killing or inactivating the pathogenic factors in the human body. In this process of studying the state changes caused by pathogenic factors acting on the human body, the body's response state described by the holistic model of the human body is taken as the frame of reference. The holistic model is a macroscopic model, and the state changes caused by pathogenic factors and the actions of interventions (such as medications) on the human body (including therapeutic actions and side effects) revealed by the model, are stable in time and space. Taking the response state of the body described by such a model as a reference frame, it is easy to find a treatment plan that can kill or inactivate pathogenic factors, correct the deviation of the human body state caused by it, and have as little side effect as possible on the human body.

On the other hand, in reality, there are undoubtedly some diseases for which the entity pathogenic factors cannot be detected or isolated, diseases caused by non-entity factors, or diseases for which the entity pathogenic factors have been isolated but no effective treatment has been found for them. In this case, by taking the state deviation caused by the pathogenic factors as a controlled quantity, we can still find ways to eliminate the pathogenic factors, correct the deviation of the body state, and carry out effective treatment. The effectiveness of TCM on virus-induced diseases is a strong demonstration of the superiority of this research method.

More importantly, the environment in which people live is extremely complex, diseases occur, develop, and change in the process of interaction with the environment. It is quite difficult to track down the various factors that have had effects during its development and grasp the various factors that affect the human body now. In this case, the method of grasping the cause from the body's response state shows its superiority. Because the state of the human body at the moment reflects the result of the accumulation and comprehensive action of various factors in the past, the treatment methods (such as medication) developed using the body state as a controlled amount have naturally the treatment effects to address these cumulative and comprehensive factors. This method is obviously simpler and more effective than finding out the factors that affect it one by one and then seeking treatment methods one by one.

(2) Research on treatment methods for single pathological changes or single clinical manifestations

In many cases, the remarkable effects of modern medicine for local pathological changes and symptomatic treatment in quickly controlling the disease, relieving pathological changes, and reducing secondary damage are beyond the reach of TCM. The focus of symptomatic medication research in modern medicine is usually the direct link or adjacent link that occurs in pathological changes. These links are often not the fundamental links in the disease causal chain, but the links where the pathological changes cause compensatory changes. For example, in patients with hypertension, the continuous increase in blood pressure is usually a compensatory response caused by some reason, and the underlying disease is often not in the heart and blood vessels closely related to blood pressure. Since the human body is a whole, that is closely related to the functions of its various parts, the pathological changes of a certain entity part and the changes in the morphology and function of other parts are interrelated and mutually cause and effect. Treatment methods (or medications) developed for the change of morphology and function of a certain link in the disease causal chain can usually temporarily eliminate or improve this part of the pathological changes. However, changes in this part often cannot eliminate the fundamental cause of the disease and the adjacent links before this part in the causal chain, pulling the entire system out of the pathological state. Therefore, due to the action of the primary link of the disease on this part along the causal chain, once the medication action disappears, the state of this part will return to the previous pathological state. This is the reason why people feel that symptomatic medications of Western medicine 'do not eliminate the roots'. Also, due to the adaptability caused by the body's self-regulation and self-adaptive ability, there will be 'rebound' phenomena which could aggravate the condition. The antihypertensive medications developed for the purpose of 'controlling blood pressure' in modern medicine, the antidiabetic medications developed for the purpose of 'controlling blood sugar', and the antiasthmatic medications that relieve bronchial smooth muscle spasm are all the results of this medication research method. These medications developed based on the concept of disease control usually have good short-term action in quickly controlling the disease and reducing secondary damage. But usually, long-term use of these medications will interfere with or change the body's compensatory response determined by its own nature, and therefore often cause some side effects that cannot be ignored. This is also the reason why many people are unwilling to use the method of control diseases of modern medicine to treat chronic diseases.

Although TCM treatment is mainly based on syndrome differentiation, it also has its disease classification system (DCS). A large part of its disease classification is defined based on the symptoms that the patient feels and the signs that can be observed, such as headache, dizziness, vomiting, abdominal distension, etc. TCM also has symptomatic medications that directly target diseases, such as CHMs that treat headaches, stop dizziness, or have hemostatic and antiperspirant effects. Since the research on symptomatic medications in TCM directly takes these diseases as controlled amounts, in principle, the effective medications found can affect any link in the causal chain of these diseases. Comparing the clinical features of symptomatic medications in modern medicine and TCM, we noticed that the specificity of TCM's symptomatic medications is not good enough, but there is less adaptation and 'rebound' phenomenon. Moreover, when the deviation of the controlled amount associated with the disease returns to normal under the action of the medication, the medication usually does not continue to act, causing excessive correction of the controlled amount, deviating from normal in the other direction. That is, Chinese medicinal herbs usually have the action of 'two-way adjustment'. This phenomenon may be related to the situation that the actual action point of CHMs in the disease causality chain is usually not close to the controlled quantity, or the causal chain associated with the controlled quantity has different branches. Usually, in the causal chain of diseases, the links that have a good correlation with the controlled quantity are the links adjacent to the controlled quantity. Because the farther away from the controlled quantity in the disease causal chain, the more middle links between them, or the more branches of the causal link, and the smaller the correlation with the controlled quantity. For example, regulating blood pressure usually reduces blood pressure by dilating blood vessels, which has a direct action and usually has better specificity. However, high vascular tension is usually not the original cause of high blood pressure. Although the expansion of blood vessels can temporarily reduce blood pressure, because the original cause of hypertension is not resolved, blood pressure usually rebounds after stopping the medication. In the case that the primary cause of the increase in blood pressure persists, the effectiveness of the original dose of antihypertensive medications will gradually decrease, and the maintenance of normal blood pressure often requires gradually increasing the dose. Taking medications that lower blood pressure by dilating blood vessels when blood pressure is normal, although restricted by the body's automatic blood pressure adjustment mechanism, usually also exerts a certain blood pressure-lowering effect, which may cause hypotension. In other words, symptomatic medications developed based on modern medical medication research mechanisms usually do not have a 'two-way regulation' mechanism.

Due to the better specificity of symptomatic medications in modern medicine and their positive effects in quickly controlling the condition and reducing secondary damage, they have an irreplaceable role in clinical treatment. How to apply them appropriately, the TCM treatment concept of 'emergency treats its symptoms, and slowness treats its roots' provides valuable enlightenment. Since they are symptomatic medications used to 'control' the disease, their application should be limited to the right period and the right link. In other words, their application should be limited to the control of disease progression, rather than uncontrolled long-term abuse. Once the disease is under control, it is necessary to treat the disease fundamentally based on the concepts of 'treatment of the root when the disease is slow' and 'pursuing the root of treatment'. Or based on the principle of 'treatment of both the symptoms and the root', apply symptomatic treatment and the method of 'curing the root cause' simultaneously. Based on this concept, it is possible to organically combine the symptomatic treatment of modern medicine with the TCM holistic treatment based on syndrome. Through the synergistic coordination of Chinese and Western medical treatment methods, it is possible to minimize the various side effects of applying modern medical symptomatic medications alone, and to greatly improve the overall level of medical treatment of diseases.

For example, in the practice of treating diabetes with integrated TCM and Western medicine, people have noticed that the method of pure TCM can improve the symptoms of patients through differentiation and treatment, but the effect of lowering blood sugar is often not ideal; while the effect of Western medicine in lowering blood sugar is quick, but improvement of the patient's symptoms are not satisfactory, and after long-term use, side effects are almost inevitable. Combining the blood sugar control of Western medicine with the fundamental treatment for the root of TCM will greatly improve the treatment effect. When blood sugar is under control, the dosage of Western medicine is gradually reduced, which minimizes the side effects of long-term use of hypoglycemic medications. The same is true for treating hypertension with the combination of Chinese and Western medicine.

In short, from the perspective of synergistic treatment of diseases with traditional Chinese and Western medicine, if modern medical symptomatic medications are used as an expedient measure for rapid control of the disease, they can be combined with holistic treatment based on syndrome to enhance its efficacy and minimize its side effects. If a treatment strategy that takes both symptoms and root causes into account is adopted, it should be regarded as a link in the overall plan to play a role in treating the symptoms, and in the matching method of treating the root causes, take into account the restriction of its adverse reactions.

(3) Research on treatment methods target to body state variables (syndromes)

Based on the model and SDS, the state variables are used as the reference system for the study of control means, and then based on the control means developed in this way, the regulation-control of the system is realized. This is the current mainstream approach for science to regulate/control various application systems. As mentioned above, the syndrome differentiation and treatment of

TCM is in the same line with the regulation-control approach of modern science. The research on medications used in TCM syndrome differentiation and treatment is based on the human body model and the corresponding SDS as the frame of reference, and the syndromes as controlled quantities. From ancient times to the present, whether there are thousands of Chinese medicinal materials or the vast array of CHM formulas, the description of the medication action in TCM is mainly based on its unique syndrome system.

Until recently, medication research in modern medicine always focused on pathogen-specific medications and symptomatic medications. In recent years, due to the rise of precision medicine, the concept of personalized medicine has begun to have substantial significance in the modern medical system. With the continuous discovery of biomarkers used to distinguish the personalized state of the human body, a biomarker system that is different from the previous DCS is being established. However, as we will see in the next part of the analysis, the personalized medical system of modern medicine based on the development of biomarkers is obviously of great significance for further distinguishing individual differences in disease conditions at the micro level. However, this method does not break out of the limitations of the reductionist method, and it is still impossible to realize the overall grasp and comprehensive regulation-control of the state of the human body based on it. TCM's theoretical models on the holistic level, and the treatment method system including medications and herbal formulas, will still play an irreplaceable role in the path of medicine towards holistic integration.

In TCM, the research on medications and treatment methods with the holistic model as the reference frame focuses on the body's state deviation, that is, clinical findings corresponding to the syndromes that appear during the disease are regarded as the controlled quantity. This group of clinical findings is a comprehensive reflection of the internal morphological and functional changes when the body is in a specific pathological state. If these clinical findings are reduced or eliminated under the intervention of a certain medication, the corresponding morphological and functional changes of the previous disease will often move in the normal direction, that is, the disease has been alleviated or cured. Using this research method and based on the corresponding relationship between symptoms/signs and syndromes (state variables), even if we do not know the entity part of the disease, the nature of the disease, the interrelationship of various entity elements, the exact pathogenesis, and the causal relationship of the development and transmission of the disease in the body, we can still find effective medications and treatment methods by observing the changes of the patient's clinical findings before and after intervention.

Compared with medications developed by using the pathological changes of a certain entity part as a controlled quantity, the medications which developed by taking the syndromes (state variables) defined of the symptoms/signs as a controlled amount are obviously more targeted to the specific body state, so not only the long-term effect is good, but the side effects are also small. This is also the reason why TCM treatment based on syndrome has an enduring vitality. In recent years, the large number of side effects exposed by the application of CHMs and TCM treatment methods in modern medicine without the support of TCM syndrome differentiation and treatment system, also illustrate the irreplaceable role of the SDS of TCM in guiding the clinical application of CHM from another aspect.

In the records about the actions of CHM and TCM herbal formulas, each CHM/herbal formulas have indications, and these indications are described mainly based on the syndrome of TCM. Improper use that does not follow these indications can often cause certain side effects. Even the foods that people often eat, such as ginseng and jujube, which are very beneficial to health, if taken for a long time, could also have the side effects of 'blocking' and 'excessive internal heat'. Since the 1990s, the United States and the United Kingdom have successively disclosed cases of the use of weight loss products containing Chinese herbal ephedra leading to a large number of side effects and even death, which led the US FDA in December 2003 to announce a complete ban on the sale of weight loss food containing ephedra (U.S. Food and Drug Administration 2003). Based on the knowledge of TCM, ephedra is a kind of CHM that has the actions of making sweating to relieve the exterior, dispersing the lungs, soothing panting, and warming and dredging blood vessels. It is usually used to treat colds (including the common cold and flu), coughs, asthma, wind-cold dampness arthralgia, tuberculous masses, gangrene and other diseases. After taking it, it usually has excitatory effects such as more sweating and faster heart rate, especially not suitable for people with weak body and poor heart function for long-term use. Taking such a violent CHM as a healthy food for weight loss for a long time is incredible for TCM practitioners. Therefore, the question is not whether a medication has side effects, but how we apply it, and how to maximize its beneficial therapeutic actions on the human body during the application process and minimize its side effects and adverse reactions. The reason why the application of CHM today shows fewer side effects is that, on the one hand, the TCM practitioners who use these CHMs clearly understand the actions and side effects of these CHMs and know under what circumstances they can be used; what's more, in their application, TCM practitioners take advantage of the synergy and antagonism among the CHMs to enhance the beneficial therapeutic actions of the formulas while minimizing their side effects.

The cases of so-called side effects of CHM disclosed in recent years are basically caused by abuse of CHM without regard to syndrome differentiation of TCM and regardless of the scope of application of the CHM. In the 1990s, the incident of the side effect of Xiao Chai Hu Decoction, which caused a strong shock in Japanese society, was also caused by similar reasons. Xiao Chai Hu Decoction is a well-respected classic herbal formula in Kampo medicine, and after long-term clinical application in Japan, its efficacy has been widely confirmed. The Ministry of Health and Welfare of Japan recognized the effect of Xiao Chai Hu Decoction in improving liver dysfunction in 1994, and officially included the herbal formula as a liver disease medication in the National Pharmacopoeia. As a result, Japan has seen a 'grand event' of tens of thousands of patients with liver disease taking this herbal formula. However, two years later, 88 patients with chronic hepatitis appeared interstitial hepatitis due to the side effects of Xiao Chai Hu Decoction, and 10 died. After this incident, the sales of Xiao Chai Hu Decoction in Japan dropped by one-third, and it also experienced the danger of expulsion from medical insurance. This is completely the bad result of improper application of Xiao Chai Hu Decoction in the absence of TCM syndrome differentiation and treatment (Dai Zhaoyu et al. 2000a).

We have compared the medication research methods of TCM and modern medicine from the three perspectives of pathogenic factors treatment, symptomatic treatment, and syndrome treatment. But whether it is in TCM or in modern medicine, clinical application requires a comprehensive understanding of the efficacy and characteristics of the medication. That is to say, medications developed for the purpose of eliminating the cause of the disease must also understand their actions on various state variables of the body and their actions on certain pathological changes, including beneficial therapeutic actions and adverse reactions. The research on symptomatic medications and the medications target to syndrome are also the same. Only on the basis of a comprehensive understanding of their actions can they be combined into effective formulas in the clinic, strengthen the therapeutic action for the disease, and limit the side effects of each single medication in the formula to the greatest extent.

4.7 The treatment based on diseases: The disease description and regulation-control method based on disease classification

The current modern medicine is basically a medical system built around diseases. The International Classification of Diseases (ICD) is an internationally unified disease classification method formulated by the WHO. It is a multi-axis DCS established according to characteristics such as the disease's etiology, location, pathology and clinical manifestations. As of the end of December 2021, the world-wide version is the 10th revised International Statistical Classification of Diseases and Related Health. The latest ICD-11 version has entered into force on January 1, 2022. ICD-10 contains more than 29,000 disease records, covering various diseases in all departments of the hospital, and ICD-11 contains approximately 55,000 codes related to injuries, diseases, and causes of death. The current classification of diseases in modern medicine is mainly based on this classification system.

Although TCM treatment is mainly based on syndrome, it also divides diseases into categories, combining the treatment based on syndrome differentiation with the treatment based on disease differentiation.

4.7.1 Classification and definitions of diseases in TCM

Similar to the DCS of modern medicine, the disease system of TCM is also classified on the basis of department. Its departments mainly include internal medicine, surgery, gynecology, pediatrics, dermatology, ENT, etc., which are roughly similar to those of modern medicine. Its classification of diseases is mainly based on the etiology and clinical manifestations. Different from modern medicine, the disease classification based on etiology in TCM is not based on differences in pathogenic factors, but on the differences in the response state of the body caused by pathogenic factors; its diseases classification based on clinical manifestations can be based on a common symptom, as well as a comprehensive syndrome described by a set of symptoms and signs.

Diseases classified based on pathogenic factors mainly refer to infectious diseases caused by bacteria, viruses or other pathogenic microorganisms, which belong to the category of colds in TCM. Cold is divided into common cold and influenza. According to the onset season, the initial clinical manifestations and the characteristics of development and transformation, TCM divided influenza into diseases such as spring-warm, summer-heat, and autumn dryness. Obviously, the classification of influenza in TCM has nothing to do with the pathogens that cause the disease, but is based on the onset season of the disease and the clinical manifestations that describe the body's response state to the pathogens. Consistent with the isofinality of organisms revealed by Bertalanffy, the same reaction state can be induced by quite different pathogens. Like the syndromes of wind-heat attacking the surface, they can be caused by different viruses, and even quite different viral and bacterial infections can cause the same or similar syndromes. The same pathogen may cause different clinical manifestations in different seasons, environments and violations of different individuals. TCM's understanding of diseases caused by such external pathogenic factors is based on the body's response state to pathogenic factors, and the regulation and control of such diseases is also based on the identification of the body's response state. Therefore, the disease classification and diagnosis based on pathogenic factors have little therapeutic significance for TCM.

At present, the word 'colds' in TCM is mainly used to distinguish the concept of exogenous and internal injury diseases, in order to define a large class of sudden diseases caused by external pathogenic factors. Of course, studying the development and transmission laws of diseases caused by different types of pathogens is of positive significance for predicting the development and evolution of diseases and taking preventive treatment measures. Today, modern medicine has conducted quite in-depth research on pathogenic micro-organisms, making it possible to classify pathogenic micro-organisms based on differences in clinical manifestations, law of development and transmission, and treatment methods at the initial stage of disease. As a result, the body's response state under the action of pathogenic factors can be refined, and then the etiology model of TCM can be refined and perfected. At the same time, some effective treatment methods (if any) for pathogens in modern medicine can be incorporated into the treatment system based on syndrome of exogenous diseases in TCM. For example, the clinical manifestations of Ebola virus infection are sudden high fever, headache, sore throat, weakness and muscle pain. If it can be confirmed that the initial response state is limited to the scope of exogenous wind-heat in TCM without individual difference, and the development and evolution of the disease shows a unique regularity (such as nausea, vomiting, diarrhea, and rapid internal and external bleeding), it can be considered as a subtype of exogenous wind-heat. If there is no diagnostic correlation between pathogens discovered by modern medicine and the response state of the body, but there are effective treatments for the pathogens, the corresponding disease classification in TCM can be refined based on these pathogens, and then treatment methods of modern medicine for the pathogens should be incorporated into TCM's treatment system combining syndrome differentiation and disease differentiation. For example, if the initial clinical manifestations of Covid-19 infection are individualized, but there are treatments for the virus (e.g., vaccines and antiviral medications), Covid-19 can be considered a subtype of exogenous disease.

The classification of diseases based on a single clinical manifestation is usually based on the symptoms and signs that people feel unwell and need to be resolved, such as headache, dizziness, cough, and bloating. The significance of listing the symptoms and signs separately as a disease classification has two aspects: on the one hand, the problem can be localized, focusing on the pathological state (syndrome) associated with this disease, so as to adopt corresponding treatment methods based on syndromes. For example, headache in TCM is divided into two categories: external feeling and internal injury. Headaches caused by external infection can be found in the syndromes of wind-cold, wind-heat and rheumatism; internal injury headaches can occur in different syndromes such as hyperactivity of liver-yang, kidney deficiency, insufficient qi and blood, and blood stasis. On the other hand, it is convenient to find effective treatment methods and medications specifically for these symptoms. In TCM, many of the descriptions about actions of CHMs are aimed at symptoms, such as, for headaches usually add Rhizoma Chuanxiong, and different parts of the headache will require different symptomatic CHMs: forehead pain plus Angelica Dahurica rhizoma, head top pain plus radix ligustici sinensis (*Ligustici rhizoma et radix*), bilateral pain plus Bupleurum radix, and headache involving neck plus Incised notopterygium (*Notopterygii rhizoma et radix*), etc. In addition, Bamboo Shavings (*Bambusae caulis in taenias*) and Pinellia (*Pinelliae rhizoma*) for relieving vomiting; loquat leaf (*Folium eriobotryae*), asters (*Asteris radix et rhizoma*), and coltsfoot flowers (*Farfarae flos*) for relieving cough; Mantis egg-case (*Mantidis Ootheca*), golden cherry seeds (*Rosae laevigatae fructus*), and raspberries (*Rubi fructus*) for relieving enuresis are all specific treatment methods of TCM for symptoms.

We have noticed that in TCM clinical textbooks, the syndromes listed under most diseases usually do not cover all the syndromes that actually appear in the clinic related to this disease. Although some syndromes are not common in this disease, they are also realistic as small probability events. In addition, in many cases, a certain disease that a patient has cannot simply be attributed to one of the listed syndromes, but is also related to other syndromes that the patient has identified as co-existing, which means that the patient's disease is associated with a combination of more than one syndrome. For example, a patient with headache has syndromes of qi deficiency, blood deficiency, and kidney deficiency in the diagnosis based on syndrome differentiation, as well as the syndrome of disturbance of wind and heat. That is to say, the patient's headache is caused by the disturbance of wind and heat in the case of insufficient qi and blood and kidney deficiency. Only when the methods of replenishing qi and blood, invigorating the kidney, and dispelling wind-heat are used at the same time, can a better therapeutic effect be obtained. Therefore, the syndrome classification of a certain disease listed in the textbook of TCM is only the basic syndrome type that can be combined. In clinical practice, the syndromes related to the disease should be determined based on the actual clinical manifestations of the patient, rather than limited to the basic syndrome types listed in the textbook.

The classification of diseases based on a comprehensive syndrome is based on a set of clinical symptoms and signs that often appear simultaneously. There is no obvious external pathogenic factor in the occurrence of this kind of disease. Its cause is either the result of a combination of multiple factors, or the accumulation of some hidden factors, or a result of the evolution of the pathological state during the development of the disease. However, once the body exhibits this syndrome, it will either maintain this pathological homeostasis afterwards, such as diabetes, menopausal syndrome, or show a certain regular evolution.

From the perspective of TCM, this type of disease is a generalization of a certain type of pathological stable state or pathological process. For such diseases, there is no symptomatic treatment method in TCM, it completely adopts the treatment based on syndrome. The focus of treatment is to make the body state gradually break away from a specific pathological homeostasis or pathological process, and return to a normal homeostasis balance. The clinical practice of TCM shows that personalized treatment based on syndrome can effectively improve the clinical findings of polydipsia, polyuria, polyphagia and weight loss in patients with diabetes. One point to be explained here is that there is not necessarily a one-to-one relationship between diseases defined by comprehensive syndromes in TCM and syndromes in modern medicine. For patients diagnosed with dispersion-thirst disease by TCM practitioners based on their symptom groups, if the blood glucose and urine glucose levels are found to be high after testing, they will be diagnosed as diabetes by modern medicine. Of course, there are also patients diagnosed with diabetes who have no symptoms of polydipsia, polyuria, polyphagia, and weight loss. Today, modern medicine can accurately diagnose and quantify diabetes by detecting blood sugar and urine sugar. The medication research using these indicators as controlled amounts has developed some effective medications that control these indicators and reduce secondary damage, such as the control of blood sugar and diabetes-related symptoms and signs by hypoglycemic medications, and the control of neurasthenia symptoms by nerve inhibitors. However, as analyzed in the previous medication research methods, the purpose of these medications is to control the disease, not to treat the disease fundamentally.

Regarding the relationship between syndrome and disease, TCM generally believes that syndrome is the root and disease is the tip. When modern medicine introduces objective indicators to refine these diseases, and finds a symptomatic treatment method that directly targets these indicators, such diseases based on comprehensive syndrome definition are actually transformed into diseases based on single symptom (detection indicator). Of course, by introducing some objective detection indicators to achieve accurate qualitative and quantitative diagnosis, diseases defined in TCM based on a single symptom can also be refined. The treatment methods targeted to these indicators discovered in modern medicine are the same as the symptomatic treatment methods of TCM from a methodological point of view. In many cases, the organic combination of TCM treatment based on syndrome and modern medical symptomatic treatment shows better clinical effects than the application of TCM methods or modern medical methods alone. Obviously, integrating the disease systems of TCM and modern medicine and combining the treatment methods of the two systems to clinical treatment has very broad prospects.

However, the classification of diseases does not mean that the finer the division and the more types of diseases, the better. The excessively detailed division will undoubtedly increase the complexity of diagnosis and grasping the condition, making the treatment plan increasingly lose its overall comprehensiveness. If further refinement of the disease based on symptoms, signs, or modern medical test indicators is of no significance in treatment, there is no need to refine it. This is precisely the ‘simplest applicable’ rule that the natural sciences apply when dealing with all kinds of complex systems, i.e., under the premise of satisfying application, the simpler the model construction and system classification, the better (Yuan Bing and Zhao Zheng 1986).

4.7.2 *The description of the disease process and the exploration of the law of development and evolution*

As mentioned earlier, the human body is a complex super-large system composed of some subsystems with self-organization, self-adaptation and self-regulation capabilities. The various state variables of each subsystem are inter-connected and mutually restricted, maintaining the subsystem in a stable dynamic balance. The inter-connection and mutual restriction between the subsystems maintain the steady-state balance of the whole system, as shown in Figure 4.4. However, the ability of any system to maintain steady-state balance is limited. If the external interference acting on a certain state variable of the system is too strong or continuous, and exceeds the ability of the subsystem to maintain a stable balance, the steady-state balance of interconnection and mutual restraint among

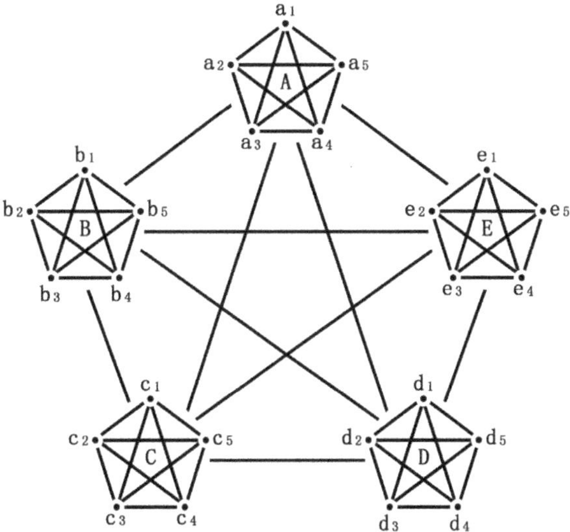

Figure 4.4. The normal steady-state balance of the human body.

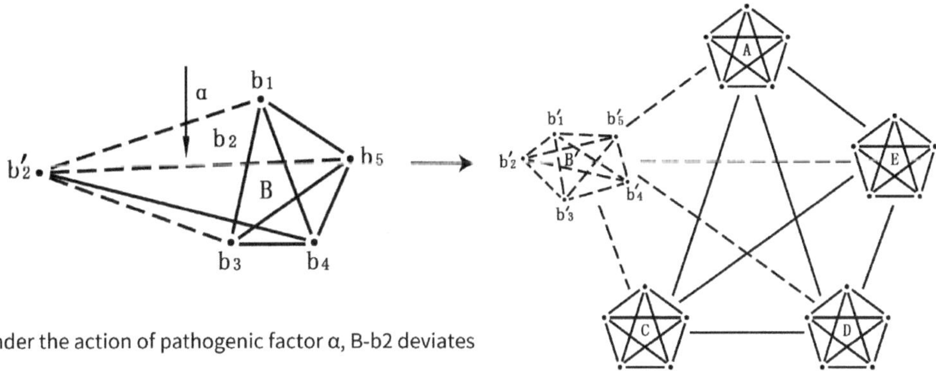

Figure 4.5. Under the influence of B-b2', B deviates to B'.

the various state variables in this subsystem will be broken. As a result, the state of this subsystem has entered a pathological state, as shown in Figure 4.5.

Due to the inter-connection and mutual influence between the various subsystems, if a not too large state deviation occurs in one subsystem, then the influence of other subsystems on it will usually bring it back to the normal state. The return of the subsystem to the normal state means that the steady-state balance of interconnection and mutual restraint among the state variables within the subsystem is rebuilt. This interaction among the subsystems is embodied in the system's repair of the normal steady-state balance of each subsystem.

However, the system's ability to repair the steady-state balance of each subsystem is also limited. If the state of a certain subsystem deviates too much or lasts too long, the action of other subsystems on it is not enough to make it return to a normal steady state, that is, the destroyed steady-state balance among the state variables cannot be restored. Then, in turn, its action on other subsystems will also cause the state of other subsystems to deviate from the normal steady state, that is, disrupt the internal steady-state balance of other subsystems. So, the disease gets extended to a larger scope (as shown in Figure 4.6).

As a self-organizing, self-adapting, and self-adjusting system, the human body's ability to maintain steady-state balance and repair stable structures should naturally be attributed to the human body's self-organization, self-adaptation, and self-adjustment capabilities. In this way, disease can

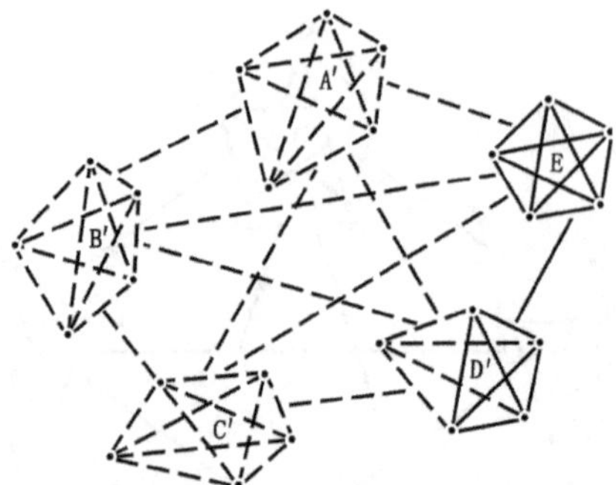

Figure 4.6. Under the influence of B', the states of multiple subsystems deviate.

be seen as a process of change in the body's state in the mutual struggle between the destructive effects of pathogenic factors and the repair action on damage by body's self-organization, self-adaptation, and self-regulation ability. A batch of state variables (or subsystems) that are separated from the normal steady state are always constantly changing in interaction with other state variables (or subsystems) to tend to a certain steady state equilibrium. In this way, its natural development trend must be in one of the following three situations:

1. Under the action of external controllable input (therapeutic intervention), the body gradually repairs the stable structure of each subsystem, corrects deviated state variables, and restores the state of the system to a normal steady state balance. As a result, the steady-state balance between the various subsystems and between the state variables is restored or repaired, which is commonly referred to as recovery from disease.
2. Under the action of pathogenic factors, the state of the system (or subsystem) deviates too far from the normal steady-state balance, or some elements of the system (or subsystem) that maintain the normal steady-state balance are damaged and hard to repair, so that the system cannot restore the previous normal steady-state balance and stabilize in a pathological state. This is commonly referred to as 'stubborn illness'.
3. Because the role of pathogenic factors exceeds the ability of the system to maintain a steady-state balance and repair a stable structure, the state of the system moves in the direction of decreasing order according to a certain procedure until the system disintegrates. This is the process where the disease gradually tends to worsen.

When we use state variables to describe the system and use the state space to analyze the trajectory and possibility of the system's state evolution, using the steady-state structure, we can deduce and predict the process of the extremely complex development and evolution of the disease in the human body. After the body state deviates from the normal steady state equilibrium or a pathological steady state, there will usually be a period of instability where the state is constantly changing. In this process, there will be more than one possibility for the evolution path of the human body state. If we can find those states with stable structures that are adjacent to the previous stable state in the state space, we can determine the limited possibility of disease development and change. Therefore, a reasonable judgment can be made on the development and outcome of the disease (Yuan Bing and Zhao Zheng 1986).

TCM uses the human body's functional model to dynamically describe the occurrence, development and evolution of diseases, and thus can effectively guide the regulation and control of the human body's state in diseases. Based on this model, it is possible to identify the state (syndrome) of state variables that characterize the human health state through the correspondence between symptoms, signs and syndromes. The combination of all the patient's state variables (syndrome) completely reflects the state of the body at a certain moment. After the state of the body is clearly identified, through model-based deduction, the possibility of the development and transformation of the state of the body in the disease can be accurately predicted. TCM treatment based on syndrome is to formulate appropriate treatment intervention programs according to the result of state identification and the prediction of disease development and transformation possibilities, to prevent the body's state from developing in the direction of the spread and deterioration of the disease, and promote the deviation of state variables to return to the normal range. As a result, step by step the body's state tends to a steady state balance relatively close to the normal steady state, and finally returns to the normal steady state balance.

The description of the occurrence, development, evolution and regulation of epidemic and infectious diseases in TCM is based on its unique human body model and state description system. In fact, whether it was SARS, avian influenza, H1N1, Ebola in the past, or COVID-19 today, this type of epidemic called the 'plague' broke out many times in ancient China. Today, the six-meridians, wei-qi-ying-blood, and tri-jiao syndrome systems of TCM are the human body models and state description systems established by TCM to deal with epidemic and infectious diseases. They were gradually developed by TCM scholars after 1500 years of long experience in the struggle against various epidemic and infectious diseases. The epidemic and infectious diseases that have been prevalent in recent years, from the perspective of TCM, the initial symptoms and corresponding treatment options are mainly divided into the following types:

1. *Wind-heat invading the exterior*: Fever, chills, headache, body pain, nasal congestion, runny nose, thirst, sore throat, thin white or thin yellow tongue coating, floating pulse.
2. *Wind-cold invading the exterior*: Aversion to cold, mild fever, headache, body pain, nasal congestion, heavy voice, sneezing, runny nose, itchy throat, cough.
3. *Wind-cold and damp stagnation*: Aversion to cold, fever, headache, chest tightness, epigastric fullness, stomach pain, abdominal pain, nausea and vomiting, bowel sounds, diarrhea, white greasy tongue coating.
4. *Wind-heat and damp stagnation*: Aversion to cold, headache, heavy body, body pain, limb burnout, pale yellow face, chest tightness, loss of appetite, hot body in the afternoon, thirst, white or yellow greasy tongue coating.

There are mainly several types in the middle stage:

1. *Exuberant heat in the qi aspect*: The patient manifests as strong fever, thirst, profuse sweating, yellow tongue coating, and large and fast pulse.
2. *Evil-heat choking lung*: The patient presents with high fever, thirst, cough, wheezing, hemoptysis or rusty sputum, red tongue, yellow coating, and large and fast pulse.
3. *Phlegm-heat choking lung*: The patient manifests as fever, thirst, cough, asthma, phlegm, yellow and thick phlegm, large amount, red tongue, yellow greasy coating, slippery and fast pulse.
4. *Spleen-stomach damp-heat*: Fever, chest tightness, fullness of the stomach, abdominal pain, diarrhea, mucous stool or loose stool, unsmooth defecation, yellow and greasy tongue coating.

There are mainly several types of extreme stages:

1. *Heat entering ying-blood*: Dark or bloody stool, coffee-like vomiting blood, red eyes, erythema, maculopapular rash, purpura and internal bleeding on the skin. Bleeding occurs from any orifice of the body, including the nose, mouth, anus, and genitals. The tongue is crimson or purple.
2. *Heat knot in the large intestine*: Clinical manifestations include abdominal pain, refusal to press, constipation, or fever, vomiting, or irritability, thirst, dry tongue, red tongue, yellow coating, and deep and powerful pulse.
3. *Heat harassing the heart spirit*: High fever, upset, insomnia, coma, delirium.
4. *Extreme heat engendering wind*: Manifested as limb twitching or neck stiffness, fast pulse like a string.

After the extreme period, some severely ill patients or patients with severe complications may develop hypotension, hypovolemia, tachycardia and/or severe damage to internal organs (especially kidney, spleen and liver), causing diffuse systemic necrosis and proteinuria. More serious illness may eventually lead to heart failure, shock, and liver and kidney failure, until death. This stage is equivalent to the process of blood deprivation and qi deprivation in TCM leading to the separation of yin and yang. Figure 4.7 shows the main path of the development and change of exogenous fever disease (epidemic and infectious disease) in the human body.

Obviously, by establishing the understanding of pathogenic factors based on the body's response state, TCM has summarized the syndrome types of epidemic and infectious diseases in the early stages and the possible syndrome evolution during their development and evolution. From the perspective of TCM, no matter what kind of virus cause the infectious diseases and how these viruses mutate, it is always nothing more than several types, such as wind-cold, wind-heat and with dampness or their compound types. The state of patients at various stages in the development process and the corresponding diagnosis, treatment methods, and even prescription and medications are described in detail in the theory of epidemic febrile disease of TCM. This is a summary of the experience of Chinese ancestors fighting against infectious diseases caused by viruses in an era when modern medicine had not yet developed. In the historical data of TCM, a large number of cases cured by this method are documented. It's just that with the rise of modern medicine based on western modern science, they were regarded as 'non-scientific' things and were sealed in the scientific 'garbage dump' (Yuan Bing 2015a).

Since the fifth round of COVID-19 outbreak in Hong Kong in February 2022, which was mainly caused by the Omicron variant, our clinical team diagnosed and treated over 1000 patients infected with COVID-19 of the infection and sequelae stages. Based on the clinical manifestations of patients, the pathological state of most patients in the early stage of infection was attributed to the syndrome of wind-heat invading the exterior, wind-cold invading the exterior, wind-cold and damp stagnation, and occasionally wind-heat and damp stagnation. Also commonly seen was the syndrome in which cold is not resolved, and internal heat has formed. Among the patients receiving treatment, there were also cases that Western medicine considers severe or with a tendency to develop into severe disease, and the syndromes involved included exuberant heat in the qi aspect, evil heat choking lung, phlegm-heat choking lung and spleen-stomach damp-heat, etc. In all these cases, some only manifested as a single syndrome, and quite a few manifested as multiple syndromes. In conclusion, the pathological state and development of the disease of patients in the early stage of infection had strong individual characteristics, which could be divided into many different situations.

Among the patients we received, except for two patients in the infectious stage who did not insist on taking traditional Chinese medicine due to lack of confidence in traditional Chinese medicine after the first video consultation and turned to Western medicine for treatment, the rest of the patients were cured. All sequelae of patients with sequelae have been improved and eliminated through TCM treatment. Among the patients we treated, the oldest is 95 years old, and the youngest is only 2 months old. Severe cases may include symptoms such as high fever (the maximum body temperature reaches 41°C), severe cough, chest tightness and chest pain, shortness of breath, loss of appetite, vomiting, and diarrhea. There are also patients who are old, frail, suffering from old illnesses, or even pregnant. Regardless of the infection stage or the sequelae stage, all the pathological states of patients could be described by the syndromes of the TCM treatment system based on syndrome. Using the treatment method based on syndrome, if only the syndromes are accurately identified, the patients all showed good responses.

Our own clinical practice, the countless successful treatment cases in mainland China during the COVID-19 epidemic, and the excellent results of TCM treatment during the epidemics of SARS, avian flu, swine flu, and H1N1, all fully prove the effectiveness of the personalized treatment based on TCM syndrome in dealing with infectious diseases caused by viruses. These clinical practices also illustrate from an empirical point of view: killing viruses and curing diseases are two completely different concepts in the treatment of infectious diseases caused by viruses. In today's world, it is not

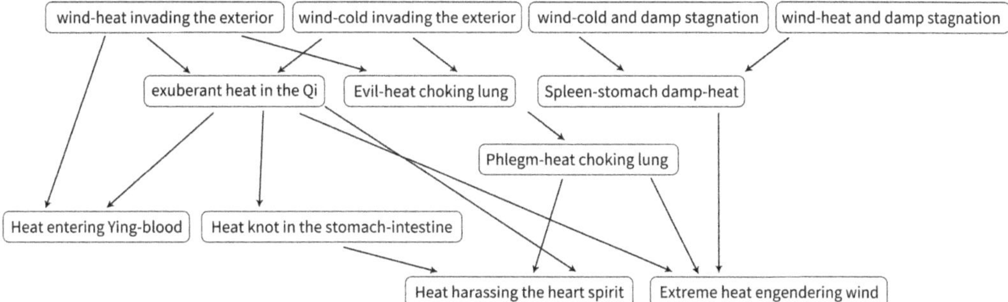

Figure 4.7. The main change path of exogenous fever disease.

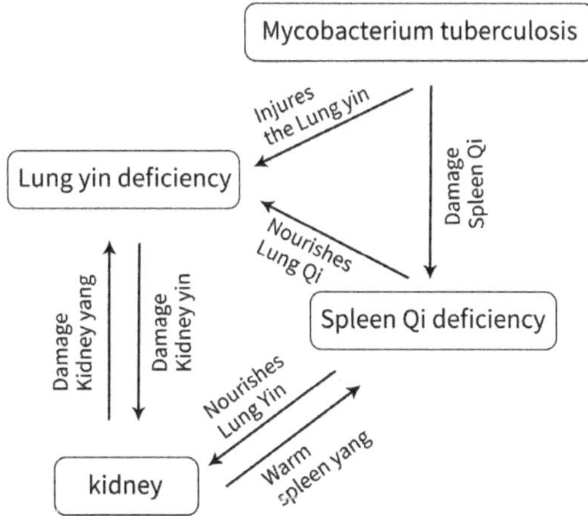

Figure 4.8. Disease evolution of tuberculosis.

that there are no effective methods against epidemic infectious diseases caused by viruses, but due to differences in philosophy, the mainstream medical community turns a blind eye to these methods. Whether it is to deal with the current COVID-19 or the epidemic infectious diseases that will occur in the future, finding out the virus that causes the disease and then developing vaccines and antiviral medications in a targeted manner is not the only feasible method. Without any preliminary research, based on the body's response state caused by pathogens, personalized TCM treatment based on syndrome can also effectively regulation-control such diseases.

Another example is the description of TCM on the development and evolution of phthisis (equivalent to tuberculosis in modern medicine). In TCM, the causative factor of phthisis is phthisis insect (equivalent to tuberculosis bacillus in modern medicine), and its main pathogenic action is to damage the lungs and spleen and consume qi and yin. The development process of the disease can be divided into three stages:

1. *Initial stage*: Tuberculosis invades the lungs and spleen, causing signs of consumption of lung yin and impaired spleen and stomach functions. Clinical manifestations include dry cough, little sputum, or blood in sputum, chest pain, hot flashes, usually heavy in the afternoon, night sweats, flushed cheeks, dry throat, thirst, fatigue, loss of appetite, weight loss, red tongue, thin and fast pulse.
2. *Mid-term*: If the initial disease is not controlled, its development mainly presents two different trends. In some patients, the loss of lung yin is relatively severe, spreading to the kidney yin, causing kidney yin loss, and inflammation on the false fire. The clinical manifestations include increased cough, dry cough with little sputum or yellow and thick sputum, dry mouth and polydipsia, hot flashes, flushed cheeks, or coughing up blood from time to time, and even a

lot of coughing up blood, night sweats, insomnia, chest tightness, chest pain, upset, irritability, dream emission in men, amenorrhea in women, red tongue, and their pulses are thin and fast. Some patients present syndromes of spleen and stomach deficiency and qi and yin deficiency. In addition to the clinical manifestations of cough, hemoptysis, hot flashes, flushed cheeks, spontaneous sweating, and night sweats, there are also pale, tired expression, shortness of breath, low voice, fatigue, loss of appetite, red tongue, less moss, and thin, fast and weak pulse, etc. Initial-term and mid-term disease associations and trends are shown in Figure 4.8.

3. *Late stage*: Based on the human body model of TCM, kidney yin is the root of yin fluid, kidney yang (the gate fire) is the root of yang qi, and kidney essence is the most fundamental substance for human life activities. Kidney yin and kidney yang (life door fire) is produced and maintained on the basis of kidney essence. When the kidney yin is consumed, it will inevitably damage the kidney essence; and the weakness of the spleen and stomach will lead to insufficient production of the kidney essence, which will cause kidney essence loss over time. When the kidney essence is lost, the kidney yang (the life fire) based on it will also decline. Therefore, whether it is yin deficiency and fire prosperity or both qi and yin deficiency, it will lead to the depletion of kidney essence and the decline of kidney yang, and eventually the lung, spleen, and kidney will be damaged, and both yin and yang will be deficient, as shown in Figure 4.9 (Yuan Bing and Zhao Zheng 1986).

For diseases defined by a single symptom and physical sign, TCM research focuses on the basic syndrome types associated with these diseases. It is to determine which syndromes are related to the symptoms of the patient according to the patient's clinical manifestations. However, diseases that occur in actual patients are often not only related to one basic syndrome type, but may be related to a combination of several basic syndromes.

Vertigo, TCM has found is related to at least three types of factors:

1. Deficiency of qi, blood, yin and yang;
2. Qi stagnation, blood stasis and phlegm obstruction;
3. Disturbance of evil qi, including actual fire, virtual fire, wind heat, phlegm heat and damp heat.

Vertigo seen clinically may be related to several factors at the same time, such as deficiency of qi and blood, stagnation of qi, and wind-heat. In this situation, the treatment plan usually needs to address several factors at the same time to achieve better results.

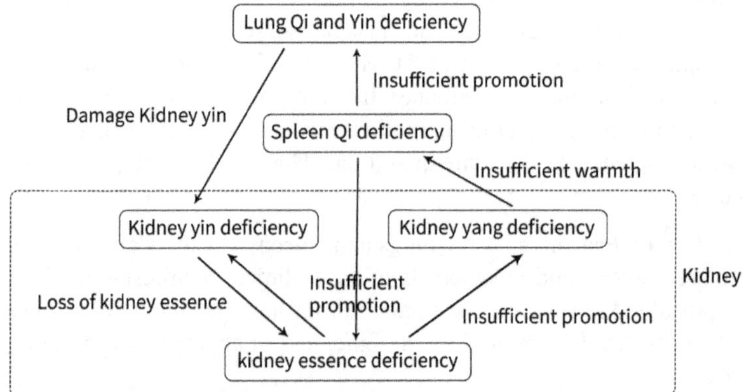

Figure 4.9. The disease-related situation of pulmonary tuberculosis.

4.7.3 The art of disease regulation and control

Based on the theoretical model and SDS of TCM, the state of the body can be accurately described and grasped. The research of CHMs and herbal formulas have revealed the properties and therapeutic actions of TCM therapeutic interventions relative to the human model. Then, the treatment process is to appropriately change the body's input, or take certain intervention measures, or directly act on the body's state variables, or mobilize the body's disease resistance factors, in order to break the stable pathological balance of the system, stop the movement of the body's state in the direction of decreasing order, and restore and rebuild the normal steady-state balance of the interconnection and mutual restraint of various subsystems and state variables.

The most direct way to treat diseases is to choose interventions that have the opposite action on the deviation of the state variables, and push the deviated state variables to move in the direction of the normal state. Treatment based on this concept can be called direct confrontation regulation.

(1) Direct confrontation method (Yuan Bing and Zhao Zheng 1986)

The direct confrontation method is to directly kill the pathogenic factor or correct the deviation of the body's state by choosing an intervention method (such as drugs) that targets the pathogenic factor or acts in the opposite direction to the deviation of the body's state. This treatment method is called 'directed control method' in 'Modern Methods of Chinese Medicine', which is currently the most widely used method in TCM and Western medicine. In modern medicine, the method of promoting thyroxine secretion or supplementing thyroxine for treating the decline of thyroid function caused by the decrease of thyroxine secretion, measures to reduce blood pressure and blood sugar for treating hypertension and hyperglycemia, and measures to kill bacteria and viruses for treating diseases caused by bacteria and viruses, all belong to this method. In TCM, the methods of replenishing qi and blood are used to treat qi deficiency and blood deficiency, and the methods of promoting qi and blood circulation are used to treat qi stagnation and blood stasis, as well as the 'cold is treated with heat' and 'heat is treated with cold' principle, all based on this concept.

However, the clinical practice of traditional Chinese and Western medicine has proved that in the human body whose functions are closely related, implementing direct confrontation for individual links does not always achieve the desired purpose. Under the action of pathogenic factors, the body faces not only the obstacles to functional activities, but also the accumulation of pathological products, and the reduction of its own self-organization and self-regulation ability. Obstacles to functional activities have led to the lack of basic substances in life activities and the accumulation of metabolites, and the accumulation of pathological products in turn affects the performance of functional activities. The body's self-organization and self-regulation capabilities are not adequately nourished and gradually decline. In turn, the decline of self-organization and self-regulation is not conducive to the recovery of body functional activities and the elimination of pathogenic and pathological products. Diseases usually develop and change under the interaction of these factors. On the other hand, diseases often involve more than one subsystem. Different subsystems have different functions, and each function occupies a different position in human life activities. Therefore, the pathological changes of various subsystems, the abnormalities of various functional activities, and the various state variables describing the pathological state have different importance in the development and recovery of the disease.

Usually, the method that can be used to break the steady state balance of pathology and stop the movement of the state of the body to the direction of decreasing order is not the only one, just like traveling to a place, you can use different means of transportation and follow different paths. This is especially true for complex disease conditions with multiple abnormal state variables. Obviously, the formulation of a treatment plan should not ignore the interrelationship of various elements in the disease, the interrelationship of the various subsystems and functional activities in the disease, and the importance of the functional activities of each subsystem in the process of human life activities and disease recovery. On this basis, treatment should be aimed at the key links and key factors of the disease, instead of blindly grasping a certain diseased subsystem or a certain deviated state variable. Otherwise, due to the stability of the pathological state maintained by the mutual connection and mutual restraint between the various parts of the disease, good curative effects are often not received. And because medications often have multiple actions, medications that have a therapeutic effect on the deviation of one state variable may aggravate the deviation of other state variables, and may produce results that are not conducive to the recovery of the body as a whole. On the contrary, based on a comprehensive and accurate grasp of the body's state and disease characteristics, and clarifying the main factors and links of the disease, the treatment can start from the link that is most conducive to the body's recovery. With the improvement of this part of the state, the secondary impact on other parts will also make other parts improve to varying degrees, resulting in a multiplier effect.

So how do we grasp the key factors and key links of disease? This requires a comprehensive and clear understanding of:

1. The importance of various subsystems and state variables in human life activities and in the recovery of various diseases.
2. The urgency of various diseases and the degree of harm to the human body.
3. The inter-relationship and mutual restriction of various parts of the human body under various physiological and pathological conditions.
4. The therapeutic actions of various medications and their adverse effects on the body.

In TCM, the above principles for disease treatment analysis are presented in the following way:

- Emergency treats its symptoms, and slowness treats its roots.
- The five internal organs are all deficient, and the spleen should be strengthened first.
- Replenish qi before nourishing blood, replenish yang before nourishing yin.
- Prevent blockage when replenishing qi, prevent stagnation when nourishing yin.

With such principles, clinically based on a comprehensive analysis of the patient's various aspects of the condition, the best treatment plan can be formulated according to the severity and urgency of the patient's various aspects of the disease.

Over thousands of years of clinical practice, ancient TCM scholars have created and developed a variety of unique treatment approaches. For example, in the Jin and Yuan dynasties, Zhang Congzheng's three methods of sweating, vomiting and diarrhea, here called 'breaking steady state method', did not pay much attention to the differences in the pathological state of patients when applying this method. Based on the concept of 'Internal injury to the spleen and stomach causes various diseases to arise', Li Dongyuan created the method that the treatment of vary diseases start from the spleen and stomach. Based on this method, although the pathological states of the patients vary greatly, the treatment is mainly to regulate the spleen and stomach and nourish qi of the spleen. Zhu Danxi put forward the concept of 'yang is often surplus, yin is often insufficient', and pays more attention to the application of nourishing yin and heat-clearing methods, even for patients with insignificant signs of internal heat.

(2) The method of breaking the steady state (Yuan Bing and Zhao Zheng 1986)

Use strong interference input, or promote the body's functional activities, or eliminate the cause and pathological products, directly act on the key links of supporting pathological homeostasis to break the pathological steady state of the patient. Once the body state leaves the previous pathological steady state, the body will automatically seek and tend to a new steady state under the action of self-organization and self-regulation. In the course of treatment, the method used to break this steady state is usually a method that is conducive to the recovery of the patient's disease resistance or weakens the pathogenic factors. Therefore, the new steady state is usually closer to the normal state of the body than the previous pathological steady state, which usually means that the patient's condition has significantly improved.

The mechanism of this method can be vividly illustrated by a simple physical experiment: take a piece of paper full of small magnetic needles and place it on a magnet. Obviously, there is a magnetic field around the magnet at this time, and the force of the magnetic field acts on each small magnetic needle to make them tend to be arranged in an orderly manner. If the magnetic field is not strong enough to overcome the friction between the small magnetic needle and the paper sheet, even though there is the force to order the small magnetic needle, the magnetic needles still remain in the original messy condition. If we gently shake the paper with our hands, we will find that the magnetic needles are arranged in an orderly manner in the direction of the magnetic field lines. Why? Obviously, the jitter will not increase the magnetic field force of the magnet, nor will it produce a driving force for orderly arrangement of each magnetic needle in different states. Because if we remove the magnet below, no matter how you shake it, it won't make the magnetic needles orderly. The secret is that in the moment of shaking, the friction between the magnetic needles and the paper are reduced, breaking the steady-state balance formed by the interaction of the friction and the magnetic force, so the system reaches an orderly structure under the action of the magnetic force. As shown in Figure 4.10.

Similar to this, the human body has the self-organization, self-adaptation, and self-regulation ability to repair damaged stable structures, and maintain and achieve a new steady-state balance. This ability still plays a role when the body is in a certain pathological steady state or moving in a direction of reduced order. Just because the disease as a whole or a certain link is too stable, this ability is not enough to take the system out of the pathological steady state or reverse the progression of the disease. Choosing some appropriate and strong intervention measures to promote the body's functional activities to remove pathogenic factors and pathological products may break the steady-state structure of the disease and make the body's state get out from the pathological steady-state balance. In this way, even though we did not impose any direct intervention against some deviated state variables, the body state will still move in the direction of normal steady state in the process of interaction between various state variables and seek a new steady-state balance.

Common diseases, especially those considered by TCM to be 'excess syndrome' or 'deficiency-excess complication', have always both functional obstacles and accumulation of metabolites. Either the system is in a very stable pathological steady state for a long time, a batch of deviated state variables are intertwined, showing some incomprehensible phenomena or illusions, it is difficult to break the stability of the pathological balance with the general direct confrontation method; or the pathogenic factors are too strong, the pathological products accumulate too much, and a certain link in the disease is too stable, which causes the overall state to evolve towards the deterioration of the disease. In this case, the method of breaking the steady state can often receive unexpected results, and the effect is much faster than the method of direct confrontation. When the patient's state deviates too far from the normal steady-state balance and the body's ability to resist disease is weak, although this method is difficult to completely restore the patient's normal steady-state balance, it can also make some deviated state variables move toward the normal direction. What is important is that strong interference stimulation can break the pathological homeostasis, completely or partially eliminate the cause and pathological products, eliminate some illusions caused by the

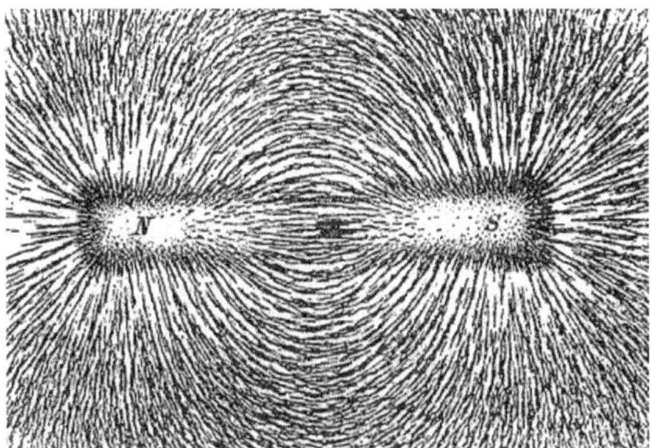

Figure 4.10. Ordered structure formed by a small magnetic needle.

intricate and dysfunctional parts of the disease, and make the condition clear. At this time, it is much easier to treat with direct confrontation methods. During the Jin and Yuan period of China, the famous medical scientist Zhang Congzheng (1156–1228) widely used the methods of sweating, vomiting, and diarrhea to treat all kinds of diseases. My grandfather, Hebei folk TCM practitioner Liu Hongyi (1904–1982), used the unique purgative method to treat various patients. They are all typical representatives of applying this method.

Zhang Congzheng's clinical treatment focused on attacking pathogens, and he was very skilled in the application of the three methods of sweating, vomiting and diarrhea. He believed that as long as the evil energy exists on the surface of the body and has not yet penetrated deep into the body, or the disease has entered inside, but the external evil has not been completely eliminated, the sweat method can be applied. The principle of using the sweat method is to induce sweat gradually, and sweat all over the body, but it should not be too much. As for the spit method, he believed that all the symptoms such as wind-phlegm, food and alcohol accumulation above the chest and diaphragm can be applied. The purgative method is not limited to laxative. Promoting birth, lactation, eliminating indigestion, removing ascites, treating amenorrhoea, and lowering qi, all belong to the scope of application of the method. Zhang Congzheng began to diagnose and treat patients in his twenties. Due to the excellent curative effects, he became a famous local TCM doctor in middle age (Chen Dashun et al. 1988).

My grandfather, Liu Hongyi, advocated purging first and then replenishing, and was good at using the purging method to attack accumulation and stagnation. In his clinical diagnosis and treatment of patients, no matter whether excess syndrome or syndrome of intermingled deficiency and excess, as long as there were signs of chest tightness, abdominal distension, abdominal pain, he would use the diarrhea method first. For many patients with intractable diseases who had been treated by many TCM practitioners without success, after a short period of the purgative, main symptoms disappeared, and they were refreshed. Even for patients with long-term illness and physical weakness or mixed deficiency and excess, after a short period of purging, the condition will be simplified, less intertwined, making subsequent adjustment and supplementation easier.

Obviously, the diseases of different individuals are very different, and the deviation of their states cannot be the same. However, regardless of Zhang Congzheng's sweating, induce vomiting, and diarrhea methods or my grandfather's unique diarrhea method, 'response to all changes with the same method', most of them can achieve unexpected results. The reason is that although the location, nature and extent of the lesions of different individuals are different, most of them have some pathological steady state or more stable pathological links, which hinders the body's efforts to restore normal homeostasis by adjusting itself. And using this strong and violent action to break this pathological homeostasis, the body's auto-adjusting ability will push the body to recover from the various pathological homeostasis toward the normal steady state.

Using the method of 'breaking the steady state', how can it be ensured that the system deviated from the pathological homeostasis will move towards normal homeostasis? First of all, the methods used to break pathological homeostasis usually act on state variables closely related to the maintenance of pathological homeostasis and have the effects of pushing the state variables to move to the normal state, so they usually have the impetus to restore the body's state to the normal steady-state balance. However, as mentioned above, Zhang Congzheng's three methods of sweating, vomiting, and diarrhea, and Dr. Liu Hongyi's unique diarrhea method, their strong medicinal power will also cause certain damage to the body's self-organization, self-adaptation, and self-regulation ability. Therefore, using this method, the body's self-organization, self-adaptation, and self-regulation capabilities, that is, the strength of the human body's righteousness, becomes the key to success or failure. If the body's self-organization, self-adaptation, and self-adjustment capabilities are strong enough to withstand the severe intervention required to break the homeostasis, the body that deviates from the previous pathological homeostasis will usually move in the direction of normal homeostasis. And if the patient's condition is severe and intertwined, and the body's self-organization, self-adaptation, and self-regulation are too weak to withstand such severe medicinal intervention, then the system that deviates from the original pathological homeostasis may also move toward the opposite direction into a worse pathological steady state, and even enter an unstable state where the condition is constantly deteriorating, leading to the disintegration of the system. Therefore, when the body's self-organization, self-adaptation, and self-adjustment capabilities are weak, this method is usually used with caution. No matter what the situation, this method is not suitable for long-term use, and the direct confrontation method is usually used as the follow-up support of this method. Of course, while adopting this method, it is also a common clinical choice to be used together with medications that replenish righteous qi, enhance the body's self-organization, self-adaptation, and self-regulation capabilities.

(3) The treatment of various diseases starting from the spleen and stomach

In TCM, qi originates from the innate, but it depends on the continuous replenishment of acquired water and valleys in order to maintain its abundance. After birth, the innate source of qi has ceased, and its only source is the acquired spleen and stomach. If the function of the spleen and stomach is healthy, the qi can be continuously replenished and full. If the spleen and stomach are deficient, the qi will not be replenished and nourished and will decline, which will also inevitably affect the functional activities of the viscera and the whole body.

Li Dongyuan's real name was Li Gao, courtesy name Mingzhi (1180–1251), a native of Zhengding, Hebei Province, and in his later years he used the name Dongyuan Old Man. He emphasized the important role of the spleen and stomach, and believed that the spleen and stomach were the foundation of vitality and the source of power for human life activities. As a result, Li Gao's diagnosis and treatment of internal injury and deficiency syndromes mostly started with the spleen and stomach. In the treatment of diseases of other viscera, he also attached great importance to the function of nourishing the spleen and stomach. In addition, Li Gao believed that the spleen and stomach are the pivots of the rise and fall of the body's qi. The absorption and distribution of essential qi in the diet depends on the rise of the spleen, and the discharge of the metabolites in the diet depends on the descending of the stomach qi. Therefore, the dysfunction of the spleen and stomach is the main pathogenesis of internal injury disease. In terms of elevating and descending, Li Gao attached great importance to the ascending function of spleen qi, and believed that as long as there is sufficient qi, the chance of suffering from various diseases would reduce, and qi deficiency is mostly caused by the lack of ascending spleen qi. Li's treatment was to invigorate the spleen and stomach, raise the yang and purify the fire, and adjust the abnormal rise and fall. Buzhong Yiqi Decoction is one of his famous prescriptions.

Li Gao also used the treatment principle based on syndrome, to replenish deficiency and reduce excess. For excess syndrome, the methods of sweating, induce vomiting and purging were also adopted. Li Gao also used the methods of purging fire with bitter cold or dispelling heat from the surface, but he was very cautious and thought that it should not be used for a long time, because if the cold was too great it could deplete the yang qi. And too much bitter cold was more likely to damage the spleen and stomach, so unless the fire is blazing, it cannot be used. When applying the method of purging fire, he usually added medicines of invigorating qi and regulating the spleen and stomach to protect the body's righteousness and the functions of the spleen and stomach.

When Li Gao was 20 years old, his mother was sick and bedridden, and later died due to the disorderly diagnosis and treatment of various TCM practitioners. Li Gao regretted the loss of his biological mother because he did not understand medicine, so he decided to study medicine, and took the famous Yishui doctor Zhang Yuansu as his teacher. With his solid and profound literary foundation, after several years of hard study and clinical practice, his medical skills became increasingly consummate, and he could diagnose and treat diseases in various disciplines. People of that time regarded him as a miracle doctor (Chen Dashun et al. 1988).

The disease states of clinical patients vary widely, and Li Gao mostly used the method of strengthening the spleen and replenishing qi. 'Responding to changes with the same', why are most patients able to achieve good results after such treatment? The reason is that the spleen and stomach are acquired foundations. If the spleen and stomach are strong, the qi and blood will be prosperous, and the self-organization, self-regulation and self-adaptive ability of the human body will naturally be strengthened. The strengthening of the function of spleen and stomach will naturally push state variables in other parts that have deviated from the normal state to restore to the normal steady state balance. Therefore, strengthening the digestion and absorption function of the spleen and stomach, and regulating the lifting and lowering function of the spleen and stomach, which are the hubs of qi, will play a positive role in the recovery of pathological states of other parts. Clinically, this method is of course applicable to patients with spleen-stomach qi deficiency and/or abnormal on lifting and lowering of qi as the main pathological state. However, patients with diseases of other viscera are often more or less accompanied by abnormal spleen and/or stomach functions. In this condition, the use/combined use of this method also has every advantage and no drawbacks. And because the CHMs used in this method are mild in nature, even for patients without abnormal spleen and stomach functions, they can strengthen the functions of the spleen and stomach and treat disease before it arises without any other side effects. Therefore, the method of treating from the spleen and stomach first is applied not only when the corresponding spleen and stomach syndromes appear, but it is a universal method that can be applied to many kinds of diseases of various organs.

(4) The methods to focus on nourishing yin and clearing heat

Zhu Danxi, whose real name was Zhu Zhenheng (1281–1358), was a famous TCM doctor in the Yuan Dynasty, a native of Yiwu, Zhejiang Province, and was named 'Danxi' because his former residence had a beautiful stream. Zhu Zhenheng was highly skilled in medicine, and his clinical efficacy was quite good. Many of his patients, after taking the medicine once, were cured without the need for follow-up consultation. Therefore, people at that time also called him 'Zhu half immortal'. He studied Confucianism first and then changed his medical practice. On the basis of his study of classical works such as Su Wen and Nan Jing, he visited famous doctors and later became a famous doctor of his generation who integrated the strengths of various schools. Zhu Zhenheng advocated the saying that 'yang is always in excess, and yin is always in deficiency' and was later called the founder of the 'School of Nourishing Yin'.

In terms of human physiology, Zhu Zhenheng attached importance to yin and blood, believing that yin essence is difficult to form but easy to lose, and proposed the famous theory of yang surplus and yin deficiency. In terms of etiology, Zhu attached importance to damp-heat and phase fire, pointing out that although normal phase fire is the dynamic energy of the human body, it can become an evil qi if the libido is too strong. In terms of treatment, Zhu focused on nourishing yin, nourishing blood and clearing heat, and opposed the misuse of warm nourishing and blindly attacking evil (Chen Dashun et al. 1988). Zhu Danxi treated fire-heat diseases separately into deficiency fire and excess fire. He proposed the principle that excess fire can be purged. For people with strong fire and weak bodies, he also believed that cool medicines should not be used directly, and the method of homeopathic or reverse should be applied. For the syndromes of internal stagnation of fire that cannot be dispersed and released, the divergent method can be used to treat it. As for the deficiency of fire, if the fire is caused by yin deficiency, it is appropriate to nourish the yin and reduce the fire.

If the four great masters of the Jin-Yuan era lived in the same period, facing same a patient suffering from multiple diseases, mixed with deficiency and excess, and involving multiple internal organs, what would happen if everyone wrote prescriptions based on their own academic opinions and medication habits? Zhang Congzheng would first expel evil qi and then support healthy qi, or focus on expelling evil qi and taking into account healthy qi. Li Dongyuan would focus on regulating the spleen and stomach and taking into account expelling evil qi. Zhu Danxi would pay more attention to the cold and heat properties of the prescription to avoid damaging yin fluid, in addition to supporting healthy qi and expelling evil qi. There is a good chance that the prescriptions they prescribe will be very different, but they will all likely achieve good clinical results. Unlike today's times, these TCM masters were general practitioners who were not limited to specific diseases or even clinical branches. But, as a generation of famous doctors, their reputations were established by their outstanding clinical efficacy.

Obviously, for patients suffering from multiple diseases and involving multiple internal organs, using different treatment ideas and treating from different parts of the patient may all achieve the curative effect expected by the TCM practitioner. In other words, TCM treats diseases, and in many cases, the effective treatment method is not unique. The art of regulating and controlling what to treat first and what to treat later, what to treat mainly and what to take into account, has great room for exploration and play for the ingenuity of TCM practitioners. In recent years, with the rapid expansion of the modern medical knowledge system, clinical divisions have become more and more detailed, and clinicians have become increasingly specialized. This trend of thought has also profoundly affected TCM. TCM is increasingly limited to the framework of 'TCM syndrome typing under Western medicine disease identification'. The treatment scope that TCM practitioners are good at is increasingly limited to narrow specialist areas, and there is almost no room for exploration and development of this art of clinical regulation of TCM. From ancient times to the present, there are famous general practitioners throughout China who are as skilled and effective as the Four Great Masters of Jin Yuan. Even in the Republican era, the four famous TCM doctors in the capital, Xiao Longyou (1870–1960), Shi Jinmo (1881–1969), Kong Bohua (1884–1955), and Wang Fengchun (1884–1949), were also widely known. However, after decades of integration of traditional Chinese and Western medicine, it has become difficult for the medical system in mainland China to produce such TCM masters.

Part II

Towards Holistic Medicine

Chapter 5

The Crises and Reform & Opening-Up of Modern Medicine

Medicine, according to the definition given in Canon of Medicine by Avicenna (980–1037), the greatest Arab medical scholar in the Middle Ages, is the art of maintaining health and restoring it when health is lost. The British 'Concise Encyclopedia Britannica' defines medicine as 'the science of studying how to maintain health and the prevention, mitigation and treatment of diseases, and the techniques used for the above-mentioned purposes' (Wen Liyang et al. 2001).

Modern medicine is an empirical knowledge system born out of ancient natural medicine since the 16th century, with the rise of modern natural science. After more than four hundred years of development, medicine has greatly promoted understanding of the human body, health, and disease, as well as the advancement of diagnostic diagnosis and treatment technology in depth and breadth, and established a huge knowledge system. However, today, in the face of many complex diseases that occur in the human body, medicine is still powerless. Moreover, the side effects of interventions used for treatment of human illness are becoming increasingly unacceptable. Today, more and more visionary scholars have begun to question the limitations of modern medical methods and turn their sights for exploring the path of medicine development to new methods and new fields.

5.1 The rise and development of modern medicine characterized by analysis

The rise of modern medicine can be traced back to the European Renaissance (approximately 14th to 17th centuries). This period experienced the re-examination of ancient medical concepts and the rise of scientific methods, which pushed medicine in the direction of science, precision and evidence. The medical progress during this period laid the foundation for modern medicine and provided important frameworks and methods for medical practice and medical research to this day.

5.1.1 The foundation of modern medicine

Ancient western natural medicine originated from ancient Greece in the 7th to 6th centuries BC. Based on the prevailing philosophy at that time, all objects in the universe were thought be made up of four elements: air (wind), water, fire, and earth, which mixed in certain quantities and proportions. Hippocrates (460–377 BC), the representative of ancient Greek medicine, and his students developed the 'four body fluid pathology theory' on this basis. Based on this theory, the four body fluids of blood, mucus, yellow bile and black bile determine the life of the organism. Everybody fluid corresponds to a certain body constitution, and each individual's body constitution depends on which body fluid is dominant in the body. When the four body fluids are balanced, the body is healthy, vice versa, imbalance among the fluids leads to sickness (Castiglioni 2019), as shown in Figure 5.1.

In ancient Greek medicine, all parts of the human body are an interconnected unity, and local diseases may cause systemic reactions. Hippocrates once described in his work: "Diseases

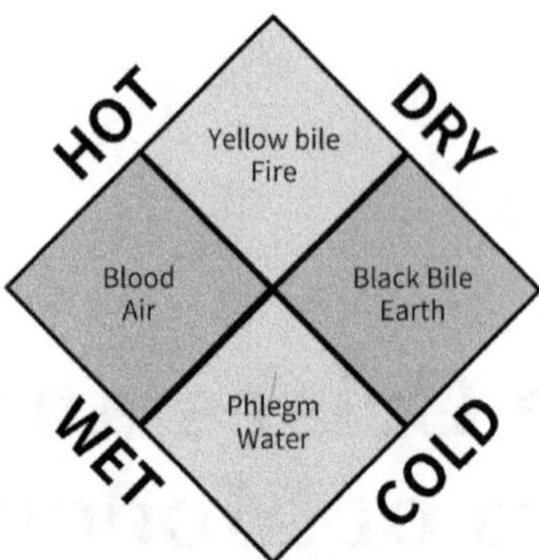

Figure 5.1. Four humoral pathology theory.

in individual parts of the body can cause diseases in other parts, waist problems can cause head diseases, and head problems can cause muscle and abdominal diseases.... And these parts are interrelated... and can spread these changes to all parts." Hippocrates also emphasized the unity of the human body and nature, and considered external factors such as climate, soil, water quality, air, living conditions, and other environmental factors will affect physical health, and had a relatively clear concept of prevention (Wen Liyang et al. 2001). Ancient natural medicine inevitably had the form of 'natural philosophy', which was based on rough and intuitive observations and experience summaries, supplemented by philosophical speculation and logical reasoning, and was influenced by myths and witchcraft. After entering the Middle Ages, it was inevitably branded with religion. Therefore, it is not surprising that its theoretical system is general, rough, and contains a certain degree of logical confusion.

In the 16th century, the advancement of astronomy and the great discoveries of navigation greatly expanded people's horizons. In the field of medicine, medical scientists broke through the barriers of more than 1,000 years in the Middle Ages, and anatomy made great progress in the research of the actual human body. In 1543, the Belgian scholar Vesalius (Vesalius 1514–1564) published his epoch-making work 'De Humani Corporis Fabrica' (Human Body Fabric), establishing human anatomy (Castiglioni 2019). In 1628, the British physician William Harvey (1578–1657) published 'An Anatomical Dissertation Upon the Movement of the Heart and Blood in Animals' and established the theory of blood circulation (Castiglioni 2019). As a result, anatomy, physiology, and even medicine was liberated from theology and became areas where science can study.

An important symbol of the rise of modern medicine is the introduction of scientific methods. Medicine began to place more emphasis on observation, experimentation, and systematic research methods, rather than relying solely on traditional experience and concepts. The application of this scientific method makes medicine a more rigorous and verifiable scientific field, rather than relying solely on traditional beliefs or personal experience. This greatly improves the accuracy and predictability of medical practice. In general, the rise of modern medicine is a critical period for medicine to become scientific and modern.

5.1.2 Establishment of modern scientific method characterized by reductive analysis

The establishment of modern scientific methods can be traced back to the period from the 17th century to the early 18th century: a series of important thinkers, mathematicians and scientists applied experiments, observations, mathematical modeling, theoretical verification and other methods to pioneeringly change the way people understand the natural world and laid the foundation for modern scientific methods. These methods emphasize the acquisition of scientific knowledge through systematic and reproducible research and the breaking down of complex problems into simpler parts to better understand natural phenomena. The establishment of this method had a profound impact on scientific progress to this day.

Modern natural science inherited the tradition of philosophical speculation and rationality from ancient Greece, and, at the same time, turned its attention to practice, forming a methodology that combines perceptual methods of observation and experiment with rational methods of hypothetical reasoning. Scientists no longer just rely on perceptual intuition and philosophical speculation as means of research and exploration, like ancient scholars used to do. For every natural phenomenon, they penetrated into the essence behind it by strict observation and measurement to establish a rigorous theoretical system accordingly. For this reason, modern scientists often need to use special instruments and equipment to extract certain natural phenomena from the whole of nature, and to study under the intervention and control of artificial forces, so as to obtain more reliable and accurate understanding than perceptual intuition. Of course, extracting the part from the connection with the whole for separate research, separates the wholeness of the thing itself, the wholeness of the thing and its environment, and the dynamic process of development and change of the part in the connection with the whole. The drawbacks and limitations of this methodology are also obvious.

Natural science in the 17th century had moved out of the form of 'natural philosophy' because it had completed the transformation from the scholastic philosophy tradition to the empirical scientific method. Mathematics, physics, chemistry, biology and other disciplines were separated from natural philosophy, became independent disciplines, and developed vigorously. Francis Bacon (1561–1626) and Rene Descartes (1596–1650), as representatives of modern natural sciences, emphasized the need to draw conclusions based on experimental observations and calculation results, thus opening up the way through rigorous observation and experiment to discover natural law (Li Meng 2016).

Bacon pointed out in The New Organon that "There are and can be only two ways of inquiry into and discovery of truth. The one flies from the senses and particulars to axioms of the most general kind, and from these principles and their (supposed) immutable truth, proceeds to judgement and to the discovery of intermediate axioms. And this is the method that is now in use. The other calls forth axioms from the senses and particulars by a gradual and continuous ascent, to arrive at the most general axioms last of all. This latter is the true but untried way" (Bacon 1878). Therefore, Bacon proposed a new inductive logic method, which is a new three-stage method of induction, summary and analysis based on observation, experiment and experience, and then discovering and verifying theories. The proposal of this method laid the foundation of scientific research methods and had a huge impact on the development of modern natural sciences.

Descartes, a French philosopher and scientist of the same era as Bacon, pioneered modern reasoning. Descartes argued that science begins with suspicion. He believed that what is believed to be true and generally regarded as truth must be doubted, nevertheless this suspicion is not the purpose, but to ensure that the foundation of knowledge is absolutely reliable and error-free (Descartes 1637). Descartes advocated deductive methods in scientific research. He mentioned in his early book 'The Rules for Guiding the Mind to Seek Truth': "There are two ways to reach an understanding of things, either experience or deduction. The deductive method must start from a few self-evident axioms and introduce other theorems step by step, until it constitutes a knowledge system that can justify itself. And every step of reasoning must be clearly understood, only in this way can the truth be reached" (Li Jianshan et al. 2002).

On the basis of observation and experiment, through induction, analysis and summary, combined with rigorous reasoning and deduction, the methodology for discovering and verifying truth in modern natural science has been formed. As Cohen (Bernard Cohen 1914–2003) said in 'Revolution in Science': "Anyone who understood the art of making experiments could test the truths of science—a factor which completely differentiated the new science from traditional knowledge, whether the old science, philosophy, or theology. Furthermore, the method could be easily learned and would then permit anyone to make discoveries or find new truths. ... No aspect of seventeenth-century science was as revolutionary as the method and its consequences" (Cohen 1987).

5.1.3 The rapid development of medicine from early modern times to modern times

The revolution of scientific concepts and the establishment of scientific methods have made science an organized activity for mankind to explore the truth and advance technological progress. In this context, medicine after the 17th century has made continuous progress and development (Wen Liyang et al. 2001, Wear and French 2011).

1. The emergence of the microscope and the discovery of microorganisms laid the foundation for the development of pathological anatomy and microbiology.
2. The application of the microscope and the discovery of capillaries have brought the medical understanding of human body structure to a deeper level.
3. The idea that cholera is transmitted through water sources laid the foundation for the development of epidemiology.
4. Disinfection theory established the foundation of microbiology, revealed the relationship between microorganisms and diseases, and laid the foundation for the modern practice of infection control and surgery.
5. The anesthetic gas ether was successfully used in surgical operations, creating modern anesthesiology and making surgical operations safer and more effective.
6. The establishment and systematization of pathological anatomy connects human understanding of diseases with the abnormalities of internal tissues and organs of the human body.
7. The discovery of cells, the establishment of cytology and cytopathology have advanced human understanding of human body structure and the occurrence and development of diseases to the level of cells.
8. The establishment and development of bacteriology opened up human research on tiny organisms that may cause diseases, and laid the foundation for the prevention and treatment of epidemic and infectious diseases caused by bacteria.

Modern medicine has experienced the foundation-laying of the 16th to 17th centuries, the systematic classification of the 18th century, and after the 19th century, it has been closely integrated with the development of modern science and technology and entered a period of rapid development. With the prosperity of the mechanistic view of nature, the reductionist method that advocates the reduction of high-level motion forms to low-level motion forms for experimental analysis has become the mainstream scientific method. The medical method characterized by anatomical decomposition and single factor analysis is regarded as a scientific method in line with the scientific concepts of the time. Analytical medicine, which was gradually developed based on this method, became a scientific medical system and occupied the status of mainstream medicine in Europe, where the level of holistic medicine was relatively weak. Traditional medicine based on the holistic approach was regarded as unscientific and was marginalized and gradually faded out of the stage of history. With the rapid progress of science and technology in the 20th century, modern medicine also entered a golden period of rapid development. The landmark events of this period include (Wen Liyang et al. 2001, Porter and Porter 1997):

1. The discovery of sulfa, penicillin and other antibiotics; and the effective treatment of infectious diseases have brought major advances in pharmacology and therapeutics, and also greatly promoted the establishment of the mainstream medical status of modern medicine.
2. The establishment and development of molecular biology marked that the decomposition of the human body by mankind has reached the most basic level of life.
3. The discovery of immune phenomena and the establishment of immunology marked the beginning of human exploration of the immune mechanism and the occurrence and development of immune-related diseases, and the use of immune mechanisms to effectively prevent and treat

diseases. This provides the basis for vaccine development, organ transplantation, and treatments for autoimmune diseases.

4. The discovery of viruses has enabled human understanding of the pathogenic factors and the development process of infectious diseases to enter a more basic molecular level than cells.
5. The widespread application of vaccines has made remarkable achievements in controlling infectious diseases, such as the development and vaccination of measles, mumps, rubella, pneumococcal and hepatitis B vaccines.
6. The advancement of medical imaging technology (X-ray, ultrasound scan, CT, MRI, fMRI, PECT, and their enhanced imaging), the development of continuous electrocardiogram and brain electricity recording equipment, and the maturity of human blood and metabolite analysis technologies have greatly advanced the human level of observation of the disease processes and disease diagnosis.
7. Long-term and detailed research on disease classification has established a huge disease classification system (DCS) which includes the current more than 55,000 codes of diseases and illness in ICD11.
8. The completion of the Human Genome Project, the beginning of various omics studies, and studies based on the correlation of genes, proteins, metabolites and other biomarkers with human diseases have revealed the essence of previously unknown diseases from a deeper level and discovered new effective treatment. At the same time, it also more profoundly demonstrated the complexity of life and disease processes and the limitations of modern medical methods.

Today, with the development of natural science and technology, modern medicine is still expanding its knowledge system in depth and breadth by continuously introducing new technologies and new methods. New pathogenic factors, new detection technologies, and new treatment methods are constantly being discovered, the DCS are constantly being refined, and diagnostic standards are going further and further on the road of accuracy and quantification. With the rise of omics research, medical research has begun to expand on a large scale at the molecular level, the lowest level that constitutes life. The method of medical research has changed from static separation research to dynamic observation; from single factor analysis to the establishment of a dynamic model reflecting the association of more than one factor; from focusing on the commonness of diseases to focusing on the individualization of the body's state. Even compared to 20 or 30 years ago, it can still be said that earth-shaking changes are taking place in medicine today. However, in confronting many diseases that are a serious threat to mankind, medical scientists have found that they are still in a situation of confusion and powerlessness in many cases.

5.2 The dilemma faced by the study of life with reductionism method

The anatomy and analysis of the human body in modern medicine started from the level of organs and tissues. However, knowledge at this level alone cannot fully reveal the mysteries of human basic physiological and pathological activities and life processes. With the advent of microscopes, the discovery of cells, and the rise of biochemistry, human understanding of life gradually deepened. The founding of molecular biology indicates that human understanding of life has reached the molecular level which is the most basic unit of life.

5.2.1 The great achievements of modern science are almost all based on reductionist methods

The natural sciences that began to rise in the 16th century actually gradually developed under the guidance of the scientific concept later called Reductionism. Descartes believed that if a thing is too complicated to be solved at once, it can be broken down into small enough problems to be analyzed separately, and then they can be combined together to obtain a complete and accurate understanding of complex things. This is the popular expression of reductionism in philosophy. The reductionist method is the core of the classical scientific method, which decomposes high-level, complex objects into lower-level, simple objects for processing. The essence of the world lies in simplicity.

The idea of reductionism is deeply ingrained in natural science. In the eyes of modern scientists, chemistry is based on physics, biology is based on chemistry, and so on. However, even in the era when reductionism was most prevalent, in the social sciences, there are still great disputes about whether psychology can be attributed to biology, whether sociology can be attributed to psychology, and whether political science can be attributed to sociology.

In the analysis based on reductionist methods the world picture is presented with unprecedented simplicity. As early as the 19th century, the German physicist Helmholtz·Von (1821–1894) believed that "Once all of natural phenomena are reduced to simple forces, and it is proved that natural phenomena can only be simplified in this way, then the task of science is completed" (Einstein and Infeld 1938). Based on the concept of reductionism, modern physics attributes the existence of the world to elementary particles and their interactions; biologists believe that research at the molecular level will reveal all the mysteries of the complexity of life. Through reduction, the complex world was clearly decomposed into simple particles and parts that can be reorganized, and knowledge about the world is also decomposed into a variety of different and complex disciplines and departments. As Fritjof Capra (1939) pointed out: "Over emphasis on the Cartesian method has led to the fragmentation that is characteristic of both our general thinking and our academic disciplines, and to the widespread attitude of reductionism in science-the belief that all aspects of complex phenomena can be understood by reducing them to their constituent parts" (Capra 1983). As a result, the reductionist method is transformed into the reductionist belief in the ontological sense.

The core idea of reductionism is that 'the world is composed of individuals (parts)', i.e., the simplicity of the world. The 18th–19th century when Newton's view of mechanics prevailed was the peak of reductionist beliefs. The ancient organic, life and spiritual view of the universe was replaced by the notion of the world as a 'watch machine'. Holders of reductionist beliefs believe that the objective world is certain, and the world is composed of 'universe bricks' similar to elementary particles in an infinitely sophisticated way. The nature and interaction of the universe bricks fundamentally determine the nature of the world, and even the most complex objects are assembled from the lowest level (and also the most fundamental) 'basic components'. From Democritus' (Democritus, 460 BC-370/356 BC) conception of atomism to Newton's objects with a certain mass and motion, through Dalton's (John Dalton 1766–1844) atomism, it eventually developed to contemporary reductionists' confirmation of the basic particles and energy inside atoms. Since the world is composed of basic units at different levels, the smallest entity that cannot be reduced in the end is the essence and origin of the universe.

As a result, the reductionist belief rooted in the reductionist method in turn strengthened the reductionist method and had a profound impact on the scientific methodology. In the eyes of modern scientists, various complex phenomena can be understood by breaking them down into basic building blocks and their interaction. Different branches of science describe different levels of objective reality, but in the end they can all be attributed to the most basic science of objective reality, physics (Liu Jinyang 2007).

The positive role of reductionism in the development of science cannot be underestimated. In fact, modern science has developed along this path. Cowan (George Arthur Cowan 1920–2012), the academic authority on complexity science and the director of the Santa Fe Institute said: "The grand road to the Nobel Prize is usually opened by reductionist methods" (Xie Shusheng 2009). But as science developed to the current stage, the limitations of reductionist methods became more and more obvious.

5.2.2 Limitations of the reductionist approach in the face of complex living organisms

The real world is a complete whole, and the various parts that make up the whole are always in a dynamic process of interconnection and interaction. Their interaction determines the nature and characteristics of the system as a whole, and often causes the system to 'emerge' at the holistic level with new attributes that none of its parts have. After any part of the system is separated from the whole, it is no longer in a state of 'interaction', and the functions it displays are not exactly the same as the actual functions in the system. This situation is common in reductionist biology research, and it often leaves scientists in a dilemma of understanding.

For example, IL-2 is a cytokine produced by activated immune cells, and it is the first FDA-approved immune medication for metastatic melanoma and kidney cancer. Purified IL-2 can stimulate T cell growth *in vitro*, so it is also called T cell growth factor. After scientists used gene knockout technology to remove the IL-2 gene in the mouse genome, the mice lost the ability to produce IL-2. The expected result should be that the number and function of such mouse T cells will be greatly reduced. But the result was completely the opposite. IL-2 knockout mice showed massive proliferation of T lymphocytes, enlarged lymph nodes, and obvious symptoms of autoimmune disease. This shows that once IL-2 is separated from the whole, the displayed function is different from the function in the *in vivo* environment. It is now known that IL-2 has the function of promoting T cell apoptosis *in vivo*, which cannot be shown *in vitro* (Xie Shusheng 2009).

For another example, in studying the actions of ginseng (including sun-dried ginseng, Korean ginseng, white ginseng, American ginseng, red ginseng, etc.) on cells, scientists have discovered that ginseng can promote or stimulate metabolism, which can enhance cell activity, including promoting the proliferation of cancer cells in cancer patients under certain conditions. In other words, under the stimulation of ginseng, the vitality of both normal and abnormal cells will increase. Obviously based on this, cancer patients should not take ginseng. However, research on the pathogenesis of cancer has shown that the human immune function is closely related to the occurrence and development of cancer. Clinical trials have shown that people who take ginseng are significantly less likely to develop cancer than those who do not. Studies have also shown that ginseng can also improve the liver function of patients and increase the function of peripheral blood white blood cells, so that the immune function of patients can be improved, so it can be used as an auxiliary medication for the treatment of tumor patients. Clinical studies on the effective ingredients of ginseng used in cancer treatment have shown that a variety of active substances contained in ginseng, such as ginsenoside Rh2 and Rg3, have anti-cancer actions. They can effectively regulate human immune function, inhibit cancer cell proliferation, and induce cancer cells to transform into normal cells. The use of ginseng during radiotherapy and chemotherapy can effectively prevent leukopenia, enhance immune function, and increase effect and reduce toxicity; it can not only improve the quality of life of patients, but also help prevent cancer recurrence and metastasis. Obviously, the direct effect of ginseng on cancer cells is significantly different from the indirect effect of ginseng in the body by enhancing immunity and the interaction of various parts of the body (Luo Linming et al. 2017).

The traditional reductionist analysis method requires that in observations and experiments, 'only one variable can be changed at a time', because if there are two or more variables that change at the same time, it is difficult to determine which variable caused the result and the causal relationship among them. But in the human body, which relates closely between various levels and among various parts, we cannot strictly control the conditions, whether it is observation or experiment. And the human intervention to control the conditions often affects the functional activities of the human body determined by its own nature, and brings great errors to observation and experimental results.

In current life science research, reductionist methods still dominate. The launch of the Human Genome Project at the end of the 20th century made people think that as long as the entire genome is sequenced, all the mysteries of life can be revealed. However, this huge project, which lasted for more than ten years, ultimately provided only the arranged sequence of 3 billion bases in the human genome. In layman's terms, what structural genomics gives is only the words that make up the book of life, and the content of the book of life is still mostly unclear. The completion of the human genome map neither reveals the mystery of life, nor does it significantly promote clinical medical practice. The work of interpreting the book of heaven will be an important task for life scientists in the 21st century (Xie Shusheng 2009).

From the level of organs, cells to molecules, human cognition of life has gradually moved away from the scope of direct perception by human senses. To establish a description of the state of the human body at the molecular level, it is not only necessary to understand the static information of which elements the human body is composed of at this level, but also to understand their structural relationships and interactions. Genomics, proteomics, lipidomics, glycomics, transcriptomics and other omics research are carried out at this level. At present, research in this area has just begun, and the road to uncovering the mysteries of life is still very long.

In the past 400 years, biology and medicine, separated from ancient empirical medicine, have established a huge knowledge system through the decomposition and analysis of the human body. Today, along the levels of organs, tissues, cells, and molecules, the understanding of life has reached the most basic level that composes life. But what puzzles scientists is that there are still many internal laws of the human body that have not been well revealed, and people are still so helpless in the face of many diseases that seriously endanger human health and life. Vibrant life phenomena that cannot be understood at the macro level lose the whole picture after reaching the micro level, and the understanding of their essence become more and more difficult (Yuan Bing and Zhao Zheng 1986). In 2000, Nature·Medicine published an editorial article titled 'Biomedical Research in Transition'. The article said: "A considerable number of biomedical scientists have now realized that reductionist biology cannot solve any problems except the simplest problems. However, there are almost no simple problems in biology" (Xie Shusheng 2009). Facing reality, scientists have gradually woken up: life cannot simply be attributed to the interaction between cells and cells, molecules and molecules, and atoms and atoms that make up life. Considering reductionist methods as the only feasible method to reveal the mystery of the essence of life is inherently problematic.

5.3 The crisis facing disease medicine

As for what is a disease, there is still no unified and clear definition in the world. As the more than 120-year-old International Classification of Diseases (ICD) has become more and more widely accepted by the medical community, the current modern clinical medicine can basically be said to be a diagnosis, treatment and prevention system established to deal with the diseases defined by this classification system.

5.3.1 A disease classification system that continues to expand as the number of diseases increases

ICD is an internationally unified DCS developed by the World Health Organization (WHO). It is based on the four main characteristics of the disease, namely the etiology, location, pathology and clinical manifestations (including: symptoms and signs, staging, classification, gender, age, acute and chronic onset time, etc.) classify diseases into categories. Each of these characteristics constitutes a classification standard, forming a classification axis, so ICD is a multi-axis classification system. As of December 31, 2021, the internationally accepted disease classification is based on ICD-10. The 11th revision of the ICD (ICD-11) has come into force on January 1, 2022, and the types of diseases have further increased and been refined (World Health Organization 2018).

However, today, clinicians find that this already large DCS is still far from enough to grasp the complex diseases that actually appear in the clinic. Certain types of medications specially developed for the treatment of certain diseases are still effective for some people and ineffective for other people due to individual differences of patients, and side effects of one kind or another cannot be avoided.

The success of the Human Genome Project has greatly broadened people's horizons. According to recent studies, the number of human protein-coding genes is about 19,000–20,000 (Ezkurdia et al. 2014). This is the result of gradual downward revisions as genome sequence quality and methods for identifying protein-coding genes improve. The human genome has about 3 billion DNA base pairs (Pray 2008), and human gene mutation refers to the changes in gene structure caused by the substitution, addition and deletion of base pairs in the DNA molecule. Therefore, the types of genetic variants associated with human disease may also be counted on a scale of hundreds of millions. Further research has shown that the association of most gene mutations with human diseases is not absolutely specific. In other words, whether a person suffers from a certain disease in his life is far from being completely determined by genes. Just like in the manufacturing process of plastic kitchenware, mold design is of course the most important, but the processes of mold manufacturing and product processing cannot also be ignored. If there is a problem with the product, it may be a defect in the mold design, the mold processing may not meet the requirements, or there may be a problem with the production process. The problem may also lie in the technical level of the workers and the management and the control of the production process. Similarly, if a person suffers from a certain disease, there may be a problem with the genes related to it. The problem may also lie in the process of protein synthesis and metabolism, or it may be due to external pathogenic factors or emotional factors, as well as diet, and living habits that have affected the body's normal physiological functions and metabolic processes at the organ and system level.

Obviously, just figuring out genes is far from enough to understand the human life process and the occurrence and evolution of diseases. In recent years, with the progress of genomics research, proteomics, metabolomics, transcriptomics, lipidomics, immunoomics, glycomics, RNAomics and other omics research have emerged and flourished. At present, research on the correlation between hundreds of millions of gene variants and human diseases is still in its infancy. However, biologists clearly know that the proteome and metabolome may have more abnormalities in the course of human life than gene mutations. And different from relatively stable gene mutations, their changes will show remarkable dynamic characteristics. If the abnormal indications found in the process of omics research are used to further classify human diseases, we will also face a huge DCS with billions or even tens of billions of diseases, even if it does not include the combination of various indications. This will be a situation that science and human intelligence cannot cope with.

Preliminary studies on the correlation between genetic variation and human diseases have shown that a human disease is often associated with more than one genetic variation, that is to say, human diseases are not determined by a certain gene alone in most cases. In this way, different combinations of gene variants related to a disease may have different effects on the disease. If we use differences in gene mutations to classify diseases, we must consider not only diseases caused by single gene mutations, but also diseases caused by different combinations of multiple gene mutations. The two-two combinations of hundreds of millions of gene mutations reached the order of trillions, plus the three-three combinations and four-four combinations, which is an unimaginable astronomical number. It is conceivable that even if only one out of ten thousand of these combinations are clinically meaningful, we still have to face trillion-scale disease types in the case of only counting the two-two combinations. What's more, the current gene-related research reveals that it is not uncommon for more than 3 or even more than 5 gene mutations to affect a disease. A study published in 'Molecular Cancer Therapeutics' in 2015 illustrates this problem. In this study, Schwaederle et al. conducted genetic testing on 439 patients with different tumors. The results showed that 96% of patients had at least one gene mutation in their tumors, with an average of 3 mutations, and most of the patients have different types of gene mutations

(Schwaederle et al. 2015). However, we will also face the proteome, metabolome, etc., which are larger than the genome.

The current DCS used to describe the complicated pathological states of the human body during the disease process has dazzled medical scientists. However, the large number of abnormal biomarkers discovered by omics research, whether used for the refinement of the current disease classification, or the establishment of a new classification system for the differentiation of the personalized characteristics of the disease, will make medicine face a much larger DCS than it is now. The expansion of disease classification and pathological state description system (DCS & SDS) means that the number of diseases or abnormal health states that can be diagnosed or found in a person has increased. At present, it is commonplace for dozens of diseases to be diagnosed from one patient. If the abnormal biomarkers found in gene and protein testing are added, the number of diagnoses that can be made will be even greater.

5.3.2 The expansion of the disease classification system makes the comprehensive treatment of diseases more and more difficult

At present, for many diseases, modern medicine has developed some symptomatic drugs for the purpose of controlling the disease, which have been applied in clinical practice. In clinical practice, more than one medication is often used to treat a disease of a patient. In order to increase the curative effect, clinicians usually choose several medications with different mechanisms of action in combination to exert a synergistic effect. As mentioned in Chapter 4 of Part I, due to the limitation of the controlled amount as the research frame of reference, the purpose of medication research in modern medicine is usually to control the direct damage and secondary damage of the disease. The actions of the therapeutic medications studied are often not on the fundamental link of the disease, but on the link where the compensatory response occurs in the disease. Faced with diagnoses of dozens or even more diseases, not to mention that one disease requires the combined use of multiple medications, even if only one medication is used for each disease, it will face the problem of combined application of dozens of medications at the same time. If dozens of medications with the same or different side effects are taken together, their mutual cooperation and antagonism are inevitable, and differences in the body's state will also bring this diversity of cooperation and antagonism. In this case, it is often uncertain what comprehensive and long-term effects will be produced on the patient.

The development of science and the progress of technology are increasing the complexity of the knowledge system describing the human body and diseases. However, the vast medical knowledge system we have established based on analytical methods cannot still describe, let alone deal freely with, the complexity caused by this interconnectedness. Today, biology, physiology, and medical research continue to make progress in depth and breadth. However, modern medicine is still powerless to deal with a large number of diseases that seriously endanger human health and life, especially chronic diseases and diseases caused by viruses. No matter what new medications are discovered, side effects and medication resistance are still a nightmare that people cannot get rid of.

The research progress of biology and medicine has increasingly revealed the integrity of the human body and the close relationship between human beings and nature. However, our clinical medicine still mainly adopts the method of classifying diseases and treating them independently. The total amount of knowledge about the human body, diseases, treatment methods and medications, and related fields is increasing rapidly. However, due to the limitation of human physical factors on intelligence, the proportion of knowledge that a person can master throughout his life relative to the total knowledge in this field is getting smaller and smaller. Yesterday we could call Mr. Zhang an expert in internal medicine and Mr. Li an expert in surgery; today, we can only call Mr. Wang an expert in cardiovascular surgery and Mr. Xu an expert in thoracic surgery. What will these people be called tomorrow?

The development of science has brought people's vision into depth, and the result of entering depth has made people lose their grasp of the whole. People are increasingly confined to narrower and narrower areas, knowing nothing or very little about the areas adjacent to them. However, the human body as an organic whole with closely related among parts, the control of the overall health status is by no means equal to the simple addition of the disease control of various levels and parts. To adjust and control the human body comprehensively and optimally requires us to deeply grasp each part and even each detail of the human body, and more importantly, we must grasp the interconnection between the various parts, that is, to integrate them into a whole (Yuan Bing and Zhao Zheng 1986).

Obviously, disease-centric medicine has come to its end. The development of modern medicine not only requires new technologies and methods, but also looks forward to the birth of new models.

5.4 Reform and opening up of modern medicine

In the face of reality, while advancing theoretical development and technological progress along the way of previous research, some visionary scholars in the medical field began to think about the limitations of medical approaches and inherent systems, and turn their exploration beyond the current medical system. As a result, natural therapies and traditional medicine have attracted more and more attention around the world, and the terms and concepts of complementary and alternative medicine, evidence-based medicine, and integrated medicine are increasingly entering people's field of vision. Thus began the 'reform and opening' era of modern medicine.

5.4.1 Return to naturopathic medicine

Naturopathic medicine is a non-mainstream medical method that originated in Europe in the 19th century. It is usually understood as a medicine that uses the natural environment and the substances existing in nature to cure diseases, or mobilize the body's own ability to restore the body to health (Smith and Logan 2002). According to the Association of Accredited Naturopathic Medical Colleges, naturopathic medicine is a distinct medical profession that combines the wisdom of nature with the rigors of modern science. It emphasizes the prevention and treatment of disease and the improvement of health using methods that increase the body's self-healing. It can use modern, traditional, scientific, and even a priori methods. Its basic principles are (Association of Accredited Naturopathic Medical Colleges 2020):

1. *Inducing the body's self-healing ability.*

   Naturopathic physicians believe that the human body possesses considerable self-healing power, and the role of the naturopathic physician is to facilitate and enhance the process of self-healing through natural, non-toxic remedies.

2. *Find and treat the cause of the disease, not just eliminate symptoms.*

   Naturopathic physicians are trained to identify the causes behind diseases, not just suppress symptoms. The causes of disease may originate from the physical, mental and emotional, as well as spiritual level.

3. *The most important thing in treatment process is not to cause harm to the patient.*

   Naturopathic physicians strive to use safe and effective natural remedies to achieve therapeutic effects without causing harm to patients.

4. *Deploy therapy from a holistic perspective.*

   Naturopathic physicians strive to view patients and health problems from a holistic perspective. They view disease as the result of the interaction of multiple factors such as physical, mental, emotional, spiritual, and social, and consider the synergistic application of various therapeutic measures from a holistic perspective.

5. *Provide proper education and guidance to patients.*

   The important role of a naturopath is to act as a teacher, educating and motivating patients to adopt a correct attitude towards disease and to adjust unhealthy lifestyles and eating habits.

6. *Pay attention to the prevention of diseases and safety hazards.*

   Naturopathic medicine believes that prevention is the best cure. Therefore, naturopathic physicians attach great importance to preventing diseases by educating correct health concepts and improving unhealthy living and eating habits.

Naturopathic medicine treats the human body as an organic whole, and believes that the human body has vitality and a certain degree of self-healing ability. It strives to use the substances existing in nature and the subjective initiative of people to prevent and treat diseases. It encourages people to minimize surgery and chemical drugs, and to use natural and harmless treatments to improve the condition, promote healing and maintain health. Practitioners of naturopathic medicine generally oppose the use of vaccines to prevent diseases. There are many methods used in naturopathic medicine, the most basic of which is diet and regular exercise. Naturopathic medicine also uses many treatment techniques derived from traditional medicine, such as acupuncture, herbal medicine, Gua Sha, massage, cupping, flower essence, hydrotherapy, etc. It also widely uses homeopathy, physical therapies, and hypnosis, Daoyin and other mind-body therapies to treat patients. In recent years, it has opposed artificial genetically modified foods and advocated the restoration of the body's natural balance through natural and balanced diet therapy. Currently, naturopathic practitioners are increasingly recognized as health care practitioners and experts of natural and preventive medicine. The positions of naturopathic practitioners exist in hospitals, multidisciplinary clinics and specialized health centers in the United States, Canada, Europe and other countries and regions.

Naturopathic medicine in a broad sense includes all kinds of natural therapies and natural medical systems, including European medicine, Chinese medicine, and Indian medicine that have a history of thousands of years. Therefore, various non-chemical synthetic medications and non-surgical medical methods used by different countries and nations in the world can be categorised as naturopathic medicine. Obviously, naturopathic medicine contains mainly effective treatment methods and medications that have been discarded with the rise of modern medicine.

Although naturopathic medicine has a long history, the popularity of this term is only in recent decades. It appeared in response to the social reality of the flood of chemical drugs and the large number of side effects of medications, a lot of addition of chemical additives to food, and the change of dietary structure that caused the increase of foodborne diseases. It embodies the human resistance to chemically synthesized medications and the return to nature. After hundreds of years of development, in the face of the helplessness of a large number of diseases endangering human health and the increasingly obvious drawbacks and limitations of their treatment methods and medication systems, modern medicine has to pick up these things that were once thrown away in the 'historical garbage dump' to re-understand them. In recent years, with the rise of the concept of complementary and alternative medicine based on the perspective of modern medicine, naturopathic medicine has been included in the category of complementary and alternative medicine.

5.4.2 The rapid development of complementary and alternative medicine

As defined by the World Health Organization, the terms 'complementary medicine' or 'alternative medicine' refer to a broad set of health care practices that are not part of that country's own tradition or conventional medicine and are not fully integrated into the dominant health-care system (World Health Organization 2023). Obviously, Complementary and Alternative Medicine (CAM) usually refers to methods of diagnosis, treatment and prevention other than mainstream medicine that can supplement the deficiencies of mainstream medicine. It includes medical knowledge that has not been taught in ordinary medical schools and medical methods that have not been widely practiced in general hospitals. Based on the concept that modern medicine is the mainstream medicine, modern medicine is 'Conventional Medicine' or 'Orthodox Medicine', and Unconventional Medicine (Unorthodox Medicine) that has not been included in modern medicine belongs to the category of complementary and alternative medicine.

In recent years, the development of complementary and alternative medicine in the United States has been promoted at the national level. In 1992, the US Congress authorized the National Institute of Health (NIH) to establish the Office of Alternative Medicine (OAM), which was later upgraded to the National Center for Alternative Medicine (NCCAM) in 1998. In 2014, NCCAM was renamed the National Center for Complementary and Comprehensive Health (NCCIH). NCCIH funds complementary and alternative medicine research, including CAM-based clinical trials (National Institutes of Health 2023). Since the establishment of OAM, scientific research in this field has developed rapidly. Ten universities, including Harvard University, Columbia University, and Stanford University have successively set up research centers, and started curriculum education for some students. Outside the United States, many medical schools in Western countries have opened complementary and alternative medicine courses. The content includes acupuncture, massage, homeopathy, herbal medicine, spiritual and body interaction, natural therapy, etc. Canadian medical workers go further. They believe that complementary and alternative medicine should be emphasized and understood as an integral part of the health care system.

The status of complementary and alternative medicine in the field of medicine has increased rapidly in recent years. The World Health Organization (WHO) classifies 65%–80% of the health care in the world as traditional medicine that currently belongs to complementary and alternative medicine. In terms of population size, more people in the world currently use complementary and alternative medicine, rather than modern medicine. The publication of literature on complementary and alternative medicine in mainstream medical journals is also an important sign that mainstream medical circles accept complementary and alternative medicine. In the 1950s, articles about complementary and alternative medicine that appeared in these journals were all articles that expressed concern about patients' use of complementary and alternative medicine, which was seen as an unscientifically proven method. The 1980s saw the emergence of literature investigating the use and knowledge of complementary and alternative medicine in patients. After the 1990s, reports of scientific research on complementary and alternative medicine began to appear in mainstream medical journals. In recent years, JAMA, New England Journal of Medicine, The Lancet, British Medical Journal and some professional journals such as Cancer and The Journal of Clinical Oncology have published scientific research articles on complementary and alternative medicine.

Although the terms 'supplement' and 'replacement' are usually combined into one category when talking about complementary and alternative medicine, it still makes sense to distinguish the meaning of these two terms. NCCIH distinguishes the difference between the two in this way: when a non-mainstream practice is used together with conventional medicine, it is regarded as a 'supplement', and when a non-mainstream practice is used instead of conventional medicine, it is regarded as 'replacement'. There may be overlap between these categories. For example, aromatherapy can sometimes be used as a complementary treatment, and in other cases as an alternative treatment (NHS Site 2022).

Complementary and alternative medicine cover all therapies and medicines other than mainstream medicine. Naturally, it also includes the aforementioned naturopathic medicine, integrated medicine, family therapy, and oriental medicine. Proposing complementary and alternative medicine within the scope of modern medicine, in a sense, means the expansion of the boundaries of modern medicine. Currently, NCCIH divides the coverage of complementary and alternative medicine into five main areas (National Cancer Institute 2022):

1. *Mind–Body Therapies*

   These combine mental focus, breathing, and body movements to help relax the body and mind. Some examples are:

   - *Meditation*: Focused breathing or repetition of words or phrases to quiet the mind.

- *Biofeedback*: Using simple machines, the patient learns how to affect certain body functions that are normally out of one's awareness (such as heart rate).
- *Hypnosis*: A state of relaxed and focused attention in which a person concentrates on a certain feeling, idea, or suggestion to aid in healing.
- *Yoga*: Systems of stretches and poses, with special attention given to breathing.
- *Tai Chi*: Involves slow, gentle movements with a focus on the breath and concentration.
- *Imagery*: Imagining scenes, pictures, or experiences to help the body heal.
- *Creative outlets*: Interests such as art, music, or dance.

2. *Food therapy*

   This type of CAM uses things found in nature. Some examples are:

   - Vitamins and dietary supplements.
   - Botanicals, which are plants or parts of plants, such as kava (Piper methysticum) or ginkgo biloba.
   - Herbs and spices such as turmeric or cinnamon.
   - Special foods or diets.

3. *Physical therapy, including*:

   - *Massage*: Massaging, tapping and stroking of soft tissues.
   - *Chiropractic therapy*: A type of manipulation of the spine, joints, and skeletal system.
   - *Acupoint pressure therapy*: Pressing specific parts of the patient's body with his hands or feet.

4. *Biofield therapy*

   Sometimes called energy medicine, it involves the belief that the body has energy fields that can be used for healing and wellness. Therapists perform the therapy by using pressure or placing their hands in or moving the body through these fields. Some examples are:

   - *Reiki*: Emitting energy either from a distance or by placing hands on or near the patient.
   - *Therapeutic touch*: Moving hands over energy fields of the body.

5. *Whole Medical Systems*

   These are systems of medicine or healing that have developed around the world during different historical periods and in different cultural contexts. Some examples are:

   - *Ayurvedic medicine*: A system from India in which the goal is to cleanse the body and restore balance to the body, mind, and spirit.
   - *Traditional Chinese Medicine (TCM)*: A medical system with a holistic concept and treatment based on syndrome differentiation as the core.
   - Acupuncture is a common practice in TCM that involves stimulating certain points on the body to promote health or to treat diseases.
   - *Homeopathy*: Uses very small doses of substances to trigger the body to heal itself.
   - *Naturopathic medicine*: Uses various natural methods to help patients heal. An example would be herbal treatments.

In recent years, there have been more and more unconventional and unorthodox therapies whose mechanism and effectiveness have been scientifically proven, and the number of patients receiving these therapies has been increasing year by year. As a result, mainstream medicine can no longer avoid unconventional and unorthodox therapies, and has begun to re-recognize them with an objective attitude. Acupuncture, which originated from TCM, has been widely accepted by mainstream medical circles. In February 2007, the US Federal Food and Drug Administration

(FDA) published the 'Alternative Medicine Product Management Guidelines', which recognized acupuncture needles as medical devices and recognized TCM as a Whole Medical System with complete theory and practice independent of the mainstream medicine (U.S. FDA 2007).

Roughly speaking, complementary and alternative medicine are mostly non-invasive therapies with almost no side effects. Especially for those patients who are recognized as suffering from intractable diseases by modern medicine, it undoubtedly provides more choices. Moreover, complementary and alternative medicine has positive significance for solving various medical problems in the 21st century, such as side effects of medications, environmental pollution, crisis of trust in medical practitioners, economic problems, and improving the overall quality of medical care.

5.4.3 Evidence-based medicine: the 'passport' to enter the modern medical clinical treatment system

Evidence-based medicine (abbreviated as EBM) is a new medical model that has developed rapidly in the international clinical field since the 1990s. It represents a more scientific and systematic approach to medical practice, emphasizing scientific evidence-based medical decision-making to improve patient care quality and safety. It is now one of the fundamental principles of modern medicine and continues to have a profound impact on medical practice.

(1) Concepts and definitions of evidence-based medicine

Evidence-based medicine (EBM) is a new medical model that has rapidly developed in the international clinical field after the 1990s. The concept of evidence-based medicine was proposed by the internationally renowned clinical epidemiologist David Sackett (1934–2015). Its definition is: Prudently, accurately, and wisely applying the best research evidence currently available, while combining clinicians' personal professional skills and long-term clinical experience, taking into account patients' values and wishes, and perfectly combining the three to formulate a specific treatment plan (Sackett et al. 1996). EBM aims to obtain more accurate and reliable diagnostic methods, more effective and safer treatment options, and ultimately enable patients to obtain the best treatment outcomes.

EBM requires clinicians to strive to find and obtain the best research evidence, combined with personal professional knowledge, including pathophysiological theories of disease occurrence and evolution, and personal clinical work experience, as well as expert opinions. On this basis, it should be based on the principle of 'patient first', respect the patient's personal wishes and actual possibilities, and then make diagnosis and treatment decisions.

Whether the clinician determines the treatment plan, or the expert determines the treatment guidelines, it must be based on the best evidence obtained in clinical scientific research. Evidence is the cornerstone of evidence-based medicine. Clinical evidence mainly comes from large samples of randomized controlled trials (RCT) and systematic reviews or meta-analysis. Using EBM's ideas to guide clinical practice, the most important thing is to conduct a systematic literature search based on clinical problems, understand the research progress of related problems, and scientifically evaluate the research results to obtain the best evidence.

EBM is a medical diagnosis and treatment method that emphasizes the application of research evidence to optimize decision-making. While based on scientific beliefs, all medicine needs to be backed by some level of experience, but EBM goes a step further. It categorizes the evidence by epistemological strength and requires that only the strongest evidence (such as meta-analysis, systematic reviews, and randomized controlled trials) can be summarized as strong evidence for recommendations; relatively weak evidence (such as case-controlled studies) can only be classified as a weak recommendation. The method of EBM was originally used in the teaching of medical diagnosis and treatment and to improve the decision-making methods of clinicians when facing different patients (Sackett et al. 1992). However, its scope of application has rapidly expanded to design guidelines and policies applicable to patient groups and the entire group (Eddy 1990), including other research fields such as education, management, law, public policy, and building safety (Li 2019).

Broadly speaking, EBM is the application of scientific methods to make medical decisions. It advocates those applications whether in medical education, individual decision-making, the formulation of guidelines/policies applicable to groups, or management decisions/policies in general health services should all be based on evidence and not solely on medical practitioners, experts, or administrators' belief. Therefore, it tries to ensure that the opinions of clinicians are based on the analysis and summary of all available data in the scientific literature to ensure the best diagnosis and treatment. The difference between evidence-based medicine and traditional medicine is that it does not rely on empiricism or past cases. It advocates the use of serious and clear methods to analyze evidence and provide it to decision-makers (Wikipedia contributors 2023a).

The development of EBM has two main branches. One is to insist on a clear assessment of the effectiveness of evidence when issuing clinical diagnosis and treatment guidelines; the other is introducing epidemiological methods into medical education and individual patient-level decision-making. In 2005, Eddie provided an overarching definition for the two branches of EBM: "Evidence-based medicine is a set of principles and methods intended to ensure that to the greatest extent possible, medical decisions, guidelines, and other types of policies are based on and consistent with good evidence of effectiveness and benefit" (Eddy 2005). The following are the five steps of implementing EBM for the purpose of medical education and individual-level decision-making, compiled by the representatives who participated in the 2003 Evidence-based Health Care Teachers and Developers Conference (Dawes et al. 2005):

1. Translation of uncertainty to an answerable question that includes critical questioning, study design and strength of evidence;
2. Systematic retrieval of the best evidence available;
3. Critical appraisal of the internal validity of the evidence from the following aspects:
    - Systematic errors due to selection bias, information bias, and Interfering factors
    - Quantitative results of diagnosis and treatment
    - The effect size and aspects regarding its precision
    - Clinical importance of results
    - External validity or generalizability
4. Outcomes of application in clinical diagnosis and treatment;
5. Evaluation of performance.

(2) Evidence from evidence-based medicine

Systematic reviews of published studies are an important part of evaluating specific treatments. Once all the best evidence has been evaluated, the evaluation of treatments can be divided into three categories: possibly beneficial, possibly harmful, or insufficient evidence to support beneficial or harmful.

Evidence-based medicine categorizes and grades different types of clinical evidence based on whether research can avoid the various biases that can affect it. For example, the strongest evidence for therapeutic interventions is provided by the systematic review of randomized, well-blinded, placebo-controlled trials with allocation concealment and complete follow-up involving a homogeneous patient population and medical condition. In contrast, patient testimonials, case reports, and even expert opinions (some critics have argued that expert opinion does not represent a form of empirical evidence and it would seem to be a separate, complex type of knowledge) have little value as proof because of the placebo effect, the biases inherent in observation and reporting of cases, difficulties in ascertaining who is an expert and more (Wikipedia contributors 2023a).

The following are the recommendations for rating the quality of evidence proposed by the US Preventive Services Task Force (USPSTF) in 1989 (Sherman et al. 2011):

- *Level I*: Evidence obtained from at least one properly designed randomized controlled trial.
- *Level II-1*: Evidence obtained from well-designed controlled trials without randomization.
- *Level II-2*: Evidence obtained from well-designed cohort studies or case-control studies, preferably from more than one center or research group.
- *Level II-3*: Evidence obtained from multiple time series studies with and without treatment intervention. Uncontrolled trials with particularly significant results might also be regarded as this type of evidence.
- *Level III*: Opinions from respected authorities, clinical experience, descriptive studies, or reports of expert committees.

In 2000, Grading of Recommendations Assessment, Development and Evaluation (GRADE) developed a rating system that considers more dimensions in addition to the quality of medical research (GRADE working group 2006). Although there are differences between the rating systems, their purpose is to guide users of clinical research information to distinguish which studies are more reliable. However, individual studies still need to undergo rigorous evaluation.

An ongoing challenge to EBM is that the conclusions of the evidence for or against current treatment methods are constantly changing, and it is practically impossible to keep abreast of each change in time. This is also an important reason why some clinicians are unwilling to conduct diagnosis and treatment based on evidence (Epstein 2017). For example, from 2003 to 2017, hundreds of empirical conclusions involving medical diagnosis and treatment changed, from whether hormone replacement therapy is safe, to whether infants should take certain vitamins, and whether antidepressants are effective for Alzheimer's disease (Herrera-Perez et al. 2019). In addition, many problems faced in clinical work cannot be answered in the existing EBM review. If urgent clinical questions need to be answered, EBM analysis is too time-consuming to be feasible.

Clinicians following EBM also realize that what kind of medical treatment is used for individual patients will be affected by various factors, including different understandings of the quality of life and the value of life, etc., which cannot be fully covered by pure scientific methods. However, evidence-based medicine tries to clarify those medical methods that can be covered by scientific methods, and tries to use scientific methods to ensure that such medical methods can bring the best treatment effect, even though there is still controversy as to which treatment effect is most desirable for people.

(3) Implementation steps of evidence-based medicine

EBM does not blindly apply recommendations gathered from recently published literature. In fact, EBM requires the use of a series of steps to effectively obtain useful information to solve patient problems. Fully integrating the EBM principle must also consider the patient's value system, including healthcare cost, the patient's religious beliefs, and the patient's autonomy. The clinical application of EBM principles usually includes the following steps (Mandell 2021):

1. Asking clinical questions

   The problem must be clear. Clear problems are easier to locate in the medical literature. A better question should include the population involved, interventions (such as diagnostic tests, treatment), comparison (such as comparison of different treatment options), and prognosis.

2. Find evidence to answer questions

   Collect relevant research extensively by reviewing the literature. Standard literature sources include MEDLINE, Cochrane Collaboration, National Guideline Clearing-house, and ACP Journal Club.

3. Assess the quality and accuracy of evidence

   Different types of research have different scientific value and rationality. Any specific research or case report has different values according to its methodology, internal validity, and universality of results (external validity). The level of evidence is usually divided into 1 to 5 levels. For EBM analysis, high-quality evidence should be selected as much as possible. However, because high-quality randomized controlled trials may be slightly different from clinical issues, lower-level evidence is often used. Low-quality evidence does not mean that EBM will not adopt it, but only shows that the conclusions drawn from the evidence are weaker.

4. Decide how to apply the evidence to a specific patient

   Since the highest-quality sources of evidence have some characteristics different from those of the patient who asked the question, whether it is applicable to a specific patient requires a clinician's judgment. In addition, it is necessary to consider the patient's tolerance for aggressive examinations and treatments, as well as the tolerance for discomfort, danger and uncertainty. For example, an aggressive chemotherapy regimen suggested by EBM can prolong the lifespan of certain cancer patients by 3 months, but different patients have different ideas about whether they are willing to obtain this extra survival time or avoid this extra pain. The cost of examination and treatment may also influence the decision of clinicians and patients, especially when some options are particularly expensive for patients.

(4) Medical guidelines are increasingly becoming an indispensable tool for clinicians in clinic

The concept of EBM is indisputable, but it is actually unrealistic for clinicians to go through the above-mentioned evidence-based process when encountering every case. So various medical guidelines came into being. Medical guidelines are also called clinical guidelines or clinical practice guidelines. They are currently used more and more clinically, and many professional committees have published guidelines. Good clinical guidelines are usually formulated based on the principles of EBM and unanimous advice from experts. The purpose of issuing clinical guidelines is to provide documents of medical decision-making and standard guidelines regarding diagnosis, management and treatment in specific areas of health care. In the history of medicine, there were documents similar to medical guidelines thousands of years ago. However, in contrast to previous approaches, which were often based on tradition or authority, modern medical guidelines are based on an examination of current evidence within the paradigm of EBM. They usually include codified medical consensus or best practices in healthcare. A healthcare provider is obliged to know the medical guidelines of his or her profession, and has to decide whether to follow the recommendations of a guideline for an individual treatment.

Modern clinical guidelines identify, summarize and evaluate the highest quality evidence and latest data on prevention, diagnosis, prognosis, and treatment, including dosage of medications, risk/benefit and cost-effectiveness. Then they define the most important questions related to clinical practice and identify all possible decision options and their outcomes, some guidelines also contain decision or computation algorithms to be followed. On this basis, they integrate the identified decision points and specific protocols for each step with the clinical judgment and experience of practitioners. Many guidelines place the treatment alternatives into classes to help clinicians in deciding which treatment to use.

Additional objectives of clinical guidelines are to standardize clinical medical process, to raise the quality of medical services, to reduce various risks and to achieve the best balance between cost and medical parameters such as effectiveness, specificity, sensitivity, etc. It has been demonstrated repeatedly that the use of guidelines by healthcare providers such as hospitals is an effective way of achieving the objectives listed above.

Guidelines are usually produced at national or international levels by medical associations or governmental bodies, such as the United States Agency for Healthcare Research and Quality. Local healthcare providers may produce their own set of guidelines or adapt them from existing top-level guidelines. Insurance companies often issue their own clinical guidelines as well. The USA and other countries maintain medical guidelines clearinghouses. In the US, the National Guideline Clearinghouse maintains a catalog of high-quality guidelines published by various health and medical associations. In the United Kingdom, clinical practice guidelines are published primarily by the National Institute for Health and Care Excellence (NICE). In The Netherlands, two bodies-the Institute for Healthcare Improvement (CBO) and the College of General Practitioners (NHG)-have published specialist and primary care guidelines, respectively. In Germany, the German Agency for Quality in Medicine (ÄZQ) coordinates a national program for disease management guidelines. All these organizations are now members of the Guidelines International Network (G-1-N), an international network of organizations and individuals involved in clinical practice guidelines (Wikipedia contributors 2023b).

Although the clinical guidelines describe standard practices, the guidelines themselves cannot be a medical plan for individual patients. Some clinical guidelines follow 'if, then' rules (e.g., if a patient is febrile and neutropenic, then institute broad-spectrum antibiotics). More complex, multistep rules may be formalized as algorithms. Guidelines and algorithms are generally straightforward and easy to use but should be applied only to patients whose clinical characteristics are similar to those of the patient group described in the guidelines. Furthermore, guidelines do not take into account the degree of uncertainty inherent in test results, the likelihood of treatment success, and the relative risks and benefits of each course of action. To incorporate uncertainty and the value of outcomes into clinical decision making, clinicians must often apply the principles of quantitative or analytical medical decision making (Mandell 2021).

With the emergence of new medical discoveries and the progress of medical research, the content of medical guidelines is being constantly enriched and updated and some content that has lost its clinical significance may be deleted. The study found that nearly 20% of the items previously strongly recommended in the medical guidelines were later cancelled (especially those based on personal opinions rather than trial results) (Neuman et al. 2014).

5.4.4 The progress brought about by 'reform and opening up' and the problems faced

Obviously, the knowledge systems of naturopathic medicine, complementary and alternative medicine, and modern medicine are completely different. These methods and medications need to have a certain basis to be recognized and applied by modern medicine. And evidence-based is the simplest way to solve this problem. 'Regardless of white cats and black cats, those who can catch mice are good cats'. No matter where these methods and medications come from, as long as there is evidence to prove their scope of application and effect in clinical practice, they can be used in modern medicine.

Naturopathic medicine, complementary and alternative medicine are actually a kind of 'reform and opening up' measures for modern medicine to make up for its own shortcomings, conditionally recognize traditional medicine and incorporate traditional medicine therapies and experience into the modern medical system. In a sense, they are an extension of the boundaries of modern medicine. EBM gives the necessary procedures to incorporate medical methods and treatment techniques outside the system into the modern medical system and get them recognized by modern medicine.

The treatment methods and techniques of traditional medicine are largely based on its own theoretical system. For example, the description of the actions of TCM treatment method and Chinese Herbal Medicine (CHM) are usually inseparable from the treatment system based on syndrome of TCM, and this theoretical system is incompatible with the modern medical system. Naturopathic medicine, complementary and alternative medicine, in the process of incorporating traditional medical treatment methods and technologies into their systems, usually ignore their actions descriptions relative to traditional medical theories, and replace them with descriptions of their actions relative to modern medical disease systems. As a result, these treatment methods and medications based on traditional medical theories have become 'water without a source' and 'wood without roots' to a certain extent, and their applications is often inappropriate according to the theory of traditional medicine.

For example, in TCM, Epimedium is a kind of aphrodisiac, which is not suitable for patients with internal heat or yin deficiency. However, according to the research on the pharmacological actions of its active ingredients, epimedium flavoniods (EF) has an immunomodulatory effect and can significantly enhance human immunity. Icariin plays a role in inducing tumor cell apoptosis, and Icariside II also has good anti-cancer activity (Guo Feng and Zhao Na 2007). Therefore, the use of Epimedium extract for immunocompromised cancer patients by naturopathic doctors is obviously justified. However, for patients with severe internal heat or yin deficiency, it often encourages internal heat or aggravates yin deficiency, leading to some adverse reactions that are not understood by naturopathic practitioners.

Part I mentioned a series of events that have occurred in recent years: the alkaloids contained in ephedra that appeared in the United States were used indiscriminately for weight loss; Xiao Chai Hu Decoction was used indiscriminately in the treatment of chronic hepatitis in Japan, with its indications in Kampo medicine ignored. These events that triggered serious adverse reactions were essentially caused by the abuse of these medications regardless of their indications in the original traditional medical system.

EBM puts aside the discussion on the basic principles of medicine, dilutes the discipline attribution of medical knowledge, experience and method sources, focusing only on their authenticity and validity. EBM uses rigorous methods to verify and analyze the evidence in order to make an objective evaluation, and then incorporate them into clinical guidelines. This opens the door for incorporating knowledge, experience and methods that are not part of the modern medical system into the clinical practice of mainstream medicine on the basis of empirical verification. Ever since medicine entered modern times hundreds of years ago, mankind has been looking forward to finally establishing a set of rigorous medical theories on the basis of observation and experiment through decomposition and analysis methods, so that medicine can be developed into a real science that can carry out clinical practice under the guidance of theory. However, today, the prevalence of the concept of EBM in the medical field marks the failure of the effort to fundamentally realize the practice guided by theory under the framework of modern medicine. The development of medicine is turning back and returning to the age of experience. The difference is that the current experience is based on evidence, rather than ancient experience with a rather subjective color.

However, the path for EBM to incorporate knowledge, experience, and methods that are not part of the modern medical system into mainstream medical clinical practice through empirical verification is not straightforward. It is clearly facing challenges from several aspects:

(1) The challenge of precision medicine to evidence-based medicine

Modern medicine attributes the treatment of the sick body to the treatment of diseases of the various parts of the human body. The evidence it is based on is classified and relatively isolated. So far, medication R&D and clinical research have been clinically evaluated with randomized controlled trials as the core. The core of the randomized controlled trial is to select the test population as uniform as possible through a series of screening criteria, and randomly assign the test population to the experimental group and the control group. Randomized controlled trials can significantly reduce trial bias and eliminate individual differences among subjects to the greatest extent by strictly controlling the experimental conditions and the random grouping of participants. Clinical guidelines based on 'high-quality' evidence such as randomized controlled trials are standardized diagnosis and treatment guidelines that exclude individual differences, this statistically confirmed treatment plan gives only a probability of effectiveness. For some specific individuals, the conclusion is inaccurate or even completely unrealistic.

According to the statistics of an American researcher published in the journal Nature in 2015, the effectiveness rate of the medications ranked in the top ten medications by sales in the United States is actually not ideal. Among them, the more effective medication is effective for 1 out of

4 users, while the less effective medication is effective for only 1 out of 25 users (Schork 2015). In 2018, a study titled 'Why All Randomized Controlled Trials Produce Biased Results' evaluated the 10 RCTs with the highest citation rate and found that the trial has a lot of deviations and limitations. These trials are only suitable for studying a small part of the problems that are easy to randomize, and can usually only evaluate the average treatment effect of the sample. Whether the results can be extrapolated to other situations remains to be further studied (Krauss 2018). In order to cope with the challenge of low accuracy brought by the non-personalized research characteristics of 'randomized controlled trials', it is also an important reason for the rise of precision medicine characterized by individualization.

As we will see later, with the rise of precision medicine and the 'basket trials' and 'umbrella trials' that are compatible with personalized medicine have been recognized by the US FDA, the dominance of 'randomized controlled trials' in clinical trials and efficacy evaluation has become a thing of the past. When clinical trials take into account different personalized characteristics, a single trial becomes a trial series consisted of multiple sub-trials. To ensure the number of cases for each sub-trial, the total number of cases required to complete it must be at least multiplied by as many sub-trials as the trial series contain. Specifically, refer to the following expressions:

The total number of cases required for a trial series ≥
*the number of cases required for a single trial * the number of sub-trials included in the trial series*

In this way, 'randomized controlled trials' conducted in limited time and concentrated locations will become increasingly difficult to operate. As shown in the aforementioned research on the correlation between human diseases and gene mutations, different combinations of gene mutations related to a disease may have different effects on this disease. If the classification is based on different combinations of biomarkers, the possible personalized types of a disease will also be a considerable number. The relationship between biomarker combinations and diseases usually does not have a one-to-one correspondence, that is to say, the same biomarker combinations may also appear in other diseases. As a result, the refinement and increased complexity of 'randomized controlled trials' will also make its operation increasingly difficult to conduct. This is also the reason why 'real world research' has attracted more and more attention from the medical community in recent years and has begun to be accepted by the US FDA as evidence.

With the rise of precision medicine in recent years, evidence-based concepts have entered the field of precision medicine based on biomarkers. However, individualized evidence is difficult to be collected by ignoring individualized 'randomized controlled trials', and the results of such trials cannot reflect the complex relationships among various elements related to the disease. How does EBM, with 'randomized controlled trials' as the 'gold standard' adapts to the emerging personalized medicine has become an important issue that urgently needs to be solved in the development of EBM. And its final solution may shake the foundation of EBM that currently uses 'randomized controlled trials' as the 'gold standard'.

(2) How can the evidence of EBM support the integration of medicine?

With the improvement and refinement of the DCS and the establishment of a new DCS based on biomarkers, more and more diagnostic conclusions can be made on a patient after a comprehensive examination. The current knowledge system of modern medicine is divided into disciplines and categories. The increase in the total amount of medical knowledge has brought about more and more detailed divisions, so that clinicians are increasingly focusing on more and more limited areas. The current EBM is usually carried out by each clinician within the scope of his or her specialty, and he or she may not be familiar with or have less experience with diseases outside the scope of his or her specialty. The human body is a whole with closely related functions of its part. An abnormality in one tissue or organ may cause abnormalities in other tissues and organs. When a medication acts on the human body, its effects are usually not limited to the organs and tissues involved in the doctor's specialty. And its effects on organs and tissues covered by other specialties are often side effects.

The disadvantages of this methodologically very low-level medical model are now recognized by more and more people by assigning the diseases of different parts of a whole with closely related functions to different medical specialists to treat. Today, 'medicine needs integration' has become a consensus in the medical community. The 'randomized controlled trials' currently based on evidence-based medicine are basically conducted for specific diseases or disease types. The trial process often needs to ignore the correlation among different diseases, or only focus on the single correlation that can be verified by the analysis method of 'changes only one variable at a time'. As mentioned earlier, the study of various complex systems in complexity science has profoundly revealed: The control of the holistic level of the complex system cannot be simply attributed to the simple addition of the control of the various levels and parts that make up the whole. Integrating EBM as the 'gold standard' evidence as the basis for combination medications, it is often difficult to predict what will happen to actual patients.

Obviously, the evidence that the current evidence-based medicine relies on is far from enough to support the combination medication required for the simultaneous occurrence of multiple diseases. Obtaining the evidence required for combination medication may shake the evidence system that is the basis of evidence-based medicine. That is to say, much evidence that are valuable for combination medication are difficult to obtain by means of 'randomized controlled trials'. The following is an actual case of combination medication in a hospital in Hong Kong that I learned about. This case shows to a certain extent that even if a rigorous treatment plan is formulated based on EBM, the combination medication in a patient with multiple diseases may have some uncertain results.

A 70-year-old male patient with diabetes and hypertension was admitted to the hospital in 2020 due to a sudden hemiplegia and speech difficulty caused by cerebral thrombosis. The treatment by the physician in charge included the following medications:

- *Norvasc*: it has the action of reducing blood pressure by dilating peripheral blood vessels
- *Plavix*: it might prevent thrombosis
- *Lipitor*: it might lower blood cholesterol and reduce the risk of cardiovascular disease. The physician suspected that the thrombus came from the heart and therefore chose it
- *Nootropil*: it might prevent brain tissue damage further into Alzheimer's due to mental decline caused by stroke
- *Glucophage, Januvia, Diamicron*: all belong to hypoglycemic medication
- *Tresiba*: a long lasting insulin medication
- *Duphalac*: a symptomatic medication for the treatment of constipation.

This treatment plan integrates the actions of stabilizing blood pressure and blood sugar, reducing the risk of secondary strokes, and preventing mental decline secondary to strokes. Based on the latest clinical guidelines, the treatment selected by the physician in charge in this plan was appropriate, and the selected medications are all mature medications that have been verified by 'randomized controlled trials' and have been used in the clinic for a long time. However, a few days later, the patient developed edema of the limbs, which is the main side effect listed in the instructions of Norvasc and Tresiba. Of course, the appearance of edema should also be related to diabetes. In the next few days, the patient developed severe depression, which is the side effect listed in the instructions of Norvasc and Nootropil. In later observation, the patient's edema was not effectively controlled and gradually worsened. After one month of hospitalization, the patient's toes turned black and showed signs of diabetic gangrene. In this treatment plan, Norvasc has the action of raising blood sugar, and Lipitor has been confirmed by the US FDA to increase the risk of diabetes. Obviously, the physician in charge had considered these factors and selected four hypoglycemic medications including insulin for blood sugar control. However, blood sugar has not been effectively controlled. The occurrence of gangrene may also be related to the uncontrolled edema for a long time.

Obviously, for patients with multiple diseases at the same time, in the face of the complex relationship among multiple concurrent diseases and the unavoidable side effects of medications, combination regimens based on valid evidence of EBM for a single disease are often difficult to produce the result that clinicians expect.

(3) The frame of reference on which EBM is based is also related to its effectiveness

Incorporating the knowledge, experience and treatment methods of traditional medicine into naturopathic medicine, complementary and alternative medicine, and incorporating them into the clinical guidelines of modern medicine on the basis of EBM, undoubtedly opens up a new channel for modern medicine to accept the effective treatment methods of traditional medicine to better deal with diseases. But we should also see that the current evidence description of EBM is based on the knowledge system of modern medicine. The action and applicable situation of treatment methods derived from traditional medicine has been described fully and in detail in traditional medicine. However, due to the limitations of the knowledge system of modern medicine and the method of 'randomized controlled trials', the verification results obtained by 'randomized controlled trials' with the knowledge system of modern medicine as the reference system are often one-sided, and cannot comprehensively reflect their actions and application situation. Separating these treatment methods from the traditional medical theories on which they are rooted, and applying them based on evidence-based evidence in the framework of modern disease medicine, the application effect will naturally be greatly reduced. And due to some inappropriate applications from the perspective of traditional medicine, it is also difficult to avoid adverse reactions.

As mentioned earlier, the effectiveness of target medications based on abnormal biomarkers is often not statistically verified in 'randomized controlled trials' based on conventional disease classification. This has led to the emergence of new clinical trial models such as 'basket trials' and 'umbrella trials', incorporating personalized biomarker-based situations into clinical trials and efficacy evaluation systems. Treatment methods of traditional medicine are also difficult to show effectiveness in 'randomized controlled trials' that are based entirely on modern medical knowledge systems. This has become common knowledge of the medical community in China, which has decades of experience in the practice of ITCMWM. If the evidence based on EBM does not include clinical trials based on the state description of traditional medicine (such as the syndrome of TCM), the effectiveness of traditional medical treatment methods is also difficult to obtain statistical verification. Therefore, in the process of incorporating the knowledge, experience, and methods of traditional medicine into the modern medical system, if EBM does not introduce the system in which they were originally rooted in some form, it will undoubtedly lose the accuracy of grasping their functions and applicable situations to a certain extent. And if the system in which these knowledge, experiences, and methods are originally rooted is incorporated into the EBM system, EBM will be no longer EBM in the current sense.

Clinical trials and efficacy evaluations based on a personalized disease description system of precision medicine have now caused a huge challenge to evidence-based medicine methods. Whether to use the personalized state description system (SDS) of traditional medicine (such as the TCM syndrome system) as the reference system for clinical trials and efficacy evaluation will be a greater challenge for EBM.

The evolution from experience based on non-statistical significance to statistical analysis based on empirical evidence is undoubtedly a kind of advancement in medicine, marking a significant improvement in its scientific degree. However, whether it is based on evidence is not the only indicator assessing the pros and cons of a medical system. What mankind needs is an evidence-based medical system that can achieve overall integration, and can describe and regulate-control the disease in a personalized way. Abandoning the holistic concept and the principle of individualization in order to follow the evidence, is obviously far from the correct path of medical development.

Chapter 6

Crises and Opportunities Brought by the Collision of Eastern and Western Medicine

Among the many traditional medicine systems in the world, traditional Chinese medicine (TCM) is the system with the longest history, the highest level of development, and the most complete. This is not only reflected in the depth and breadth of the understanding of health and disease, but more importantly, based on the ancient Chinese philosophy of holism, it has established a complete system from multiple aspects including theories, treatments, prescriptions, and medications. It has gone through thousands of years without decline, and has made great contributions to the reproduction and prosperity of the Chinese nation. Whether it is for many difficult diseases that make modern medicine helpless or epidemic infectious diseases caused by pathogens such as bacteria and viruses, TCM has shown amazing effects.

Due to its effectiveness in dealing with various diseases, TCM was the mainstream medicine in China until the beginning of the 20th century. In recent hundreds of years, with the rise of modern medicine, two medical systems with very different understandings and treatments of diseases have emerged in the East and the West. After the Opium War and towards the end of the Qing Dynasty, with the invasion of Western powers and the opening of the country, the Qing government initiated the Westernization Movement in order to strengthen its national power. With the rise of the Westernization Movement, China began to introduce Western industry and commerce, and sent students to study in the West to study science and technology, biology, and medicine. Scholars who received Western education also gradually returned to China to open schools and universities. As a result, modern medicine, as a vassal of modern science, began to enter China on a large scale, colliding with TCM, and at the same time starting a century-old course of debate between traditional Chinese and Western medicine and 'integration of traditional Chinese and Western medicine (ITCMWM)'.

In Japan, which is located in China's immediate neighborhood, medicine was entirely derived from the ancient TCM. The success of Japan's 'Meiji Restoration', which began almost at the same time as China's Westernization Movement, enabled Japan to begin the process of 'exiting Asia and entering Europe'. Amidst the background of 'total westernization' at that time, traditional medicine was abolished by the official order, and Kampo medicine in Japan embarked on another path of 'abandoning medicine and keeping herbal medicine'.

6.1 The integration of traditional Chinese and Western medicine and the debate on the continuation and abolition of TCM caused by the introduction of modern medicine

The integration of Chinese and Western medicine caused by the introduction of modern medicine refers to the process of communication, and integration between traditional Chinese medicine and Western medicine in modern times with the introduction of modern Western medicine. This process involves multiple levels of knowledge, technology, concepts, and practice, and has had a profound impact on China's medical system and medical education.

6.1.1 Thoughts on the integration of traditional Chinese and Western medicine caused by the introduction of modern medicine to China

During the late Ming and early Qing dynasties in China, when modern medicine was in the early stages of development in Europe, some Jesuit priests who came to China brought some knowledge of Western medicine. For example, German Jesuit missionary Johann Schreck (1576–1630) compiled 'Great West Human Body Theory', 'Human Body Diagram' and other anatomical and physiology works appeared. During this period, some TCM scholars began to understand and study Western medicine theories, such as the famous scholar Bi Gongchen (died in 1644), Jin Jongxi (died in 1645) in the late Ming dynasty, and the famous doctor Wang Honghan (1648–1700) in the early Qing Dynasty. In their writings, there are accounts of the brain being in charge of people's thinking and memory (Li Zhijun 2004, Anhui Cultural History Editorial Committee 2000, Li Xisuo 2002).

Wang Qingren (1768–1731), a famous doctor in the Qing Dynasty, wrote the book 'Medical Forest Correction' based on the concept of 'diagnosing and treating diseases, one must first understand the viscera'. Based on his own observations of the internal organs of those who died of illness or were sentenced to death, and the knowledge gained from asking others, the book 'corrected' a large number of 'fallacies' in the description of human internal organs in TCM (Shaanxi Provincial Institute of TCM 1976).

After 1840, due to the failure of several wars with Western powers, China lost its territory, compensated for silver, and was forced to open the door of 'closed self-defense'. The reality of 'lagging behind will be beaten' forced the Chinese to seriously think about their own weaknesses and the crises facing the country. Subsequently, under the slogans of 'learning advanced Western science and technology for strengthening oneself' and 'learning advanced Western science and technology for prosperity', China began the Westernization Movement that lasted 35 years from 1861. The Westernization Movement was the first national-scale industrialization movement in modern China. While introducing Western industry and commerce, the Qing government vigorously introduced Western science and technology, translated various Western books and documents, opened new schools, and sent groups of students to study in the West. Western medicine, as an adjunct to Western civilization and science, also began to enter China on a large scale. The missionaries coming to China, along with the translation of Western medicine books, the establishment of Western medicine schools and hospitals, and the absorption of overseas students returning home, all impacted the dominance of TCM in China for thousands of years. This also led to the comparative study of traditional Chinese and Western medicine and the early attempts to integrate traditional Chinese and Western medicine.

The core concept of the Westernization Movement in the Qing Dynasty was the 'improvement' of 'taking Chinese traditional thinking as the main body, and Western science and technology as technical means'. Therefore, the ITCMWM in this period was inevitably marked with this idea. Moreover, during this period, the main body of research on ITCMWM was by TCM scholars, so the ITCMWM was mainly carried out from the perspective of TCM and based on the concepts and methodology of TCM.

Tang Rongchuan (1846–1897), also known as Tang Zonghai, a TCM scholar in the Qing Dynasty was one of the early representatives of research on ITCMWM. He authored two volumes of 'Essence of traditional Chinese and Western Medicine Huitong' (1892). He believed that the principles of traditional Chinese and Western medicine are interlinked. Western medicine is good at understanding the human body shape and structure, while TCM is good at understanding the body's gasification and function. In terms of internal medicine treatment, the superiority of TCM is indeed beyond the reach of Western medicine. He believed that 'Western medicine also has its strengths, but how can TCM have no shortcomings?' He tried to use knowledge of Western medicine anatomy and physiology to confirm the theory of TCM, thus proving that TCM is not unscientific (Cai Jingfeng et al. 2000).

Zhu Peiwen (born in 1805), another representative figure in the early research of integrated TCM and Western medicine, was born in the Lingnan area, which is the place where Western medicine was widely spread and prosperous in China. He had extensively read ancient and current TCM books and Western medicine books translated into Chinese at that time, and personally observed autopsies in Western hospitals. His main work is 'Compilation of Images of Chinese and Western Viscera' (1893). He believed that TCM is 'good at reasoning and speculation, while the observation and description of the structure and function of the entity is poor', but 'too much emphasis on reasoning and speculation may lead to divorce from the practice'. Western medicine, he felt, 'is good at studying entities, and is poor in theoretical thinking', but 'excessive pursuit of the description of the entity structure may lead to ossification of understanding'. He advocated that the confluence of traditional Chinese and Western medicine should be based on clinical verification as the standard: 'unify the content that can be unified, and retain the different content separately'. Obviously, as far back as the late Qing Dynasty 100 years ago, Zhu realized the difference between the theory of visceral outward manifestation in TCM and the anatomical structure of the human body, and proposed a way of ITCMWM, which is very similar to today's evidence-based medicine (Cai Jingfeng et al. 2000).

The opening of the country broadened the horizons of the Chinese people. They saw the tremendous power of science and technology, and also realized the gap between China and the West in science and technology. After the Sino-Japanese War of 1894–1895 in which China was defeated by Japan, the Chinese redoubled their efforts to make the country stronger. With the rise of industry and commerce, the students studying overseas sent by the Westernization Movement returned to China, and the concepts of Western science and technology and Western civilization began to take root in the country. TCM, which has nothing to do with modern Western scientific concepts, was also hit unprecedentedly.

6.1.2 The controversy on the survival and abolishment of TCM and the proposal of the slogan of 'scientificization of TCM'

In TCM, the explanation of functions related to the human body and diseases is based on organs such as five Zang, six Fu, and basic vital substances such as essence, qi, blood, body fluid. With the introduction of Western medicine focusing on anatomy, physiology, pathology, bacteriology, and clinical diagnostics, people discovered that TCM's understanding of these visceral organs and basic vital substances is far from the actual situation of the organs and vital substances of the same name under the scalpel. In other words, the description about them in TCM is far from 'true and detailed'. Obviously, based on the concept of empirical science, the theory of TCM was not considered scientific.

In the first 20–30 years of the 20th century, with the advancement of medical observation and experimental technology brought about by the development of science and technology, as well as the huge success of antibiotics on infectious diseases, modern medicine was in full swing. So many people believed that with the development of science and the advancement of technology, modern medicine would be omnipotent. Therefore, when Western medicine was imported into China, especially after Western medicine was incorporated into the new education system as an important subject of 'new knowledge', the number of people receiving Western medicine education and the number of Western medicine practitioners grew rapidly, and the mainstream medical status of TCM suffered an unprecedented impact. In January 1913, the Ministry of Education of the Republic of China announced the 'University Regulations' for the two courses of medicine and pharmacy, which were completely set up in accordance with Western medical subjects. TCM was not included in the modern education system (Cai Jingfeng et al. 2000).

During this period, the voice of intellectuals criticizing the ignorance and backwardness of TCM kept increasing. A group of people with overseas experience, represented by Yu Yunxiu, believed that TCM was unscientific and should belong to the category of witchcraft, and advocated the abolition of TCM. The Western medical profession also broke openly with the TCM profession, and the medical profession formed two distinct confrontation camps. Yu Yunxiu, who graduated from Osaka Medical University in Japan in 1916, was deeply influenced by the abolition of Kampo medicine during the Meiji Restoration in Japan. He was the first to systematically criticize the theory of TCM, which triggered a debate on the scientific nature of TCM theory in the early 1920s. Yu Yunxiu believed that TCM is based on the philosophical utopia of yin and yang and the Five Elements, and therefore is 'unscientific', but he also acknowledged the clinical efficacy of TCM.

Under the ideological background of prevailing scientism at that time, Chinese medicine, which was difficult to be proved by modern science, naturally lacked the legitimacy of existence in reality. Therefore, the abolition of TCM became logical and in line with the trend of the times. In February 1929, at the first meeting of the Central Health Committee convened by the National Government, the proposal put forward by Yu Yunxiu, Chu Minyi and others on 'abolishing the old medicine to remove the obstacles to medical hygiene' was passed, and the activities of abolishing and eliminating TCM reached a climax.

However, with the strong opposition of the TCM community across the country and the solidarity of all sectors of society, the 'abolition of TCM' was finally cancelled, and TCM regained the opportunity and space for survival. But in fact, it was a huge blow to TCM. Yu Yunxiu's denial of the basic theories of TCM was almost accepted by public opinion at the time, so the survival crisis of TCM was not eliminated. For a period after the 1930s, the Western medical community and the government's policy of scorning, discriminating, rejecting, and attacking TCM has not fundamentally changed, and the suppression has not weakened. The TCM community's efforts to incorporate TCM into school education have also been unsuccessful (Cai Jingfeng et al. 2000).

In early modern times when analytical scientific ideas dominated mainstream science, Western medicine, which is in line with Western scientific ideas, naturally replaced TCM with a history of more than two thousand years in China, and became the mainstream medicine that continues to this day. Although TCM is the most powerful traditional medical system in the world created by five thousand years of Chinese civilization, and has created countless admiring medical miracles, it still cannot escape the destiny of being suppressed, marginalized, and even on the verge of being 'abolished'.

The most important reason why Western medical community represented by Yu Yunxiu advocated the abolition of TCM is that TCM does not conform to modern science. In the torrent of the times when scientism prevailed at that time, as the theory of TCM was unfounded in science, and could not be explained by science, it could not be recognized by science. In addition, TCM treatment of diseases is mainly based on its unique treatment system based on syndrome, and its treatment methods usually have significant individual characteristics. Conducting statistical analysis based on the DCS established in accordance with modern scientific concepts as a reference system, its effectiveness could not be verified by scientific recognition.

In the controversy surrounding the retention and abolition of TCM, the characteristics of TCM that do not conform to the concept of analytical science have been fully revealed. The Nationalist Government and the Western medical community used this as an excuse to exert strong pressure on the TCM community to force TCM to make scientific improvements. The TCM community also recognized the scientific value and existing shortcomings of TCM. In order to survive, it had to start to innovate and improve the theory of TCM, and consciously try to make TCM scientific. At the same time, the Western medical community also began to pay attention to research TCM, trying to clarify TCM with scientific theories, so as to improve the treatment level of modern medicine. The rise of the 'Scientificization of TCM' movement in the 1930s was not the result of the unilateral efforts of the TCM community, but was also closely related to the promotion of the Western medical community and public opinion (Cai Jingfeng et al. 2000).

When the argument for the abolition and elimination of TCM was rampant, and the debate between traditional Chinese and Western medicine became increasingly fierce, Yun Tieqiao (1878–1935) was the first person in the TCM field to stand up to challenge Yu Yunxiu, and played a crucial role in this debate. Yun Tieqiao was admitted to Nanyang Public School in 1903, studying foreign languages and literature, and was the first person in the TCM community to receive the new school system education systematically. As a medical scholar familiar with Chinese and Western cultures, he believed that due to the difference of Chinese and Western cultures and the different medical foundations, two different systems of traditional Chinese and Western medicine have been formed. "Western science is not the only way for academic development. Oriental medicine has its own foothold." He believed that TCM and Western medicine have their own advantages. TCM pays attention to the influence of yin and yang changes in the natural environment over time on the human body; while Western medicine pays attention to the anatomical structure in physiology and local lesions in pathology. The two medicines should communicate with each other and learn from each other's strengths. But he also emphasized that "TCM must not be assimilated by Western medicine, and only the theory of Western medicine can be used to supplement TCM."

Yun Tieqiao kept a clear head on how to develop TCM. He believed that TCM has a long history and should be sorted out and improved to make it develop and progress, and should not end with the 'Huangdi Neijing, The Yellow Emperor's Inner Canon'. Reform of TCM should 'combine new knowledge' on the basis of inheriting the academic thoughts of predecessors, gradually breaking away from ancient theories, and achieving a medicine system that is neither TCM nor Western medicine, but the integration of Chinese and Western medicine. He emphasized that "TCM has the value of evolution and development. It must absorb the strengths of Western medicine and combine with it to produce new TCM. This is the track that TCM must follow in the future." He believed that "today's medical reform, apart from dealing with Western medicine, there is no second way." But he also said that "scientificization must not be regarded as fashionable, focusing on superficial things and discarding the fundamentals of TCM" (Cai Jingfeng et al. 2000).

Among the people who made outstanding contributions to the ITCMWM in the Republic of China, the most worth mentioning is Zhang Xichun (1860–1933), a famous doctor in Hebei, who was a TCM clinical expert with a profound understanding of modern experimental scientific methods. He carefully studied and researched the theory of Western medicine based on the specific conditions of TCM, and tried hard to combine the application of traditional Chinese and Western medicine methods in clinical practice. He put forward the concept of ITCMWM of 'taking TCM as the foundation, and Western medicine as reference and supplement' and advocated 'taking the strengths of the Western medicine to make up for the shortcomings of TCM'. He authored the book 'Medical practice based on TCM and with reference to Western medicine'. Zhang Xichun's practical spirit is highlighted in the practical research on medications, careful clinical observation, and detailed and reliable medical records.

Zhang Xichun believed that: "Regarding the treatment of diseases using medications, Western medicine focuses on the local parts of the disease, that is, the branches and leaves of the disease; TCM focuses on the cause of the disease, that is, the root of the disease." Treatment of diseases should be aimed at both the roots and the branches and leaves. "Therefore, TCM and Western medicine actually have complementary effects and complement each other." He applied this idea to the clinic and achieved good results, such as gypsum aspirin soup. Zhang Xichun believed: "The nature of gypsum is most suitable for use with Western medication aspirin. Although gypsum has a strong heat-clearing power, it has a slightly less power to induce sweat and dispelling exogenous evils. Aspirin is acidic and cool, and is good at inducing sweat and dispelling exogenous evils. It cooperates with gypsum, and has the effect of complementing each other" (Zhang Xichun 2009). In addition, in the treatment of epilepsy, Western medication sedatives potassium bromide or chloral hydrate are applied with Chinese herbal medicines (CHM) that have the actions of clearing fire, purging phlegm, and regulating qi, so as to increase the effect of calming the mind (Zhang Xichun 2009). Zhang Xichun used medications of modern medicine in TCM treatment based on syndrome, and the methodology contained therein has enlightenment significance for the development of TCM even now.

The early ITCMWM began in the era when TCM was still the mainstream medicine. The conflict of ideas brought about by the introduction of Western medicine triggered the comparative study of traditional Chinese and Western medicine and in-depth thinking on medical methodology in TCM profession. Although the level of scientific development at that time was far from reaching the level of revealing the scientific nature of TCM methods, some visionary scholars in the TCM field clearly realized the differences between Chinese and Western medical methodology. They firmly believed that the effectiveness of TCM in the treatment of many diseases was far beyond the reach of modern medicine, which is a strong proof of its inherent scientific nature. Therefore, TCM cannot be abolished, nor can it be reformed and 'innovated' completely in accordance with the ideas and methods of Western medicine. The development of TCM should be based on adhering to its own fundamental ideas and methods, and absorb the valuable things in modern medicine for TCM use. Even today, after decades of westernization of TCM, this kind of thinking still has methodological enlightenment.

The controversy over the retention and abolition of TCM was triggered by the belief that TCM is unscientific, and this controversy continues until today, 100 years later. However, whether Yu Yunxiu, who advocated the abolition of TCM 100 years ago, or Fang Zhouzi, a representative who believed that TCM was 'pseudoscience' in recent years, they all based their reasoning on the scientific method characterized by experiments, analysis and reasoning laid down by Bacon and Descartes 400 years ago. In an environment where reductionism dominates mainstream science, the scientific nature of TCM is difficult to be scientifically proven. Today, when the mainstream of scientific method is transforming from simple science to complex science, this issue can no longer be viewed based on the original scientific concept.

6.2 How did national support become a 'soft knife' to 'eliminate' TCM?

Since the founding of the PRC, the government has always attached great importance to the development and promotion of TCM and has adopted a number of policies and measures to support the cause of TCM. Including: formulating a series of laws and regulations to clarify the status of TCM and standardizing the management of TCM; investing a lot of resources in TCM education and training; supporting scientific research and innovation of TCM, and encouraging research on the efficacy and mechanism of CHM, as well as the treatment methods of TCM; take measures to protect TCM knowledge and cultural heritage; actively promote the internationalization of TCM through international cooperation, exchanges and promotions; encourage medical services that combine traditional Chinese and Western medicine, and integrate TCM into the national medical system.

However, what effect has the strong support from the government level had on the survival and development of TCM?

6.2.1 Since the founding of the PRC, the national level has given great support to the cause of TCM

Since the founding of the PRC in 1949, with the support and advocacy of Chairman Mao Zedong (1893–1976), the Chinese government has given great support to TCM. In November 1954, the Department of TCM was established by the Ministry of Health. In December 1955, the state established the TCM Research Academy, and then some provinces, municipalities, and autonomous regions successively established TCM Research Institutes. In 1956, Beijing, Shanghai, Guangzhou, and Chengdu took the lead in establishing TCM colleges with a 6-year school system and an enrollment of 2,400 students. TCM education was incorporated into the formal academic education system. From 1962 to 1965, there were more than 5,600 graduates from TCM colleges across the country. In addition, through the method of 'teaching apprentices', as of 1965, more than 59,000 TCM apprentices had been trained nationwide. According to statistics, in 1952, there were 19 TCM hospitals nationwide with 224 beds; in 1960, the number of TCM hospitals nationwide increased to 330 with 14,119 beds. Most general hospitals and specialty hospitals had established TCM departments (Ouyang Xuemei 2020).

In 1958, China set off an upsurge of Western medical practitioners to learn TCM, and the TCM college opened full-time TCM training courses for Western medical practitioners. Chairman Mao Zedong affirmed this practice, and instructed: "TCM is a great treasure house, which should be explored and improved" (CPC Central Literature Research Office 1987–1998). From 1955 to 1966, a total of more than 4,700 modern medical practitioners were trained through the training courses, many of whom later became pioneers and authoritative figures in the study of ITCMWM across the country. Tu Youyou (1930), the Nobel Prize winner in medicine, is a typical representative among them. She participated in the third TCM training course organized by the Ministry of Health for Western medical practitioners and researchers across the country from 1959 to 1962. In October 1958, Mao Zedong expressed his unique opinion on the ITCMWM: some of those who have studied Western medicine should learn TCM again in order to use modern scientific knowledge and methods to sort out and study TCM and CHM. Finally, integrate the knowledge of TCM and CHM with the knowledge of Western medicine and Western medication, to create a unified new medicine and new pharmacy in China.

During the 'Cultural Revolution', the development of TCM was seriously affected. After the 'Cultural Revolution' ended in 1980, the Ministry of Health convened the National Working Conference on TCM and ITCMWM. The conference clearly put forward the policy of parallel development and long-term coexistence of the three forces, namely the TCM, Western medicine, and ITCMWM. Article 21 of the 'Constitution of the People's Republic of China' promulgated in 1982 stipulates: 'Develop modern medicine and TCM', thus establishing the legal status of traditional medicine such as TCM. In 1986, the Fourth Session of the Seventh National People's Congress listed 'equal emphasis on traditional Chinese and Western medicine' as the main content of the new era's health work policy. The Ministry of Health formulated the 'Seventh Five-Year Development Plan for TCM', which puts forward the idea of development 'based on institution building, focusing on talent training, and relying on academic improvement'. In the same year, the State Administration of TCM was established. In 1991, the 'Ten-Year Plan for National Economic and Social Development and the Outline of the Eighth Five-Year Plan' listed 'equal emphasis on traditional Chinese and Western medicine' as one of the basic guidelines for health work. In 1997, the 'Decision of the CPC Central Committee and the State Council on Health Reform and Development' further clarified the policy of 'equal emphasis on traditional Chinese and Western medicine'. In September 2001, the first 'Five-Year Plan for TCM' was promulgated. In 2003, the 'Regulations of the PRC on TCM' was promulgated. The regulations clearly state that it is necessary to protect, support and develop the cause of TCM, implement the policy of equal emphasis on traditional Chinese and Western medicine, encourage traditional Chinese and Western medicine to learn from each other, complement each other, and improve together, promote the organic integration of the two medical systems of traditional Chinese and Western medicine, and comprehensively develop the cause of TCM. In 2007, the principles and policies such as 'equal emphasis on traditional Chinese and Western medicine' and 'supporting the development of Chinese medicine and ethnic medicine' were included in the report of the National Congress of the CPC for the first time.

Since Xi Jinping (1953) assumed the Chairman of the State in 2013, he has emphasized that TCM is a treasure of the Chinese nation and increased the country's support for TCM. In 2016, the State Council issued the 'Outline of the Strategic Plan for the Development of TCM (2016–2030)', elevating the development of TCM to a national strategy. In 2017, the 'TCM Law of the PRC' was implemented, establishing a legal basis for inheriting and promoting TCM, supporting and promoting the development of TCM. In Oct. 2019, Xi Jinping gave important instructions to the work of TCM, "(We) should follow the law of development of TCM, inherit the essence, innovate on the basis of maintaining essential characteristics; accelerate the modernization and industrialization of TCM; adhere to the equal emphasis on traditional Chinese and Western medicine, and promote the mutual complement and coordinated development of TCM and Western medicine; promote the high-quality development of TCM cause and industry; promote TCM to the world, and give full play to the unique advantages and functions of TCM in disease prevention and treatment" (Ouyang Xuemei 2020).

Since the founding of the PRC, China has always adhered to the principle of parallel development and long-term coexistence of the 'troika' of TCM, Western medicine, and ITCMWM, which reflects the country's great attention and support to TCM. Under the guidance of this policy, TCM has entered the university education system and national scientific research sequence, and TCM medical services have entered the national medical system. All provinces, municipalities, and autonomous regions have established TCM college and TCM research institutes, and all provinces, cities, and counties have established TCM hospitals and/or ITCMWM hospitals. As of the end of 2020, there were 683,000 licensed (including assistant) physicians in TCM across the country, and the number of patients receiving TCM diagnosis and treatment reached 920 million that year; the total number of TCM medical institutions nationwide reached 72,355, of which 5,482 were TCM hospitals, with a total of 1,148 thousand beds; there are 43 research institutions of TCM (National Health Commission 2020). As of the end of 2017, there were 43 higher TCM colleges or universities nationwide, with a total of 858,000 students majoring in TCM (China TCM News 2019).

6.2.2 A centennial review of the quantity and quality of TCM practitioners

In the early years of the Republic of China, China had a population of about 400 million (Mi Hong et al. 1997). In 1949, it was about 540 million. The current population has exceeded 1.4 billion (Office for the 7th National Census of the State Council 2021, Wikipedia contributors 2022). However, according to the survey data of the TCM Strategic Research Group of the Chinese Institute of Science and Technology Information, in the early years of the Republic of China, there were 800,000 TCM practitioners nationwide; in 1949, there were 500,000; in 2005, there were only 270,000. According to a 2005 survey of some regional and county-level TCM hospitals, only 10% of TCM practitioners prescribe decoction for treatment. In other words, there were no more than 30,000 TCM practitioners who can really treat diseases with TCM ideas at that time. The number of TCM practitioners has rebounded in recent years due to measures taken in recognition of the seriousness of the situation. However, calculated in proportion to the population, it is still less than half of that at the beginning of the PRC, and less than 1/4 of that at the beginning of the Republic of China. In contrast, the number of Western medicine practitioners has grown from approximately 87,000 in 1949 to 4.086 million at present. Even excluding medical technical department personnel and rural medical practitioners (including hygienists), the number of practicing (assistant) physicians is

Table 6.1. List of changes in the number of TCM physicians since the early years of the Republic of China.

Year	Population (100 million)	Number of TCM physicians	Number of TCM physicians per 10,000 people
The early years of the ROC	4	80	20
1949	5.4	50	9.26
2005	13.2	27	2
2020	14	68.3	4.78

nearly 6 times that of TCM (China TCM News 2019). And behind this string of distressing numbers, what is even more distressing is the quality of TCM practitioners.

6.2.3 Regulation based on the Western medicine model restricts and squeezes the concepts and thinking of TCM

In the current national medical and health system, medical institutions consist of general hospitals that focus on Western medicine, TCM hospitals that focus on TCM, and ITCMWM hospitals that place equal emphasis on traditional Chinese and Western medicine. The current general hospitals are required by authority for the establishment of TCM departments within each of them. TCM hospitals should be equipped with the general examination and treatment facilities of Western medicine, and the configuration of Western medicine examination and treatment methods in ITCMWM hospitals is usually more perfect than that of TCM hospitals. The sub-department setting and management standards of TCM Hospitals and ITCMWM Hospitals are basically formulated with reference to general hospitals, and of course some characteristics of TCM would be taken into consideration. TCM hospitals and ITCMWM hospitals should also set up emergency departments, abdominal surgery, thoracic surgery, interventional treatment rooms, intensive care units (ICU), coronary care units (CCU) and other departments with distinctive Western medicine characteristics to improve the hospital's comprehensive service capacity.

The state implements hierarchical management for TCM medical institutions just like general medical institutions. The basis for the hierarchical management of hospitals is the comprehensive level of hospital functions, tasks, facilities, technical construction, medical service quality, and scientific management (National Ministry of Health 1994).

1. *First-level hospitals*: primary hospitals and health centers that directly provide prevention, medical care, health care, and rehabilitation services to a community of a certain population.
2. *Secondary-level hospital*: a district hospital that provides comprehensive medical and health services to multiple communities and undertakes certain teaching and scientific research tasks.
3. *Tertiary-level hospitals*: hospitals of high standard/level that provide high-level specialist medical and health services and perform higher education and scientific research tasks to several regions, most of them are affiliated to medical schools/universities.

After review, hospitals at all levels are classified into grades A, B, and C in accordance with the 'Hospital Classification Management Standards'. Among them, third-level hospitals have added special grades, so hospitals are divided into three levels and ten grades. The standards and indicators for hospital grading mainly include five aspects (National Ministry of Health 1989).

1. *Medical facilities and hardware conditions*: This includes the hospital's buildings, equipment and infrastructure. Indicators may include the hospital's building area, number of beds, operating room equipment, clinical laboratories, availability and quality of imaging equipment (such as X-ray machines, CT scanners, etc.), etc.
2. *Qualifications and quantity of medical personnel*: This involves the hospital's medical team, including doctors, nurses, technicians, etc. Indicators may include the number, education, and professional background of health care workers.

3. *Medical quality and safety*: Hospitals need to take measures to ensure medical quality and patient safety. Indicators may include a hospital's medical quality management system, medical error reporting mechanisms, infection control measures, surgical safety protocols, etc.
4. *Scope and characteristics of medical services*: Hospitals at different levels may provide different scopes and characteristics of medical services. Indicators may include whether a hospital provides care in specific areas such as emergency response, surgery, obstetrics and gynecology, pediatrics, cardiology, neurology, etc.
5. *Patient satisfaction and complaint handling*: Patient satisfaction and complaint handling are one of the important indicators for evaluating hospital quality. Hospitals need to establish a patient satisfaction survey mechanism to regularly assess patient satisfaction with medical services and take measures to improve services.

The operation and management of TCM hospitals and ITCMWM hospitals are basically the same as the requirements of modern management of Western medicine hospitals, including back-office finance, human resources, administrative logistics management and front-end medical quality management, hospital infections control, emergency and intensive care, and hospital HIS, etc. In a large TCM hospital with more than 500 beds, the medical setting is basically the same as that of Western medicine hospitals except for the special TCM departments such as acupuncture, massage, and anorectal departments. Some departments may adopt the traditional title, such as the digestion department of Western medicine is named the spleen and stomach disease department in TCM. The setting of medical technical departments is similar to that of Western medicine hospitals, which also have clinical laboratory, ultrasound, CT, MRI, etc. In the medical staffing of TCM hospitals, 70–80% are graduates from TCM colleges and universities, and 20–30% are graduates from Western medicine colleges and universities. Generally, graduates of all levels of the Western medical colleges need to participate in the two-year 'Western medicine learning TCM' training course, while the clinicians who graduated from TCM colleges have to go to the Western medicine hospital for further specialty training. Medical practitioners of TCM and ITCMWM can prescribe Western medications in addition to decoctions of CHM and proprietary Chinese medicines. Of course, Western medical practitioners can also prescribe proprietary Chinese medicines; Western medical practitioners trained in 'TCM Course' can prescribe CHM decoctions.

The hospital level is the comprehensive symbol of the hospital's function, scale, management level, quality level, technical level and service level, and is the 'golden signboard' of its comprehensive competitiveness. The 'Hospital Classification Management Measures' clearly stipulated that medical fees should be linked to the hospital level. It is reported that the fee standard of tertiary hospitals is about 30% higher than that of secondary hospitals. Therefore, as long as a hospital is promoted to 'Level 3', its charging standard can be set higher. 'Level 3' means better policies, more high-end equipment can be purchased, and more scientific research project resources can be enjoyed. For example, the Ministry of Health clearly requires that the hospital applying for the national key clinical specialty construction project should be a tertiary hospital. 'Level 3' also means a larger platform and reputation, which can attract more medical talent and patient sources. Patients who want to see a doctor prefer to seek a hospital of higher level. The label of a tertiary hospital will naturally become an important basis for patients to choose a hospital. Such a hospital has a sufficient source of patients, and there is a foundation for further development. Therefore, the hospital grade review and recheck are usually regarded as 'top priority' by hospitals. It might need to be clarified that in China, all patients have the right to choose their medical service without a referral of their local medical practitioner and they can simply pop into any hospital and find an appointment to see a doctor there if available (Xiao Sansan 2020).

After 1979, with the economic boom brought about by reform and opening up, the Chinese government increased its support for TCM. This kind of support is mainly reflected in the support of funds and the improvement of management level requirements. TCM medical services have relatively simple requirements for the clinical setting and equipment. Therefore, medical institutions naturally use state support funds for the construction of medical facilities, the purchase of advanced inspection and treatment equipment, and the improvement of management software and hardware. As a result, not only is the difference between TCM hospitals and ITCMWM hospitals shrinking, but also the differences between TCM hospitals, ITCMWM hospitals and general hospitals are also getting smaller and smaller. The level of medical equipment in some TCM medical institutions is even as good as that of general hospitals of the same level. As a result, TCM is increasingly restricted to the narrow space defined by the division setting, disease classification, examination methods, diagnostic requirements, and management standards of modern medicine. TCM practitioners have to adopt the Western medicine way of thinking for clinical diagnosis and treatment, and the proportion of TCM in the medical system is naturally getting smaller and smaller. In other words, the actual effect of the state's aid funds for TCM is to accelerate the marginalization of TCM in TCM hospitals, and to make TCM more rapidly exist only in name.

6.2.4 How TCM has been marginalized in TCM medical institutions from the comparison of Chinese and Western medicine medical record writing standards

The in-patient medical records of TCM and ITCMWM hospitals implement dual diagnosis, namely Western medicine diagnosis (for example: duodenal ulcer with upper gastrointestinal bleeding) and TCM diagnosis (for example: blood in the stool, spleen deficiency, stomach heat). Western medical diagnosis is necessary, to ensure legal protection when encountering medical disputes. After the 'troika' has kept pace with each other for 60 years, China's medical and health management agencies have successively issued medical record writing specifications for Western medicine and TCM. Comparing the medical record specifications of traditional Chinese and Western medicine, it can be seen that there are 38 clauses in medical record specifications for Western medicine, totaling 8266 words, and 39 clauses in medical record specifications for TCM, totaling 8692 words. One extra item in the TCM medical record specification is Article 10, which is about the content of 'TCM diagnosis includes disease diagnosis and syndrome diagnosis'. The requirements of Western medicine medical records are completely copied into the specification of TCM medical records, and there is no omission, including the relevant contents of surgery and anesthesia. That is to say, TCM medical records are based on the complete Western medical records with the addition of contents about TCM diagnosis and treatment (National Ministry of Health 2010, National Administration of TCM 2010).

"A medical record refers to the sum of words, symbols, charts, images, slices and other data formed by medical personnel in the process of medical activities" (National Ministry of Health 2010). The specification of writing medical records is a legally binding code of conduct that physicians should follow in the process of diagnosing and treating patients. Judging from the specification requirements for writing medical records of TCM, TCM practitioners who have not undergone systematic training in Western medicine, and do not understand the diagnosis and treatment methods, techniques and specifications of Western medicine will inevitably be marginalized even in a TCM hospital, and at best can only be barely qualified for outpatient work. This also determines that if the graduates of TCM colleges and universities want to be able to work as physicians in TCM hospitals in the future, they must comprehensively and systematically learn the theory, methods, techniques, and management norms of Western medicine. The rapid growth of the total amount of modern medical knowledge has resulted in an increasingly narrow and specialized field of physicians' diagnosis and treatment, which is already a helpless choice for medical practitioners who are fully trained in Western medicine. For students majoring in TCM who have also to systematically study TCM theory and treatment based on syndrome in their five or six years of university career, specialization will be even more compelled. And a medical practitioner, who is limited to a certain disease field for a lifetime, such as cardiovascular and renal diseases, how can he learn to conduct overall analysis and comprehensive treatment of multiple diseases belonging to different specialties? The diagnosis and treatment of diseases by medical practitioners is divided into categories. Unfortunately, the multiple diseases that a patient suffers from are often not limited by specialties. Clinically, there are many patients suffering from multiple departments and categories of diseases at the same time. How then can today's medical system, develop masters of TCM like Zhang Xichun and Qin Bowei who can treat various diseases of different departments?

The thirty-seventh article of the specifications of medical records of TCM clearly stipulates that 'the writing of medical records of ITCMWM shall be implemented with reference to this specification'. Obviously, after the 'troika' has kept pace with each other for decades, China's medical and health structure has changed from 'troika' to 'two-horse carriage' at least at the hospital level. There is basically no difference between TCM hospitals and ITCMWM hospitals, and TCM has been completely 'marginalized' even in TCM hospitals. As a result, in the medical and health system of mainland China, TCM has been reduced to complementary and alternative therapies to make up for the deficiencies of modern medicine. Modern medicine moving towards 'reform and opening up' no longer needs to explore how to incorporate TCM treatments and techniques into 'complementary and alternative medicine'. According to the TCM hospital model in mainland China, it is easy to achieve the highest state of 'complementary and alternative medicine' by grafting TCM diagnosis and treatment technology into existing Western medicine hospitals.

6.2.5 Failed education in TCM

In the education system involving TCM, the curriculum of medical colleges and universities must include a course of TCM, while the students of the TCM colleges and universities have to study modern medicine courses systematically. The ratio of traditional Chinese and Western medicine courses is about 6:4 or 7:3 in colleges and universities of TCM. Although the content of Western medicine occupies a smaller proportion of class hours, for students who have been receiving modern science education from elementary school to high school, the concepts and knowledge of Western medicine are easy to accept and logical. And TCM based on yin-yang and the Five Elements and modern science seem to be 'incompatible with each other', and not only is it not easy to understand, but it is also difficult to accept in terms of concepts. As Qian Xuesen said: "After modern high school graduates go to the TCM college, it is very difficult to read the classic works of TCM, not to mention the ancient Chinese characters, the whole idea is all not consistent". Therefore, after five or six years of university studies, few people have truly learned to use TCM ideas to diagnose diseases and prescribe prescriptions. It is no wonder that in the national standards for higher education in TCM promulgated in December 2012, the cultivation of TCM thinking ability was included in the overall goal (The Ministry of Education 2016). In other words, it is not surprising that graduates who have studied at the University of TCM for five or six years have not even mastered the thinking way of TCM.

In this regard, Professor Liu Lihong of Guangxi College of TCM said: Our education took five to eight years, even eleven years. However, quite a lot of graduates still do not have a basic understanding of TCM, that is to say, many of our undergraduates, masters and doctors have not yet entered the door of TCM (Hao Guangming 2004). A TCM practitioner at a Beijing TCM hospital said, "I graduated from a seven-year TCM major at the University of TCM. The curriculum is 60% TCM and 40% Western medicine. In the end, TCM was not fully understood by me, and Western medicine was also not proficient". Currently, less than 20 of his 110 classmates are still engaged in TCM, and some of them still do not know how to prescribe TCM prescriptions to date (Li Bin et al. 2015). Many college graduates of TCM dare not take up the profession of a TCM practitioner because they have not learned how to diagnose and treat diseases, so they have to become drug companies' medical representatives. Therefore, some people even call the University of TCM the 'Medicine Representative University'. Westernized TCM education cannot train TCM practitioners who can diagnose and treat patients. They eventually become 'gravediggers of TCM' (Wang Junping 2014).

As early as 2001, famous TCM experts such as Jiao Shude (1922–2008) and Deng Tietao (1915–2019) published an article in 'Modern Education News' entitled 'China Has Not Cultivated Real TCM practitioners for Decades', reflecting deeply on the education of TCM. Qiu Peiran (1916-2010), a TCM grandmaster, once said publicly: "The domestic TCM colleges cannot train qualified TCM practitioners, and the TCM practitioners trained are not even up to the level of mediocre doctors; mediocre doctors also understand some TCM decoctions and prescriptions. But the students who are now trained don't understand the theory, method, prescription and CHM of TCM at all. This is a big failure in TCM education. TCM education has gone astray" (TCM Heritage Network 2019).

At present, undergraduate students majoring in TCM at the University of TCM must master the basic knowledge, theories, and skills of TCM, as well as the basic knowledge, theories and skills of Western medicine. For example, physicians who are engaged in gastroenterology specialty at a hospital of TCM after graduating from a university should master the basic knowledge, theories and skills of Western medicine and gastroenterology (such as abdominal puncture, gastrointestinal endoscopy, etc.), and require an understanding of gastric mucosal pathology. To get the professional title, the attending physician needs to go to the Western medicine hospital for specialist training. They are required not only to be able to undertake the outpatient work of the gastroenterology department, but also to undertake the daily work, the work on duty and critical illness rescue of the gastroenterology ward.

The westernization of TCM education and of TCM clinic works have led to the deterioration of TCM clinical thinking based on syndrome differentiation. A TCM practitioner who has been engaged in clinical work for decades said with a sigh: "TCM practitioners who are currently engaged in clinical work are often led by the disease name of Western medicine. When seeing inflammation, they immediately think of clearing heat and detoxification, and meeting cardiac infarction, one immediately thinks of replenishing qi and nourishing yin, promoting blood circulation and removing blood stasis" (The Medical Education Network 2005). This kind of treatment based on 'symptom' identification naturally loses the most fundamental elements of TCM. At present, more and more people have lost confidence in the TCM practitioners trained by the modern TCM education system and believe that 'the real TCM practitioners are in the folk'. However, since the state began to implement the physician practice registration system in 1999, TCM physicians must pass the physician qualification examination and obtain the qualifications of practicing practitioners or practicing assistant physicians before they can legally practice medicine. Under the impact of the rise of modern medicine on the living space of TCM and the strict regulation of the qualifications of TCM practitioners, how many folk TCM practitioners can survive and pass on the examination? It has now been more than 70 years since the founding of the PRC, how many of the 500,000 TCM practitioners in 1949 have else survived, or have their medical skills been inherited?

In April 2020, the Guangming Daily published an article titled 'The Protection, Inheritance and Development of TCM by the CPC', which also admitted: "In this period, equal emphasis on traditional Chinese and Western medicine has not been well realized to a large extent. Compared with Western medicine's diagnostic methods such as radiography, laboratory tests, and ultrasound, as well as the precise theoretical analysis of diseases, TCM cannot produce 'human body data'. However, relying on the accumulation of individual case experience is difficult to gain broad recognition. These 'shortcomings' have caused the slower development of TCM and a large loss of TCM talents. At present, even in some large TCM hospitals, Western medicine is the main method of diagnosing and treating patients. The problems of westernization of TCM education, weak TCM thinking, and lack of TCM skills exist widely in college education" (Ouyang Xuemei 2020).

Obviously, the TCM policy since the founding of the PRC has prevented TCM from being artificially abolished, and TCM hospitals, TCM colleges, and TCM practitioners have been formally protected. However, for decades, the research on the methodology of TCM has not received enough attention from the management and decision-making institutions of TCM, and become the important basis for formulating TCM policies. Ignoring the significant methodological differences between traditional Chinese and Western medicine, there is a blind adoption of the concepts, methods, and management models of modern medicine to 'elevate' the level of TCM, thereby making TCM 'westernized' comprehensively. As a result, the national level of support for TCM has eventually become a 'soft knife' to stifle TCM.

6.3 Research on TCM, ITCMWM and modernization of TCM

Different from the evolution of the status of TCM in the medical and health system, since the founding of the PRC, with the development of science and technology and modern medicine gradually occupying the status of mainstream medicine in mainland China, the research involving TCM is basically divided into three aspects: ITCMWM, using the methods of modern science (including modern medicine) to study TCM and the modernization of TCM.

6.3.1 ITCMWM for clinical research

The concept of ITCMWM usually covers the unity and cooperation of traditional Chinese and Western medical practitioners, the combined use of traditional Chinese and Western medical technologies, the combined use of CHMs and Western medications, the mutual evidence of the theory of traditional Chinese and Western medicine, and the integrated research of traditional Chinese and Western medicine. According to the description of Professor Shen Ziyin (1928–2019) of Fudan University, a well-known expert on the integration of traditional Chinese and Western medicine, the integration of traditional Chinese and Western medicine includes two aspects: first, use modern science, mainly modern medical methods, to study the theory and clinical practice of traditional Chinese medicine, explore its theoretical essence, and clarify its mechanism of action; the second is to give full play to the respective advantages and strengths of traditional Chinese and Western medicine, and organically combine the two methods in clinical practice to achieve the purpose of improving medical standards, improving patients' quality of life, and reducing medical costs (Shen Ziyin and Wang Wenjian 2004). Chen Keji (born in 1930), a clinical expert on ITCMWM and a researcher from the China Academy of Chinese Medical Sciences, divided the ITCMWM into three levels: the first is to unite practitioners of Chinese and Western medicine, which is the original meaning of the integration of Chinese and Western medicine; the second is that the treatment methods of Chinese and Western medicine complement each other in terms of clinical services; the third is the organic combination of theory, that is, the medical treatment of 'combination of disease and syndrome' of Chinese and Western medicine model.

(1) The advancement of treatment methods and the improvement of clinical efficacy brought about by the ITCMWM

The ITCMWM in clinical diagnosis and treatment includes the combination of disease and syndrome in diagnosis, integrated use in clinic treatment, and mutual reference in theory. The combination of disease and syndrome is based on the DCS of Western medicine, using Western medicine diagnostic methods to make a disease diagnosis; at the same time, based on the TCM treatment system based on syndrome, a judgment is made on the type of TCM syndrome to which it belongs. In this way, clinicians can not only understand the etiology and local pathological changes from the perspective of Western medicine, but also grasp the health status of patients described by TCM syndromes, which can be used as the basis in formulating treatment plans. Comprehensive coordination during treatment refers to selecting the optimal treatment plan respectively and applying it simultaneously according to the respective theories of Chinese and Western medicine. This is not a simple combination of CHM and Western medication, but organic cooperation and mutual supplementation, which can often achieve better clinical outcomes. Theoretical mutual reference is based on the specific conditions and treatment needs of the disease, or focus on guiding treatment with TCM theory, or focusing on guiding treatment with Western medicine theory, or guiding treatment according to new theories formed after the ITCMWM.

At present, in mainland China, diagnosis and treatment of common, frequently-occurring, and refractory diseases with ITCMWM have be very common. After the 1950s, the ITCMWM has achieved remarkable results in treating certain diseases. For example, treatment of cardiovascular and cerebrovascular diseases, aplastic anemia, irregular menstruation, viral pneumonia, Japanese encephalitis, dysentery, arthritis, chronic nephritis, anorectal disease, fractures, small and medium burns, thrombotic occlusive vessels inflammation, scleroderma, lupus erythematosus and a variety of skin diseases have significant effects. In the treatment of certain acute abdomen conditions, the traditional treatment principle has been changed and a new treatment with Chinese characteristics has been developed. It not only improves the cure rate, but also saves some patients from the pain of surgical treatment and reduces comorbidities and side effects of Western medicine. Treatment of emergencies of internal medicine, such as respiratory distress syndrome, acute myocardial infarction, shock, acute diffuse intravascular coagulation, etc., also have a good effect. In the treatment of fractures, the ITCMWM forms a new reduction and fixation method, which can shorten the time of fracture fixation and functional recovery, and maintain better joint function. ITCMWM also pay attention to the use of non-invasive therapies to treat diseases. The combination of Western medicine's diagnosis and treatment methods with TCM's Qigong, acupuncture, and massage has received wide attention for its non-invasive, simple and easy operation, and reliable curative effect. For example, Qigong treatment of hypertension, acupuncture treatment of neurological diseases and coronary heart diseases, orthopedic adjustment and repair techniques to treat soft tissue injuries, etc., through clinical observation, have achieved good results.

Looking into a detailed example of how the two system work together, during severe attacks of chronic ulcerative colitis, adrenal cortex hormones are generally used to control the condition. However, at this time, the use of CHMs that have the action of clearing the intestines and removing dampness to perform a retention enema, so that the medication action can directly reach the disease site, can often significantly increase the efficacy and shorten the disease course. Especially for patients who are not satisfied with hormone control, the ITCMWM in treatment methods is very necessary. In the remission period, TCM treatment usually focuses on replenishing the spleen and replenishing qi, and it has all the obvious effects in preventing relapse and improving symptoms (Li Qiangou et al. 2001). For another example, for malignant tumors or cancers, surgery, chemotherapy, and radiotherapy are usually impossible to eliminate all cancer cells, and the unbearable side effects of these therapies determine that they cannot be used as sustainable long-term treatments. After completing the course of Western medicine, there are surgical trauma, toxic and side effects of chemotherapy and radiotherapy, low immune function, tumor thrombus formation, and potential cancer cell escape, all of which leave hidden dangers for future recurrence and metastasis. TCM has considerable advantages in reducing the side effects of chemotherapy and radiotherapy, enhancing the immune function of the body, promoting blood circulation, dissolving latent microscopic tumor thrombi, and reducing the risk of tumor recurrence and metastasis. At present, the advantages of ITCMWM treatment for different stages of major diseases such as tumors have been recognized by the medical community in China and many eastern Asian areas (Zhou Yiqiang 2006).

In addition, intestinal polyps, especially adenomatous hyperplasia with mild to moderate atypical mutations, currently have a high incidence and these patients are a high-risk population for intestinal cancer. Colonoscopy combined with endoscopic surgical removal is the treatment of choice. However, many intestinal polyps are multiple and have a high recurrence rate, especially polyps below 0.5 cm are often flat, and are difficult to eliminate. In this regard, current modern medicine lacks corresponding treatment medications, and TCM has a good effect on eliminating multiple intestinal polyps after endoscopic treatment. The principle of 'six-fu organs is better to keep unobstructed' also has a significant effect on improving intestinal symptoms (Liu Shenlin 2014).

Yet another example: the current interventional therapy for coronary heart disease, coronary balloon dilation and stent installation are very popular, but a considerable number of patients have restenosis of the coronary arteries after half a year of treatment. On the basis of conventional

Western medication treatment, the use of CHM for promoting blood circulation and removing blood stasis has a good effect on preventing restenosis. The clinical practice of Beijing Anzhen Hospital, Tongren Hospital, Beijing Third Hospital, China-Japan Friendship Hospital, and Guangdong Provincial Hospital of TCM have proved that the use of CHM can reduce restenosis by half (Pan Feng and Chen Keji 2005).

(2) The concept of 'classification of TCM syndromes under the diagnosis of Western medicine diseases' squeezes the holistic concept and clinical thinking mode of TCM

The ITCMWM clinical research characterized by classification of TCM syndromes under diseases diagnosis of Western medicine reveals the basic situation of the distribution of TCM syndromes under common diseases in modern medicine, and promotes the organic combination and comprehensive application of Chinese and Western medical treatment methods. This also lays the foundation for integrating the treatment methods of TCM into the clinical guidelines of modern medicine and applying them to the clinical practice based on the concept of evidence-based medicine. Studying TCM syndrome classification under modern medical diseases has profound methodological consistency with today's precision medicine that introduces biomarkers to distinguish personalized characteristics under modern medical diseases. This is undoubtedly of positive significance for modern medicine to formulate treatment plans based on the individual characteristics of patients to improve the effectiveness of treatment. However, the application of this method, especially the supervision and management on TCM based on this concept, has obvious negative effects on the maintenance and development of the core concepts and thinking modes of TCM:

1. Research on TCM syndrome classification based on modern medical diseases usually lists only the common syndrome types within the disease. In the clinical process, experienced TCM physicians who can uphold the concept of TCM often would find syndrome types that are not listed. For example, the TCM syndrome types of diabetes are usually divided into five types: Qi and yin deficiency, yin deficiency and fire-strength, blood dryness, blood stasis, and both yin and yang deficiency. However, in clinical practice, some patients mainly show the characteristics of damp-heat syndrome, and by adopting the method of clearing damp-heat, the blood sugar level of patients can also be effectively controlled. In other words, 'classification of TCM syndromes under the diagnosis of Western medicine diseases' is far from exhausting all the syndrome types related to the specific disease. The 'preconceived' syndrome classification framework will undoubtedly limit the thinking of TCM practitioners in clinical practice and often mislead the treatment plan.
2. In clinical practice, it is often found that there are multiple types of syndromes related to a certain disease in patients. For example, dizziness is related to insufficient qi, blood, yin, and yang, qi stagnation and blood stasis caused by poor circulation of qi and blood, and also related to the disturbance of damp heat, wind heat, and fiery heat. In clinical practice, many patients actually are a certain combination of the multiple syndrome types. However, in the research on the TCM classification of modern medical diseases, most of the listed syndromes are single syndrome types such as qi deficiency and liver-qi stagnation, or simple combined syndrome types such as spleen and kidney yang deficiency, liver and kidney yin deficiency. Such clinical guidelines will also mislead the clinical diagnosis and treatment of TCM practitioners.
3. When multiple diseases are diagnosed based on the disease classification of modern medicine, and multiple syndromes are identified based on the syndrome differentiation system of TCM, each of these diseases may relate to multiple syndrome types, and at the same time each of these syndromes may relate to more than one disease. How can a practitioner coordinate the relevance of treatment of multiple diseases and multiple syndromes? The method by classification of TCM syndromes under the diagnosis of Western medicine diseases brings more complexity, which makes it more difficult to determine an appropriate treatment plan based on the severity of the disease, as well as the relationship among various diseases and syndromes.

Obviously, the method of classification of TCM syndromes under the diagnosis of Western medicine diseases may have more advantages for simple and single disease treatment, but for patients with multiple diseases and complicated conditions, it greatly increases complexity of comprehensive analysis and overall treatment, and reduces the level of TCM treatment. The most influenced by this concept are the medical practitioners of TCM and of ITCMWM who work in the regular TCM hospitals and the hospitals of ITCMWM, as well as the graduates trained from the colleges of TCM. As the detection methods and treatment techniques of modern medicine have been introduced into TCM hospitals in large quantities, as graduates of TCM colleges have learned a lot of knowledge and skills of modern medicine, the theories and methods of TCM are increasingly being rejected and penetrated by modern medicine, and there are fewer and fewer things that truly have the characteristics of TCM.

The concept of classification of TCM syndromes under the diagnosis of Western medicine diseases prevailed in TCM hospitals and ITCMWM hospitals, and strengthened by hospital administration and assessment, has made TCM more and more confined under the DCS of modern medicine, and has made it a vassal of disease medicine. Here, the holistic concept of TCM is gone, and the dynamic process of the development and change of the body's state in the disease is no longer considered. In the process of formulating treatment plans based on this concept, the interrelationship of various parts of the disease and the influence of the development trend of the disease on the overall layout of treatment measures can no longer show rationality and necessity. Here, the 'preventive treatment' of TCM based on the occurrence and development of the disease is interpreted as a treatment for a disease condition that is not established by modern medical diagnosis, which is completely distorted. As a result, TCM's overall and dynamic grasp and regulation-control of the disease process has become the static treatment for TCM syndrome under the DCS of modern medicine. The holistic concept of TCM has been dismembered, and the thinking mode of TCM basically has no room for application in the clinical environment of ITCMWM. As we will see in later chapters, the problems and limitations faced by such concepts and methods are or will be encountered in the process of establishing its personalized DCS based on biomarkers in precision medicine.

6.3.2 Use modern medical methods to study TCM

Studying TCM with modern medical methods mainly includes the study of the 'essence' of the viscera, qi and blood, and syndromes of TCM from the perspective of modern medicine, as well as the study of the mechanism of TCM treatment methods. In 1955, the Research Academy of TCM was established in Beijing with the participation of Western medicine clinical experts and researchers whose specialties are physiology, biochemistry, and pharmacology. With the enrichment of corresponding personnel in universities, research institutes and clinical institutions of TCM, research on TCM using modern scientific (mainly modern medicine) analytical methods and technologies has gradually begun.

(1) The positive significance of using modern medical methods to study TCM to the development of modern medicine

The method of studying the 'essence' of viscera, qi and blood in TCM is seeking the theoretical combination point of TCM and Western medicine by establishing the animal model of TCM theory or animal disease model based on clinical practice. Therefore, this type of research is also attributed to research on the essence of syndromes. For example, the research on the spleen focuses on the syndromes of spleen-qi deficiency and spleen-yang deficiency, while the research on qi and blood focuses on the syndromes of qi deficiency, blood deficiency, and blood stasis. The research on the 'essence' of syndromes focuses on specific syndromes, through experiments and statistical analysis based on modern medicine, to find the pathological changes in the body corresponding to the syndromes at various levels of organs, tissues, cells, and molecules. Modern research on the treatment principles of TCM mainly focuses on the research on specific treatment principles such as promoting blood circulation and removing blood stasis, clearing away heat and detoxification, tonifying qi and nourishing blood. The method is based on affirming the curative effect, finding out the rules of medication, screening formulas and CHMs, and then conducting experimental research on the pharmacological actions, ingredients, and action mechanisms of the relevant herbal formulas applicable to the treatment principle, and finally putting the acquired knowledge to clinical practice for verification (Baidu Encyclopedia 2023a). The characteristics of this type of research are the combination of medicine and medication, clinical practice and experimental research, and single-CHM research and compound CHMs research, to strive to explain the treatment principles and methods of TCM by modern medical theories.

Due to the ethical restrictions of scientific research, modern medical and biological research always conduct limited human experiments on the basis of animal experiments. Animal experimentation refers to scientific research conducted with animals in a laboratory in order to obtain new knowledge about biology, medicine, etc., or to solve specific problems. The research on TCM based on modern medical methods is naturally conducted with animal experiments as the forerunner. Establishing an animal model combining disease and syndrome is an important foundation of basic theories and clinical research of TCM. The animal model of combination of disease and syndrome usually use separately (or simultaneously) TCM etiology to replicate syndrome animal models and modern medical etiology to replicate disease animal models, so that animal models have both the characteristics of disease and syndrome. Carrying out research on animal models combining disease and syndrome has important academic value in explaining the scientific nature of the basic theories of TCM (Lv Aiping 2013). Obviously, animal models replicated based on this method are more suitable for syndromes or diseases with clear etiology and relatively objective and generally accepted diagnostic criteria. However, for most syndromes, TCM has not yet formed a relatively objective and generally accepted diagnostic standard. Therefore, for animals living in biological and sociological environments that are very different from human beings, how much similarity can be claimed between the animal model established by some artificial intervention and the human syndromes to be simulated? How consistent are the pathological laws and responses to specific treatment methods revealed based on such models with those in the human body under corresponding conditions? These issues have been questioned and criticized by the TCM community.

Using modern medical methods to study TCM syndromes and treatment methods, such as the study of kidney-yang deficiency and spleen deficiency, and the study of promoting blood circulation and removing blood stasis, the initial research results have been exciting. However, due to the difficulty of standardization of syndromes and the establishment of syndrome models, the subsequent research subsequent research progress has been far less than people expected.

The significance of adopting the methods of modern medicine to study TCM is firstly that it has been clarified that there is no one-to-one correspondence between the viscera and vital substances in TCM and the solid organs and substances with the same name in anatomy. The scope of the viscera of TCM usually involves functional activities of multiple tissues and organs. The emergence of a syndrome in TCM will also involve abnormalities in the functional activities of multiple tissues and organs. For example, in the related research of spleen qi deficiency, several functions and indicators with higher sensitivity or/and specificity to spleen deficiency syndrome have been discovered. Experiments have shown that in the spleen deficiency syndrome, the following pathological changes have a higher occurrence rate: decreased cellular immune function; hyper-parasympathetic nerve function, lower stress ability of parasympathetic and sympathetic nerve; pathological manifestations are chronic inflammation and parenchymal organ atrophy, degeneration, some tissue cells are immature; plasma and tissue cAMP levels are reduced; D-xylose excretion rate is reduced; salivary amylase activity is decreased before and after acid stimulation; body surface gastric electric wave amplitude is reduced; digestive tract emptying speed is accelerated; physiological functions reserve are reduced, etc. The main difference between spleen qi deficiency and spleen yang deficiency lies in the degree of disease and energy metabolism (Zheng Chaowei et al. 2013). The research on the pharmacological actions and mechanism of CHM for promoting blood circulation and removing blood stasis mainly focuses on improving hemodynamics, improving hemorheology, improving microcirculation disorders, anti-thrombotic, anti-atherosclerosis and myocardial ischemia, inhibiting Inflammation, abnormal tissue proliferation and tumors (Gao Chong et al. 2013).

The use of modern medical methods and means to study TCM has advanced the understanding of TCM basic concepts, basic theories and syndromes from the perspective of modern medicine. More importantly, on this basis, some internal mechanisms of the effectiveness of CHM treatment have been discovered, and treatment plans ITCMWM for many diseases have been developed, which have enriched modern medical diagnosis, treatment technology and disease monitoring methods, and improved clinical efficacy. For example, according to the ancient TCM record of treating malaria with Artemisia annua, artemisinin was discovered, which has enriched the treatment method of modern medicine for malaria. The 'trigger point therapy' developed on the acupuncture therapy of TCM has now become a treatment method that medical practitioners and physiotherapists trained in modern medicine can legally apply, and has been recognized by the medical community of Europe, America, South Korea, and Japan, etc.

(2) The use of modern medical methods to study TCM has not played a substantial role in promoting the development of TCM

The use of modern medical methods to study TCM essentially belongs to the research carried out within the scope of modern medicine, and its results can naturally be incorporated into the treatment system of modern medicine, to enrich the 'complementary and alternative therapies' of modern medicine. Although the contents of the research are all related to TCM's human body, diseases, treatments, and prescriptions, the research results cannot be used by TCM to improve the level of clinical treatment of TCM. The reason is that the frame of reference used in the study, the description about pathological phenomena and pharmacological actions, etc., are all based on modern medicine, such as alleviating smooth muscle, dilating coronary arteries, and suppressing immune responses. In recent years, the analysis and research on the chemical composition and pharmacological actions of CHM by chemical analysis and pharmacological experimental methods have been quite in-depth. However, the research results can neither touch the theoretical system of TCM, and reveal more deeply its mystery in essence, nor use these results to further improve the clinical use of TCM on these CHM. Therefore, the practical significance of this kind of research for the development of TCM has been questioning in the TCM field.

1. In the research of applying single factor analysis method to CHM compound formulas, the technical difficulties brought about by complexity have not yet seen the prospect of solution, while the 'toxic side effects' caused by the 'variation' of the westernization of CHM have caused widespread concern in the world time and time again. The wide application of the extraction technology of the effective ingredients of CHM has kept CHM away from the appearance of 'rough, big, and dark', but its application has been seriously out of touch with the treatment based on syndrome of TCM. Therefore, the results of the research are nothing more than adding a new medication to the pharmacopoeia.
2. There has never been a consensus in the TCM community about which standards should be used to conduct research on TCM and CHM. At present, the experimental system of CHM with animal models as the core and the evaluation system of clinical efficacy and scientific research results of TCM are all derived from Western medicine. How can such an evaluation system ensure the practical value of research results to TCM?

Professor Jiao Shude, a well-known TCM scholar in China, from the China-Japan Friendship Hospital, pointed out the dilemma of this type of research sharply: "The current policy orientation is to emphasize the use of modern scientific methods to study TCM, but in fact it is to study TCM using Western medicine methods." This kind of research has been carried out "for decades, and there is no real valuable result. Such a large investment in financial, human, and material resources has been spent in vain and time has been delayed" (Hao Guangming 2004).

A doctor of Shanghai University of TCM told reporters: "Master students of the TCM colleges do experiments at the cellular level (in lab), and doctoral students do experiments at the genetic level. Is this type of TCM practitioner still a TCM practitioner? Can this type of TCM practitioner cure patients in the clinic? Can the rats in the laboratory test the effects of TCM? For example, to purify CHM that are originally non-toxic and have good therapeutic effects, usually the purer and finer the purification, the more obvious the toxicity and the lower the curative effect. So should the performance of the medication be judged by its curative effect, or should it be judged by its purity? The so-called modernization of TCM now uses purity as the standard to determine the performance of medications, which just shows that the standards of Western medicine cannot be used to evaluate TCM. However, much of the funding for TCM research has been wasted on mice. The so-called scientific research results are actually bubbles. The TCM community should seriously reflect on whether this kind of scientific research is developing TCM or eliminating TCM?" (World Health Organization 2013).

The use of modern medical methods and means to study TCM and the ITCMWM in clinical diagnosis and treatment methods has opened up a new research field of ITCMWM. In 1992, the national standard 'Discipline Classification and Code' (a national standard for classifying scholars and university degree categories) set Integrated traditional Chinese and Western Medicine (ITCMWM) as a new discipline (National Standardization Administration 2008). The original intention of this type of research was to clarify the essence of TCM theories, concepts and syndromes from the perspective of modern medicine, and reveal the mechanism of TCM treatment methods. This type of research, which has continued for more than 60 years, is obviously still far from this goal. However, research progress in this area has verified from the side that the TCM theory is not based on the internal anatomy of the body's solid organs and tissues, but a theoretical model, and has revealed some static relationships between the concepts and syndromes in the theoretical model of TCM and the structures of tissues and organs of the human body, including the cellular and molecular levels. Some medications and treatment methods developed in this kind of research can be incorporated into the treatment system of modern medicine if the mechanism is clarified; if the mechanism is not clarified, they can also be used as complementary and alternative therapies for modern medicine. However, under the efficacy evaluation system of modern disease medicine, there are very few TCM treatment methods and CHMs that can show effectiveness. Abuse in the absence of a comprehensive understanding of its actions will inevitably lead to adverse reactions similar to the incident of Japanese Xiao Chai Hu Decoction. Regarding this method as the main method of developing TCM and ignoring the research based on TCM methodology also made TCM lose the opportunity to develop based on its own methodology and unique system.

6.3.3 Research on the scientific and modernization of TCM

Research in this aspect includes adopting methods of modern science or modern medicine to standardize the TCM treatment system based on syndrome and studying the four examination methods of TCM, or creating new diagnosis methods. The slogan of the scientificization of TCM was put forward in the 1920s and 1930s, but substantive promotion materialized after the 1980s.

(1) Exploration of the standardization of the TCM syndrome system

The standardized research on TCM treatment system based on syndrome mainly includes the standardization of TCM syndromes and TCM symptoms and signs, which was started in the early 1980s under the promotion of academician Wang Yongyan (1938) of the Chinese Academy of Engineering. The modern research of TCM diagnostic equipment mainly focuses on pulse diagnosis, and tongue diagnosis. The pulse imager tries to express the sensation of the TCM practitioner's finger when he touches the pulse in the form of objective ways such as wave shape curves, frequency and rhythm, etc.; the tongue imager tries to objectively reflect the changes in the tongue coating and tongue morphology that the TCM practitioner sees when looking at the tongue through the methods of pathomorphology, cytology, biochemistry, hemorheology, and optics. This type of research is conducive to the realization of instrumentation, objectification and standardization of the four examinations of TCM.

In the 1980s, some visionary scholars in the TCM field realized the difference between the research based on the theory and methods of TCM and the use of methods of modern science/modern medicine to study TCM. Professor Lu Guangxin (1927–2014) from the Institute of Basic Theories of the Chinese Academy of Chinese Medical Sciences deeply analyzed the huge difference in the connotation of the two concepts of 'the research on TCM' and 'the research based on TCM', and pointed out that they are researches with different contents using different theories and methods, which cannot replace each other, but can only complement each other and refer to each other. Professor Lu Guangxin vigorously advocates the promotion of 'the research based on TCM' that conforms to the characteristics of TCM. However, in order to promote the research, we must first clarify what are the characteristics of TCM from the methodological level, and understand the research methods and theoretical framework as the core of TCM. In recent decades, in the world, we have seen modern medicine continue to draw nourishment from the treasure house of TCM knowledge to develop its own knowledge system and complementary/alternative therapies, such as artemisinin and trigger point therapy as mentioned earlier. In the period of the Republic of China, we can still see Zhang Xichun's 'Gypsum Aspirin Soup' which incorporated medication of modern medicine into the system of TCM. Today, even in China where government policies strongly support TCM, apart from the limited achievements of ITCMWM, it is almost impossible to see that TCM absorbs modern science and modern medical methods and techniques, and promotes the development of TCM theory and treatment techniques. In addition to the restrictions of regulatory policies, the lack of a consensus on the core concepts of TCM and the lack of a scientifically standardized description system of the human state that can be used as a reference frame are the biggest technical obstacles.

The reason why modern medicine can continue to develop some new diagnostic techniques and treatment methods with the advancement of science and technology is that modern medicine has established a relatively clear and objective disease diagnosis standard system on the basis of anatomy, histology, physiology and biochemistry, cytology, and molecular biology. Statistical analysis using this system as the reference frame can determine the correlation between the indicators of new testing equipment and the disease through empirical methods, and determine the effectiveness of new treatment methods and medications. Obviously, if TCM wants to be scientific and use scientific methods and technical means to promote its own development, it must also have such a relatively clear and objective syndrome diagnosis standard system as the reference frame. The syndrome system of TCM formed spontaneously in ancient medical practice naturally cannot be compared with the DCS of modern medicine established based on strict observation and experiment. Compared with modern medical diagnosis, which mainly relies on objective detection indicators, the diagnosis of syndromes in TCM relies more on symptoms based on the subjective perception of the patient and signs with subjective factors of the TCM practitioner. And the association of these symptoms and signs with the syndromes related to them often lacks specificity. In addition, in the process of the formation and development of the TCM syndrome system, the description of symptoms and signs, and the diagnostic criteria of syndromes and diseases have never been systematically standardized in accordance with scientific principles. Different regions, different schools, and even different TCM practitioners have great differences in understanding and expression. Obviously, the scientization of TCM is first and foremost the standardization and objectification of the TCM syndrome and symptom system. It was precisely because of this awareness that in the early 1980s, under the advocacy of Professor Wang Yongyan and the organization of the State Administration of TCM, research on the standardization of TCM syndromes, diseases, and symptoms began. The most important results of the research are the successive publication of three books edited by the Chinese Academy of Chinese Medical Sciences: 'Differential Diagnosis of TCM Syndromes',

'Differential Diagnosis of TCM Symptoms', and 'Differential Diagnosis of TCM Diseases' (Liang Maoxin and Sun Dehua 1991).

(2) Methodological research on the standardization of TCM syndrome differentiation and treatment system

As mentioned earlier, there are multiple incompatible syndrome systems in TCM, such as viscera-qi-blood-body fluid syndromes, six-meridians syndromes, wei-qi-ying-xue syndromes, and Tri-jiao syndromes. So far, the principle of syndrome standardization research is to receive all syndromes under different syndrome systems in different periods of TCM as comprehensively as possible. No structured arrangement has been made for the overall structure of the syndrome system and the relationship among syndromes. Therefore, it is inevitable that there are some overlaps, structural contradictions among syndromes of different systems. For example, some syndrome names in two systems are the same, but the connotation is different, or syndrome names are different, but the connotation is the same. In the initial standardization work, the syndromes of each part were assigned to different teams to sort them out, which exacerbated the structural confusion of the standardization system. In response to this problem, the article 'Establishing a Clear Theoretical Structure' published in the Journal of Beijing University of Traditional Chinese Medicine in 1985 (Yuan Bing 1985) clearly points out: The scientification of the theoretical system of TCM is not limited to the unification of names, the standardization of connotations, and the determination of qualitative and quantitative criteria for syndrome diagnosis, but also includes the structured problem of the syndrome system. This structured problem and the completeness and independence of state variables mentioned in this book are the same problems raised from different angles. It will be discussed in detail later.

The results of the initial standardization work were unsatisfactory, and the TCM profession gradually strengthened the methodological research on standardization. The unification of naming, the standardization of connotation, the qualitative and quantitative determination of the diagnostic significance of symptoms and signs in the syndrome diagnostic criteria, etc., the significance of these in the standardization work is obvious, and it was easy to make limited progress. After 2010, the importance of 'structured' in establishing a clear state description system (SDS) gradually attracted the attention of some visionary scholars in the TCM field. However, the structuring of the syndrome system is based on theoretical models, and the theoretical models themselves also have structured problems, which involve the methodology establishing the theoretical models and the unification of the current several major syndrome systems of TCM. That is to say, to fundamentally solve the problem of syndrome standardization may involve restructuring the overall structure of the TCM theoretical model according to the structured concept, and the top-level design of the theoretical model that can unify the several major syndrome systems of TCM. Due to the complexity of structured work, until now, the TCM profession still believes that the conditions for the standardization of syndromes are not yet available.

The standardization of syndromes involves deep-layer problems of structuring of the theoretical system, which is relatively complex, while the standardization of symptoms is supposed to be much easier. In principle, the description of symptoms and signs in TCM to reach the level equivalent to that of modern medicine have no technical obstacles. However, when looking for a more scientific and clearer description, or decomposing traditional descriptions, problems of idea and presentation structure are often encountered. There has been a saying in China since ancient times, "Since I can't be a good minister, I would like to be an excellent doctor." Under such a cultural background, many TCM scholars had profound literary attainments. In ancient books of TCM, the naming and description of symptoms and signs usually pay more attention to the beauty and elegance of the text, unlike the description of modern medicine, which pays more attention to accurate expression, easy identification, and less ambiguity. For example, plum core-qi is a symptom of TCM, and it can also be said to be a TCM disease like headache and dizziness. Ancient literature records usually say 'the pharynx seems to be blocked by the plum nucleus, which cannot be vomited out, and cannot be swallowed down' as a further description. If using modern medical terms, the symptom can be described as 'abnormal sensation or foreign body sensation in the throat, but it does not affect breathing and eating'. Obviously, the expression of modern medicine is straightforward, easy to understand, not easy to produce ambiguity, and covers a wider range. For some relatively complex and diverse clinical manifestations, such as headaches, dizziness, fever, etc., how to refine them so that specific performance in different individuals can be comprehensively, objectively and clearly expressed, there are structural problems of refinement and characteristics description. There is also the wider structural question of whether certain further descriptions of these symptoms are classified as features or as independent symptoms. For example, headaches have different pain locations, such as the whole head, inside the brain, top of the head, both sides, left and right sides, forehead, back pillow, and neck, etc.; there are different pain characteristics, such as swelling pain, throbbing pain, stinging pain, empty pain, severe pain, referred pain to different surrounding parts or routes, etc.; there are also different influencing factors, such as aggravation in the afternoon or night, aggravation after fatigue, aggravation after thinking, aggravation after exposure to cold or heat, etc. Obviously, the standardization of symptoms is not a simple explanation of the nouns and terms of symptoms and signs in TCM. In order to complete the collection of patients' clinical manifestations and characteristics conveniently and simply, it is often indispensable to carry out structural refinement of symptom descriptions and clear and unambiguous expressions. Therefore, the corresponding methodological research has shown its importance.

In addition, most diseases in modern medicine are defined based on specific etiology, location, and pathological changes (including organic and functional). Some diseases defined based on clinical manifestations will gradually introduce some objective detection indicators as their research deepens. Therefore, the diagnosis of a disease is highly dependent on the detection indicators. In TCM, a syndrome is accepted, usually after discovering effective treatments and herbal formulas for it. Therefore, the differentiation of syndromes is usually based on a specific treatment method. When the syndrome is diagnosed clinically, the treatment method has been determined accordingly. For example, in TCM, no matter which part of the tumor, nodule, or pain characterized by tingling and fixed pain, it is regarded as a blood stasis syndrome, and the treatment method is to promote blood circulation and remove blood stasis, and the prescriptions are similar. In TCM, the basis for distinguishing syndromes is mostly symptoms and signs which are lack of specificity. Symptoms are usually the patient's expression of feelings, while physical signs are the information collected by TCM practitioners through methods such as looking, smelling, and touching. In the process of collecting symptoms and signs, the subjective factors of patients and TCM practitioners are difficult to rule out. Therefore, it is difficult for symptoms and signs to have the objectivity of modern medical testing indicators. On the other hand, due to the non-specificity of the symptoms and signs contained in the syndromes, it is often difficult to judge whether the diagnosis of the syndromes is established based on the qualitative determination of symptoms and signs. Therefore, it is inevitable to introduce quantification of the diagnostic significance of the symptoms and signs. In recent years, a 'scale' that quantifies the significance of symptoms and signs in the diagnosis of syndromes has been introduced into the process of establishing diagnostic criteria for single syndrome in TCM (Li Xiantao et al. 2019). However, the application of the current scale is still in its infancy, and the quantification of the diagnostic significance of symptoms and signs basically remains at the empirical scoring stage. The mechanism for adjustment based on the statistical analysis of standardized clinical cases has yet to be established. In addition, the accuracy of diagnosing syndromes based on symptoms and signs is obviously related to the following factors:

1. The degree of symptoms and signs and their relationship with syndromes also have problems that need to be described quantitatively. It is manifested as (1) the severity of a symptom/sign (such as headache) is different in different patients or at different times of the same patient; (2) the diagnostic significance of a symptom/sign to a syndrome needs to be quantitatively described; (3) the degree of symptoms/signs will also affect the quantitative determination of

the diagnostic significance of syndromes. Obviously, it is difficult to accurately describe the relatively independent quantitative characteristics of the above three aspects with a scale that usually sets a value for a symptom/sign.

2. The correlation among symptoms and signs related to a syndrome will affect the accuracy of diagnosis. For example, the signs of a pale face, pale lips, and pale nails, which are diagnostic for blood deficiency syndrome, obviously have a certain degree of correlation. If the points accumulation of blood deficiency syndromes does not consider the correlation among them, it will definitely cause a certain degree of repeated scoring, thereby affecting the accuracy of the diagnosis.
3. The introduction of the scale has taken a step forward in the quantitative description of the diagnostic significance of symptoms/signs. However, completely adopting the quantitative description of the scale and ignoring the structural relationship between the symptoms/signs under the same syndrome often weakens the more important qualitative problems in the diagnosis process, and also affects the accuracy of the diagnosis. For example, the diagnosis of heart-yang deficiency must have two factors: heart performance and yang deficiency performance. Using the scale method, when only multiple heart symptoms appear and are more serious, the points may also exceed the cumulative value required for the diagnosis of heart-yang deficiency. Therefore, it will also cause misdiagnosis.

It can be seen that there are not only qualitative problems in the identification of syndromes, but also quantitative factors. It also needs to deal with the structural relationship between the qualitative and quantitative. The current deficiencies of TCM diagnostic indicators in terms of objectivity, diagnostic rules, and algorithms make them not have the credibility required as a reference system for clinical observation and efficacy evaluation. In other words, TCM has not yet formed a reference system comparable to the classification system of modern medical diseases that can be used for clinical observation and efficacy evaluation. Therefore, the conclusions drawn from the research on the diagnosis and treatment methods of TCM based on the current evaluation system often also do not have sufficient credibility and are difficult to be accepted by modern medicine. The diagnostic methods such as pulse image analyzers and tongue image analyzers of TCM, and CHM research to syndromes all need this frame of reference in order to carry out extensive statistical analysis of correlation, so as to provide a scientific basis for their clinical application. Without such a frame of reference, these studies are basically impossible to carry out. Obviously, the standardization of diagnostic criteria for syndromes, symptoms, and signs is an important issue that must be solved first on the road of development of TCM.

6.4 Kampo Medicine in Japan: From 'abandoning medicine (Theory) and keeping herbal medicine' to evidence-based medicine

TCM was introduced to Japan in the Sui and Tang Dynasties. For more than 1,000 years since then, Japanese medical professionals have been continuously absorbing new theories and clinical experience from TCM to enrich their knowledge system called Kampo medicine.

6.4.1 Japanese Kampo medicine survived under the policy of 'abandoning medicine and keeping herbal medicine'

In 1868, Japan entered the Meiji period, and the government implemented the policy of prospering the country and strengthening the military. Under the background of advocating Western culture, mainstream medicine has turned to Western medicine which is in line with Western science in terms of methodology. The government clearly stated that medical practitioners must study Western medicine and obtain the qualifications of medical practitioners before they can practice medicine. This law is still in effect today. As a result, a large number of Kampo Medical Centers were forced to close because Kampo medical practitioners lost their practice licenses. The Ministry of Health and Welfare of Japan, which was run by people with a background in Western medicine education, vigorously advocated Western medicine and popularized Western medicine education, thus training a large number of Western medical practitioners. Traditional medicine was regarded as unscientific and was abolished, and the Kampo schools were also forced to close. After the 8th year of Meiji, Kampo medicine even fell into the dilemma of legal existence and was close to being extinct. However, because Japanese law does not prohibit the study and research of Kampo medicine, medical practitioners who have obtained the qualifications of medical practitioners in Western medicine can still use Kampo medication to treat diseases (Otsuka Kyoo 2000).

Under the policy of 'abandoning (TCM/Kampo) medicine and keeping herbal medicine', only the 'Ancient Formula School' that respects Zhang Zhongjing's 'Shanghan Lun, Treatise on Cold Pathogenic Diseases' and attaches importance to the correspondence between herbal formula and syndromes had the possibility of survival and room for development. Therefore, the ancient formulas school, which has less conflicts with Western medicine in terms of ideas, eliminated the few theoretical and speculative components of 'Shanghan Lun, Treatise on Cold Pathogenic Diseases', and focused on applying the 'correspondence of herbal formulas and syndromes' of the original herbal formulas of 'Shanghan Lun, Treatise on Cold Pathogenic Diseases', and thus developed the 'Kampo medicine' which Japan is today proud of.

Since the 1970s, on the one hand, there have been more and more intractable diseases, mainly among the elderly, and modern medicine has been more and more helpless; and the side effects of chemical medications on the human body have been widely known. This has gradually increased Japanese interest in Kampo medicine. On the other hand, due to the restoration of diplomatic relations between China and Japan, cultural exchanges between the two countries have become more frequent. Encouraged by the success of Chinese acupuncture anesthesia, there is an upsurge in learning TCM across Japan. A large number of TCM textbooks, clinical experiences of old TCM experts and other excellent TCM books have been translated and published. Japanese students brought the knowledge and experience of TCM back to Japan after studying in China, and the exchanges and immigration of TCM practitioners in China to Japan greatly promoted Japan's understanding and acceptance of TCM. Under this background, Kampo medicine has aroused people's interest again and developed rapidly. Under the promotion of the Japanese medical profession, in 1976, the Ministry of Health, Labour and Welfare of Japan established the Kampo medication medical insurance applicable system. Subsequently, 210 ancient herbal formulas recorded in 'Shanghan Lun, Treatise on Cold Pathogenic Diseases' or other ancient books were successively approved to be used in the production of Kampo medications. As a result, the Kampo medication industry developed rapidly. Many scientific research institutions and pharmaceutical companies are committed to the extraction, separation, identification of active ingredients of Kampo medications and CHMs and modernization of Chinese medicine preparations. Proprietary Chinese medicines based on Kampo have been widely used in Japan and occupy a large share of the world's natural medicine market, which has also brought huge economic benefits to Japan. At present, the Ministry of Education, Culture, Sports, Science and Technology of Japan has included Kampo medicine in the curricula of medical universities. Eighty medical universities across the country have made Kampo medicine a compulsory course, and 80% to 90% of the 290,000 Western medical practitioners are using or have used Kampo medication to treat patients. Japanese Kampo medications are basically stereotyped products, and they are used in accordance with the product instructions. Generally, there is no need for treatment based on syndrome differentiation, nor is there need to increase or decrease herbal medicines according to the symptoms. At present, there are no pure Kampo medical practitioners in Japan, and most of the people who use Kampo medication for treatment are Western medical practitioners who have received modern medical education. The way they use Kampo medicine to treat diseases is basically the same as that used by Western medical practitioners in China to treat diseases with proprietary Chinese medicines (Dai Zhaoyu et al. 2000b).

6.4.2 The difference and relationship between 'formula-syndrome' and TCM 'syndrome'

The syndromes based on the herbal formula (formula-syndromes) in Kampo medicine are different from the syndromes of TCM treatment system based on syndrome. As mentioned earlier, TCM syndromes are state variables which are introduced to describe the state of the holistic body level and the sub-system level. TCM regards state variables (syndrome) as the controlled quantity, so that the identification of the body state can be combined through the identification of state variables, and the regulation-control of the body state can be combined through the regulation-control of state variables. The formula-syndromes of Kampo medicine usually correspond to the individualized state of the human body. So, what is the relationship between the formula-syndromes of Kampo medicine and the syndromes of TCM treatment system based on syndrome?

As mentioned earlier, a TCM syndrome usually corresponds to a set of symptoms and signs. However, to determine the establishment of the syndrome usually does not require all symptoms and signs in this set to appear, only a part of them is sufficient. Therefore, a simple formula-syndrome may be for a specific state with a state variable is abnormal. Usually, its diagnosis only requires the appearance of a group of symptoms/signs in the set of clinical manifestations that characterize the state variable. For example, Mahuang Decoction is aimed at a condition with severe cold and no sweating as the main clinical manifestations under the syndrome of exogenous wind-cold. Complex formula-syndromes usually involve states in which more than one state variable are abnormal. For example, Huoxiang Zhengqi San addresses the state in which the two state variables of exogenous wind-cold and spleen-stomach dampness are abnormal at the same time. Shiquan Dabu Decoction is aimed at the state where the two state variables of qi deficiency and blood deficiency are abnormal at the same time (formula-syndromes). More complex formula-syndromes may involve a combination of multiple syndromes at the same time, such as the simultaneous appearance of qi and blood deficiency and qi stagnation and blood stasis syndrome. From the perspective of the state space, a syndrome of the TCM syndrome system is equivalent to a dimension of the state space, and a formula-syndrome is equivalent to a state point in the state space.

In terms of complexity and workload, it is obviously much simpler to decompose the control of a pathological state into the control of the state variables that determine the state. In fact, for patients with a wider range of diseases and more complex conditions, experienced TCM practitioners usually adopt this approach to regulate/control. Of course, due to the complexity of the human body system, there may be exceptions where the control of the pathological state cannot be attributed to the control of abnormal state variables associated with the pathological state. As revealed by the science of complexity, the regulation-control of the complex system as a whole cannot be attributed to the simple addition of the regulation-control of the various parts of the system. But when we abstract a complex system into a simple functional model, the regulation-control to state variables becomes a planned and step-by-step regulation-control to all aspects and parts of the system. In this case, formulating a comprehensive regulation-control plan based on the importance of various aspects and parts in the life activities of the human body and the relationship among them will often achieve good results. Only in exceptional cases should the treatment plan be formulated based on 'formula and syndrome correspondence'. In the face of complex and individualized pathological states, this approach is obviously much simpler than using the formula and syndrome correspondence approach to regulate and control all pathological states (Yuan Bing 2017a).

Although the treatment based on formulas-syndromes also belongs to the treatment based on syndrome in TCM, the range of individualized states covered by a formula-syndrome is usually much narrower than a general TCM syndrome, which also means that formula-syndromes have better certainty than general TCM syndromes. Usually, to realize the regulation-control based on an SDS, the controlled quantity can be placed on the state variable (general TCM syndrome), or it can be placed on the state point (formula-syndrome) formed by the combination of state variables. But the number of controlled quantities corresponding to these two methods will be vastly different.

Under the scientific concept characterized by certainty, the formulas-syndromes of Kampo medicine are easier to attribute to clinical trials based on evidence than the general TCM syndromes. But every combination of symptoms and signs of each syndrome is a formula-syndrome, and every combination of syndromes is also a formula-syndrome. For a TCM treatment system based on syndrome with about 100 syndromes, the number of formula-syndromes that can be combined out will be an unimaginable astronomical number. Therefore, to comprehensively and effectively deal with various possible disease states of the human body, medicine based on formula-syndromes will be more than trillion times more complicated than medicine based on TCM syndromes.

6.4.3 The splendor and limitations of Japanese Kampo medicine

In the process of the modernization of traditional medicine, Japanese Kampo medicine has made remarkable contributions in at least two aspects, which has also made Kampo medicine today's glory. One is that when Chinese scientific workers engaged in TCM research are keen to use modern analytical methods such as pharmacology and medicinal chemistry to study the mechanism of action of CHM, the Japanese Kampo Medical Circle with world-class instruments and equipment thinks: What kind of curative effect Kampo medication has on any disease cannot be answered by laboratories or research institutes, but must be understood in clinical application. Therefore, Japan only conducts general pharmacodynamic experiments on the curative effect of Kampo medication, and basically does not conduct or rarely conducts experimental research on other modern scientific and technological means. In Japan, many housewives and elderly people usually learn to use herbs to treat or prevent common diseases by watching TV Kampo lectures and participating in various Kampo workshops. Second, Japan took the lead in introducing modern advanced technology into the industrial manufacturing of Kampo medication. In some specific links of the large-scale production of Kampo medication, such as washing, crushing, drying, mixing, decoction, concentration, filtration, molding, aseptic packaging, etc., the introduction of mature advanced technology from other industries has raised the manufacturing of Kampo medication in Japan an unprecedented level.

Since the Meiji Restoration, the policy of 'abandoning medicine and keeping herbal medicine' has caused Japanese Kampo medicine to lose its relationship with the TCM theoretical system, which is its origin, and become 'water without a source and a tree without roots'. In the application of Kampo medicine under the modern medical system without the support of the theory, the indications and contraindications clearly stated in the original theoretical system were ignored, and some undesirable consequences caused by improper use of herbs appeared. The occurrence of huge side effects of 'Xiao Chai Hu Decoction' in the 1990s caused strong shocks in Japanese society.

6.4.4 Evidence-based medicine: the 'opportunities' and problems faced by the development of Japanese Kampo medicine

In recent years, the rise of evidence-based medicine seems to have brought new opportunities for the development of Kampo medicine. With an open attitude, modern medicine no longer rejects medical knowledge systems that do not conform to the concepts of analytical science. After scientific clinical trials, the treatment methods and herbal formulas of Kampo medicine can be evaluated and applied using the method of EBM, and then incorporated into the modern medical system in a new form. In order to promote the EBM research of Kampo medicine, the Japanese Oriental Medical Association established the EBM Special Committee in 2001. The committee has done a lot of work on EBM, and obtained a lot of evidence of the practicality of Kampo through research. In 2002, the committee issued the first EBM report on Kampo therapy 'EBM in Kampo 2002, Interim Report'. The report adopted the 10th edition of the 'International Classification of Diseases' criteria for the diagnosis and classification of diseases, and included and evaluated 833 clinical treatment reports of single herbal medicine or fixed herbal formulas published after the implementation of the Japanese new Kampo preparation standards from 1986 to November 2001, including 12 randomized double-blind controlled trials and 621 controlled trials. The report gave a detailed description of the clinical research purpose, research type, research location, inclusion and exclusion criteria, intervention and control measures, main outcome indicators and measurement methods, safety evaluation, and research conclusions, and made a comprehensive evaluation and recommend. Since then, the committee has successively released 'Interim Report 2007', 'EKAT 2009', 'EKAT 2010', 'EKAT 2013' and 2014 and 2015 supplementary reports. A total of 445 RCT and 2 meta-analysis were included and evaluated in the 2015 Supplementary Kampo Therapy Evidence-Based Report released in 2017 (Yuan Zhongyu et al. 2021).

In the 2002 interim report, the evaluation of clinical evidence included research design, syndrome diagnosis, and efficacy judgment standards. Data extraction includes results, conclusions, and side effects. Evidence classification is divided into three categories according to the strength of evidence recommendation. The first category is A and B evidence that is strongly recommended for clinical use or is recommended for clinical use; the second category is C-level evidence that has no clear evidence to recommend clinical use; the third category is Level D evidence for clinical use is not recommended. Examples of recommended results according to ICD-10 disease classification are as follows:

1. Digestive system diseases—esophagus, stomach and duodenal diseases-dyspepsia

   Recommendation: Liujunzi Decoction.

   Evidence: The research on Liujunzi Decoction shows that it is effective for patients with sagging internal organs, decreased energy, and low physical strength as shown by X-ray examination. This is consistent with the traditional indications of Liujunzi Decoction, and there is also more reliable research on experimental design.

   Double-blind RCT example:

   Literature 1: Yuan Zemao and others published in 1998.

   Subjects: 296 cases of functional dyspepsia belonging to dysmotility type.

   Trial design, treatment, course of treatment and others: double-blind RCT, 7.5 grams of Liujunzitang granules, used for two consecutive weeks.

   Results: Judgment of the syndrome: decreased abdominal wall tension, self-feeling or audible liquid vibrating sound of liquid, gastroptosis on X-ray examination, decreased mental and physical strength. It has been confirmed by a double-blind randomized controlled trial that Liujunzi Decoction is an effective and safe Kampo preparation for functional dyspepsia belonging to dysmotility type.

   Evaluation: First class (strongly recommended).

2. Other bowel diseases-irritable bowel syndrome

   Recommendation: Guizhi plus Shaoyao Decoction.

   Evidence: According to a randomized controlled trial in 76 branch centers, Guizhi plus Shaoyao Decoction was effective for 286 cases of irritable bowel syndrome patients with abdominal pain and diarrhea.

   Trial design, treatment method, course of treatment and others: Double-blind RCT, granules containing 16 g of Guizhi plus Shaoyaotang extract powder, three times a day for 8 weeks.

   Efficacy evaluation: 4 or 5 grade scoring method, improvement of laboratory test results and symptoms after 8 weeks.

   Results: Compared with the control group, the Guizhi plus Shaoyao Decoction group had statistically significant improvement in abdominal pain and diarrhea.

   Evaluation: First class (strongly recommended).

The 2002 interim report also discussed the characteristics of Kampo medicine and its development direction, mainly in two aspects:

1. First, the research and implementation of Kampo evidence-based medicine was conducted in three stages. The first stage was to collate the literature to obtain preliminary data on the indications and effectiveness of Kampo medication; the second stage was to further demonstrate the feasibility of clinical trials in accordance with the requirements of modern clinical trials based on the results of the first stage; the third stage was to establish a case database and regularly analyze the accumulated data.
2. The second was to advocate the establishment of single case database containing detailed patient information, and to strengthen single-case control trials.

The two evidence-based medical evidence reports in 2002 and 2005 were not methodologically rigorous, and there were also many criticisms in Japan. For example, there was no clear inclusion and exclusion criteria, and the evaluators and methods were not strict enough, and so on. Therefore, the second phase of the Kampo evidence-based medicine research that began in 2005 was improved in many aspects, such as search scope, inclusion and exclusion criteria, evaluation methods and transparency, improvement of structured abstracts, evaluators' conclusions and readers feedback, etc. The evaluation was made more stringent (Guo Xinfeng and Lai Shilong 2010).

The EBM of Kampo medicine in Japan is mainly based on ancient herbal formulas such as 'Shanghan Lun; Treatise on Cold Pathogenic Diseases'. The production is carried out by a few large pharmaceutical factories, with reliable quality. Defining the formula-syndrome with all the symptoms and signs corresponding to a certain formula, the form is relatively standardized, and it is easier to classify and compare. On the basis of clinical epidemiology, through the scientific evaluation of individual clinical evidence, the overall systematic evaluation of multiple clinical evidence, and the preparation of clinical practice guidelines on the basis of evidence evaluation and recommendation, EBM promotes the progress of evidence-based medical practice, and improves the quality and efficiency of medical services.

However, the RCT trial of Japanese Kampo EBM as evidence is still based on the DCS of modern medicine, and the description of Kampo indications is quite rough. The distinction among different types of the same disease is not as good as the 'TCM classification under the Western medicine disease diagnoses' of the ITCMWM in China. Obviously, the attention and reflection on the Xiao Chai Hu Decoction incident by the Kampo medical circles in Japan is far from enough. Herbal formulas and medications recommended through strict evidence-based procedures still cannot ensure that events similar to 'Xiao Chaihu Decoction Adverse Reactions' will not occur again in the process of clinical application.

The result of Japanese Kampo medicine going along the path of EBM will be that, on the basis of empirical evidence, the treatment methods and medications of Kampo medicine will be completely incorporated into the complementary and alternative medicine system of modern medicine. At that time, it would not be necessary for Kampo medicine to exist independently as a system. Learning the lessons of the 'Xiao Chai Hu Decoction incident' under the concept of EBM means that diseases under the modern medical DCS will be refined based on formulas-syndrome. As a result, Kampo medicine will enter the Chinese ITCMWM mode of 'TCM classification based on Western medicine disease diagnosis', but it will also face the problems currently faced by the Chinese mode of ITCMWM.

Kampo medicine in Japan is the product of abnormal growth under the policy of 'abandoning medicine and keeping herbal medicine'. The separation from TCM caused by 'abandoning medicine' made it lose its theoretical foundation and the integrity of its knowledge as a scientific system. 'Evidence-based' is meaningful in the experience accumulation stage of a science development process, but as a science, it is impossible to stay at the level of experience forever. The reconstruction of Kampo medicine's theoretical system on the basis of formulas-syndrome means to 'recognize ancestors and returning to the clan', that is, restoring the relationship with traditional theories of TCM.

If so, Kampo medicine will be completely integrated into the TCM system, and it will also lose the meaning of independent existence. Therefore, in any case, the development of Kampo medicine is heading towards a narrower and narrower 'dead end'.

6.5 Where will evidence-based medicine bring TCM to?

After the founding of the PRC in 1949, a series of small-scale clinical observations of TCM were organized, which opened a precedent for clinical research to test the efficacy of TCM. In 1982, the results of the first randomized controlled trial involving TCM were published. Since then, until the end of the 20th century, there have been more than 10,000 randomized controlled trials of TCM that have been retrieved and published, but the quality of the research is generally not high. Although the effects of some TCM methods have been confirmed within a certain range, they lack systematic guidance and standardization, and can neither be used as a basis for clinical decision-making, nor can comprehensively promote the development of the industry of CHM (Wang Yongyan and Huang Luqi 2019).

The emergence of EBM provides reasonable ideas and methods for solving scientific decision-making problems in medical practice. In 1992, an article published by the EBM Working Group on The Journal of the American Medical Association (JAMA) marked the official birth of EBM. Since then, the evidence-based thinking model has been widely recognized, and the practice field of EBM has expanded from the initial field of clinical medicine to the field of public health. After more than 20 years of development, EBM has become an important milestone in the development of medicine in the past century, which has profoundly affected all aspects of the field of health. The concept and method of EBM of 'demand-oriented, the practice based on evidence, pay attention to the results of practice, evaluate the after-effect, and stop at perfection' is gradually applied to non-medical fields. This concept and method mainly focus on the scientificity and cost-effectiveness of decision-making, and attaches importance to the evaluation of the quality and effect of decision-making by a third party. From evidence-based medicine to evidence-based science is not only the expansion of application fields, but also the innovation of thinking and decision-making modes (Zhang Junhua et al. 2019).

6.5.1 The research history of the application of evidence-based medicine methods in TCM

EBM was introduced to China by the former West China University of Medical Sciences in 1996, and quickly received the attention and support of the former Ministry of Health, Ministry of Education, National Natural Science Foundation of China and other national ministries and related societies (Li Youping et al. 2016), EBM not only brings about the methodology of evidence production, but also establishes a feasible model for the rational flow and transformation of clinical efficacy research results. Soon after evidence-based medicine was introduced into China, some visionary scholars in the TCM academic circle began to explore the use of EBM in TCM.

In 1999, Li Youping (1949) and Liu Ming (1957) published an article titled 'EBM and Modernization of TCM', which pointed out: Use internationally recognized methodologies and standards to re-understand and interpret TCM, evaluate the efficacy of TCM; use internationally recognized academic language and theory help TCM go abroad, clinical epidemiology and EBM should be one of the best methods currently (Li Youping and Liu Ming 1999). This view was basically agreed upon by the TCM academic community. Experts and scholars such as Wang Yongyan, Chen Keji, Zhang Boli (1948), Liu Baoyan (1955), etc., all expressed their views (Wang Yongyan et al. 2003, Chen Keji and Song Jun 1999, Zhang Boli 2002), discussing the importance, feasibility and tasks of introducing EBM in promoting the development of TCM, and formed the guiding ideology of "First, learn, second, use, third, know the limitations, fourth, innovate in the evidence-based evaluation methods of TCM." Since then, the academic community of TCM has carefully accepted the concepts, methods and standards of EBM into TCM research.

Since 2002, universities in Beijing, Shanghai, Tianjin, Chengdu, Guangzhou, Jiangxi, Henan and other places have successively established Evidence Based Medicine Research Centers in TCM (Wang Yongyan and Huang Luqi 2019), forming a professional platform for EBM research in TCM covering the whole country. In 2009, the State Administration of TCM established 16 national-level TCM clinical research bases based on the key dominant diseases of TCM. TCM clinical research bases are mostly based on large-scale comprehensive TCM hospitals at or above the provincial level, and have established a domestic first-class clinical research network for key disciplines of TCM. These bases are mainly responsible for the design, implementation and quality control of major clinical research projects of TCM, carry out the evaluation and standardization of TCM clinical diagnosis and treatment technology, and cultivate high-level TCM evidence-based research talents.

In 2007, Li Youping published an article titled 'Strategies for China EBM Center to Promote the Modernization of TCM', which clarified four major measures: (1) Implement a clinical trial registration system in China and establish a Chinese clinical trial registration and publication mechanism; (2) Formulate TCM Clinical Trial Reporting Standard (CONSORT for TCM); (3) Carry out a comprehensive Cochrane system evaluation of TCM, and evaluate the efficacy of TCM according to international standards; (4) Carry out EBM education for TCM practitioners, and the establishment, and promote a TCM clinical practice model that can be widely accepted by the international community (Li Youping et al. 2007). With the support of the State Administration of TCM and societies of all levels, these measures have been gradually implemented in practice.

In order to promote research in the TCM field based on EBM, in 2018, the Evidence-based TCM Research Consortium (RC-EBCM) was established. The alliance is jointly initiated and established by the China EBM Center, the EBM Center of Tianjin University of TCM, the EBM Center of Beijing University, the EBM Center of Lanzhou University, the EBM Center of Beijing University of TCM, the Center for Health Economics Research of Beijing University, the TCM Research Center of Oxford University, the School of Statistics and Data Science of Nankai University, and the Medical Policy and Economic Research Center of the School of Pharmacy of Tianjin University. Institutions such as the Shanghai EBM Research Center of TCM, Sichuan Province EBM Research Center of TCM, Henan EBM Center and other institutions were also invited to join the alliance (He Ziyuan 2020).

In March 2019, the China Center for EBM in TCM was established, which is responsible for cooperating with other relevant domestic research institutions in China to provide clinical evidence for the effectiveness and safety of TCM. In addition, the World Federation of TCM Societies, the Chinese Integrative Medicine Society, and the Chinese Acupuncture and Moxibustion Society have established professional committees for EBM to promote clinical research on TCM in related disciplines (Tian Xiaohang 2019).

As of December 2018, the Evidence-Based Evaluation Evidence Database (EVDS) of TCM, established by the EBM Center of Tianjin University of TCM and the China EBM Center has collected more than 80,000 clinical trials of proprietary Chinese medicines and more than 4,000 systematic reviews/meta-analyses, covering more than 1,700 varieties of proprietary Chinese medicines. In accordance with international norms and standards, the EVDS system systematically examines literature, strictly screens, implements dual entry and dual verification, and strictly evaluates quality, aiming to provide evidence support for evidence-based decision-making in TCM. EVDS includes a database of evidence of clinical effectiveness of CHM, a database of evidence of clinical safety of CHM, a database of evidence of clinical evaluation of acupuncture and moxibustion, and a database of clinical guidelines for CHM. Among them, the clinical effectiveness evidence base of CHM collects RCTs and systematic reviews/Meta analyses of CHM published in both Chinese and English; the clinical safety evidence base of CHM collects safety reports related to CHM, including single case reports, case series, cohort studies, cross-sectional surveys, centralized monitoring research and national adverse medication reaction annual notifications and other data; and the CHM clinical guidelines database collects and organizes clinical diagnosis and treatment guidelines and clinical pathways issued by various societies and branches (Zhang Junhua et al. 2019). EVDS has eight functions:

1. Improve the efficiency and quality of systematic evaluation research;
2. Compare the strength of evidence for proprietary Chinese medicines;
3. Publish the annual report on the research of proprietary Chinese medicines;
4. Provide evidence support for the formulation of TCM guidelines;
5. Provide services for the revision of essential medications or medical insurance catalogues;
6. Comprehensive and continuous evaluation of clinical research methodology and report quality;
7. Support the research on the core index set of clinical trials of TCM;
8. Support the international sharing of TCM data.

On January 9, 2020, the 12th Healthy China Forum opened in Beijing, and the Evidence-Based TCM Forum was held at the same time. The Evidence-based TCM Research Alliance released the evidence index for clinical evidence-based evaluation of proprietary Chinese medicines, and determined the roadmap for evidence-based research in TCM. The goal of the alliance is to strive to complete the eight tasks within 10 years and form a collaborative mechanism for universities, research institutions, pharmaceutical companies, medical staff, researchers, contract research organizations, journals and magazines, aiming to accelerate the production and transformation of evidence-based evaluation evidence for TCM (He Ziyuan 2020).

The evidence index released this time is evaluated based on the quantity and quality of clinical research. Focusing on three types of diseases such as angina pectoris of coronary heart disease, stroke, and tumors, the current available clinical research evidence was analyzed, and the TOP 10 proprietary Chinese medicines for each type of disease were selected based on these evidence indexes. These include oral proprietary Chinese medicine compound Danshen Dripping Pills, Shexiang Baoxin Pills, Tongxinluo Capsules, Suxiao Jiuxin Pills, Naoxintong Capsules, Xuefu Zhuyu Capsules, Angong Niuhuang Pills, Yangxue Qingnao Granules, Dengzhan Shengmai Capsules, Ping Xiao Capsules, Zhenqi Fuzheng Capsules, etc. (Ren Zhuang 2020).

The eight tasks proposed by this forum to be completed within 10 years include the following six:

1. Complete the systematic evaluation of the clinical effectiveness evidence of 50 diseases;
2. Complete the systematic evaluation of the clinical safety of 100 kinds of proprietary Chinese medicines;
3. Completion of 10 annual reports on clinical trials of proprietary Chinese medicines;
4. Complete 10 high-quality clinical trials or systematic review studies;
5. Complete the study of the core index set (COS) of clinical trials for 30 diseases and their subtypes, and promote their application;
6. Complete 5 clinical evaluation index studies that meet the characteristics of TCM and apply them to clinical research on TCM.

At this forum, the academic committee of the conference drafted and issued the 'Evidence-based TCM Research Beijing Declaration', and called for 'doing clinically needed research, doing scientific and standardized research, doing transparent and usable research, and doing efficient transformation research', to provide high-quality evidence of effectiveness and safety for TCM (He Ziyuan 2020).

EBM entered China and merged with the scientific and modernization of TCM and the trend of TCM going to the world, forming the research field of TCM evidence-based medicine. This is another round of new attempts by Chine to develop TCM with the support of the national level after 50 years effort of ITCMWM. Since 1999, TCM evidence-based medicine research has gone through 20 years.

Unlike the EBM research of Kampo medicine in Japan that originated from the free exploration of the academic community, the EBM research system of TCM in China is coordinated and designed by the national level. The clinical evidence production and evaluation system of TCM is based on the concepts and methods of clinical epidemiology, integrating medical institutions, clinical research bases, relevant evidence-based medicine centers, and academic organizations to form a unified integrated system. This system includes personnel training, multi-level clinical research centers, and cooperation networks, and its scale and momentum are far greater than those of Kampo medicine in Japan. As of 2019, there have been more than 100,000 clinical studies based on randomized controlled trials, and more than 5,000 secondary studies based on systematic reviews/meta-analysis, forming a preliminary evidence base for clinical diagnosis, treatment and management decisions (Ren Zhuang 2020) Such as:

1. The publication of clinical randomized controlled trial reports of compound proprietary Chinese medicines has enabled TCM clinical trials to have internationally recognized reporting standards, and at the same time introduced theories, syndromes, treatment methods, prescriptions and other TCM concepts into international standards.
2. Acupuncture in the treatment of female stress urinary incontinence, migraine, and functional constipation has adopted rigorous randomized controlled trial methods, effectively verifying the efficacy of acupuncture.
3. The clinical research of Maxing Shigan Decoction combined with Yinqiao Powder in the treatment of H_1N_1 influenza has demonstrated the advantages of TCM in the treatment of infectious diseases.
4. The high-quality clinical studies of multiple compound proprietary Chinese medicines have strongly supported the long-term and short-term therapeutic effects of TCM on chronic or severe diseases (Wang Yongyan and Huang Luqi 2019).

6.5.2 Methodological reflection on evidence-based research of TCM

At the beginning of the introduction of evidence-based medicine into the clinical research of TCM, people thought that EBM was a ruler and a tool, which can be used by both Western medicine and TCM. The practice of EBM in TCM in the past 20 years has obtained some clinical evidence and exposed some problems simultaneously, which has triggered the TCM profession to reflect on the EBM methods used in the field of TCM. The article 'Based on High Vision and Building China's TCM Evidence-Based Medicine Center' published by academicians Wang Yongyan and Huang Luqi (1968) in 2019 clearly pointed out the problems existing in the current evidence-based research of TCM (Wang Yongyan and Huang Luqi 2019):

1. TCM treatment based on syndrome has not been included in clinical research. Therefore, the evidence obtained is difficult to guide the clinical practice of TCM treatment based on syndrome.
2. According to the current clinical trial design, the effectiveness of treatment methods that are commonly used in clinical and indeed effective has not been confirmed by statistics, and no evidence of curative effect can be produced.
3. In the clinical application of Chinese medicine preparations with evidence of effectiveness, safety disputes are often caused by individual cases, and the final solution usually is to stop the medication.
4. The quality of clinical research on TCM is generally low and it is difficult to build confidence.
5. The synergistic advantages of TCM in the treatment of infectious diseases and critical illnesses cannot be effectively evaluated and are difficult to promote.

6. Epidemiological investigations with the characteristics of TCM usually operate independently, with lack of overall coordination, and it is difficult to integrate respective advantages to support upper-level research.
7. The TCM concept is misappropriated by illegal medical activities, but it is difficult to be falsified.
8. Due to the design of TCM curative effect research and the quality of research results, the clinical evidence obtained is insufficient to fully confirm the curative effect of TCM, so it is weak in supporting TCM health management and clinical diagnosis and treatment decision-making.

This article analyzed the reasons for the above problems from a methodological point of view:

1. EBM and TCM have different views on clinical issues, rehabilitation definitions, intervention ideas, and treatment expectations. The two different cognitive methods of traditional Chinese and Western medicine have led to the differences in their measurement and evaluation concepts. When the EBM method suitable for evaluating diagnosis and treatment of Western medicine is applied to the field of TCM, it is difficult to cross over the transformation of the ideas such as from precise to vague disease definition and evaluation methods, single to comprehensive intervention ideas, and from group to individual view change. It may not be feasible to copy the evaluation methods of EBM to verify and interpret the clinical efficacy of TCM.
2. In the explorations of the past efficacy evaluation of TCM, researchers should not only respect the clinical diagnosis and treatment characteristics of TCM theories, methods, prescriptions, and medication, but also follow the requirements of evidence-based methodology, and bridge the missing standard links at the combination of the two. The missing standard links include: TCM related syndrome distribution, diagnostic criteria, evaluation indicators, judgment criteria, measurement tools, prescriptions and medications specification, quality control standards, etc., as well as CHM simulation agents and fake needles designed to achieve random double-blind control and so on. Due to the existence of different syndrome systems and academic schools, as well as the significant differences in the resulting academic opinions, it is difficult to form a consensus on the formulation of TCM standards.
3. Evidence-based clinical research is an important link connecting the development of medical academic and industry. The effectiveness and safety evidence generated under the support of a solid methodological and standard foundation is a strong guarantee to support the development of medicinal products from the laboratory to the clinic, from CHM to the medication products. Due to the differences in the methodology of traditional Chinese and Western medicine, the application of EBM to TCM lacks sufficient methodological research, and there is no internationally recognized standard system, so there are many problems in the method design and the quality of the results of the clinical research of TCM, which directly affect the recognition of the clinical efficacy of TCM.

In recent years, at the national level, although industry management departments and academic organizations have done a lot of work on efficacy evaluation methodology, data basis and normative standards, there has never been an evidence-based method suitable for evaluating the efficacy of TCM that can rise to the level of international consensus. In terms of clinical definition, design principles, quality management, reporting standards, evaluation standards, etc., most of the evidence-based efficacy studies of TCM still follow the thinking mode and standard system of modern medicine. Only a limited number of TCM treatment methods can obtain well-designed evidence-based evaluation opportunities, which are not enough to fully reflect the diagnosis and treatment practice of TCM treatment based on syndrome. Therefore, the conversion rate of research results from clinical trial to clinical practice is relatively low (Wang Yongyan and Huang Luqi 2019).

In the long history of thousands of years, the efficacy evaluation of TCM has always lacked research data that conforms to modern scientific norms, and there are problems such as unclear expression or low quality of evidence. The introduction of the concept of EBM is undoubtedly of positive significance for providing credible proof of the effectiveness of TCM and improving the safety of clinical application of TCM. But how to base the clinical application of TCM on reliable evidence? Are current EBM methods suitable for evidence-based research of TCM? After 20 years of EBM in China, the issue of TCM evidence-based methodology has provoked the attention and thinking of more and more insightful scholars in the academic field.

First of all, judging from the development status of evidence-based research of TCM and the roadmap for the next 10 years of TCM evidence-based research released by the Evidence-based TCM Research Alliance, the current evidence-based research in TCM is limited to proprietary Chinese medicines, which is basically consistent with the scope of evidence-based research in Japanese Kampo medicine. Throughout the ages, TCM has always been treating diseases by CHM decoctions based on syndrome differentiation and added or subtracted CHMs according to the symptoms. Proprietary Chinese medicine has never been the mainstream of TCM clinical practice, but is mainly used as a supplement to TCM decoctions. Based on the DCS of modern medicine only conducts evidence-based research on proprietary Chinese medicines, and uses the research results as evidence to formulate clinical guidelines and clinical pathways for specific diseases that only include these proprietary Chinese medicines. In this process, the holistic concept as the core of TCM and the treatment based on syndromes that reflect individual characteristics have completely disappeared. If EBM becomes the mainstream of the national strategy to promote the development of TCM, and evidence-based evaluation is based on the current framework of modern medicine, the living space of the mainstream of TCM will be further narrowed to the level near Japanese Kampo medicine. The difference is that Japan's 'abandoning medicines and keeping herbal medicines' was implemented through explicitly prohibited policies and regulations, while China's 'abandoning medicines and keeping herbal medicines' was implemented through technical measures under the banner of supporting TCM, promoting the development of TCM, and making TCM scientific and modern. The specific method is to gradually marginalize the core concept of TCM, and make the clinical thinking mode of TCM lose the space for existence and development.

At present, the EBM research in TCM is carried out based on the modern medical DCS as the frame of reference. The research results are incorporated into the clinical guidelines of modern medicine as complementary and alternative medicine methods, it will undoubtedly greatly enrich the treatment methods of modern medicine. This enables medical practitioners trained in modern medicine to use proprietary Chinese medicines to treat diseases more conveniently and effectively. Just like using artemisinin for malaria, and using 'trigger point therapy (dry needle)' for clinical treatment in modern medicine, these results and their transformation are undoubtedly of great significance for modern medicine to improve the level of clinical treatment. This is a deepening of the 50-year-old 'ITCMWM' research, and at the same time it has opened a new round of 'strangulation' on the concepts and ways of thinking of TCM.

The Chinese elites of the TCM profession, by introducing cutting-edge concepts and methods of modern medicine, and taking advantage of their familiarity with TCM, have dismantled the TCM, the ancient 'building' which was constructed on unique methods and structure, into 'bricks and tiles' of complementary and alternative therapies, so as to fill the 'new building' of modern disease medicine. However, they did not expect that in such 'bricks and tiles', the holistic concept of traditional Chinese medicine would disappear, and the treatment based on syndromes has also been limited to a narrow range under the category of modern medical diseases. When modern medicine uses these 'bricks and tiles' to fill the 'building' of disease medicine, it is unable to fully understand and describe the performance of these 'timber bricks' due to the limitations of the concept and structure of building this building. Therefore, it is inevitable that there will be some deviations in use, which may even lead to unforeseen adverse consequences. As academicians Wang Yongyan and Huang Luqi pointed out: In the clinical application of Chinese medicine preparations with evidence of effectiveness, safety disputes are often caused by individual cases, and the final solution usually is to stop the medication.

6.5.3 Evidence-based research of TCM needs to take the TCM treatment system on syndrome differentiation as the reference system

Although Tu Youyou's discovery of 'artemisinin' is derived from the treatment experience of TCM, it is not regarded as the research result of TCM. Similarly, research on EBM in TCM, which uses modern medicine as a reference system, does not strictly speaking belong to TCM. So, what is the development research of TCM itself, i.e., the development research based on the core concept of TCM?

The current EBM research of TCM is carried out entirely with the DCS of modern medicine as the reference frame, and the TCM syndromes have not been included in it, the evidence obtained cannot be naturally used to improve the level of TCM treatment based on syndromes. As we all know, tens of thousands of proprietary Chinese medicines have been developed based on the herbal formulas accumulated by thousands of years of clinical practice of TCM. If individualized factors are not taken into account, very few of them have significant effects on diseases diagnosed by modern medicine. At present, only a dozen proprietary Chinese medicines have submitted registration applications to the US FDA, which is enough to prove this point. The evidence involved in the introduction of TCM treatment methods and medications into the treatment system of modern medicine is often not comprehensive, and cannot cover all aspects of the treatment methods and medication applications. Therefore, it is inevitable to be biased like a 'blind man touching an elephant'. If ignore the syndromes for which Chinese patent medicines are suitable and only abuse them based on the diseases used as a reference in clinical trials, it is inevitable to have adverse reactions similar to Japan's 'Xiao Chai Hu Decoction Incident'.

Why, according to the current clinical trial design, could the effectiveness of treatments that are commonly used in TCM and are indeed effective, usually not be confirmed by statistics, and could no evidence of curative effect be provided? The question raised by Wang Yongyan and Huang Luqi's article is not unique to the clinical trials of TCM. As mentioned earlier, many target medications currently developed in modern medicine for the treatment of tumors cannot be statistically effective similarly under current DCS. Therefore, the introduction of 'basket trial' and 'umbrella trial' in modern medicine, and the establishment of a personalized DCS different from the traditional DCS showed the inevitability and necessity. Regarding this point, the TCM community has already had a clear understanding. Zhang Junhua (1960), director of the Evidence-Based Medicine Center of Tianjin University of TCM, clearly stated: the current research has many limitations, such as not following the subtype of the disease, not comparing index dimensions, not independent analysis of dosage forms, not analyzing the efficacy characteristics of varieties, etc. The follow-up will be gradually refined, deepened, and personalized. Academician Zhang Boli further pointed out that TCM "has its own characteristics, such as intervention prescriptions of CHM often vary from person to person, upon time and place." Therefore, under the framework of EBM, it is also necessary to create an individualized evaluation method of EBM that conforms to TCM treatment based on syndrome (Xinhuanet 2021). The use of clinical trial models and curative effects evaluation methods that are more suitable for TCM, like the 'basket trial' and 'umbrella trial' that are compatible with personalized medicine, is bound to have a fundamental impact on the EBM method based on 'randomized controlled clinical trials' as the gold standard.

With the introduction of TCM syndrome classification under the modern medical DCS, the EBM of TCM will be promoted to the ITCMWM level of 'TCM classification based on Western medicine disease diagnosis'. However, even based on the 29,000 diseases (including injuries and causes of death) included in the 10th edition of the International Classification of Diseases (ICD10), if a set of TCM syndrome differentiation is introduced for each disease, it will also mean a considerable amount of knowledge. As for ICD-11, which was implemented in January 2022, the number of diseases has increased to 55,000, and with the development of medicine, there is no end to the growth of this number. The TCM syndrome types corresponding to a disease include not only a single syndrome, but also a combination of syndromes, such as diabetes with deficiency of both qi and yin, and chronic gastroenteritis with spleen and kidney yang deficiency. Using a randomized controlled method to conduct clinical trials based on rigorous statistical analysis for each classification of each disease will mean that the workload of clinical trials is to be calculated in astronomical numbers, and the specific organization and implementation will not be feasible.

In TCM, if the repetition is eliminated, the total number of syndromes contained in several major syndrome systems does not exceed 100. In other words, if a TCM practitioner masters the identification and treatment of these hundred syndromes, in principle, he/she can cope with most human diseases, including the unhealthy state that cannot be diagnosed in modern medicine. The practice of ITCMWM for 50 years tells us that a variety of diseases diagnosed in a patient will ultimately be attributed to a few limited basic syndrome types or their combination. In other words, the syndrome types of different diseases suffered by a patient are often repeated. However, the treatment methods for different diseases corresponding to the same syndrome type are not much different, and the treatment methods for syndromes are not much related to the disease in most cases. If we want to explore all the syndrome types corresponding to each disease, the clinical medicine of TCM developed on the EBM method will be a larger knowledge system than the disease medicine system of modern medicine. Obviously, in the face of the continuous increase in the number of disease classifications in modern medicine, in the long run, the model of 'differentiating diseases in Western medicine and typing in TCM' is not the path that TCM should follow.

6.5.4 Evidence-based treatment based on the TCM syndrome differentiation and treatment system is bound to subvert the existing evidence-based medicine model

From the perspective of the development trajectory of precision medicine, the association between the conventional DCS and the biomarkers that describe the personalized characteristics of patients often appears 'same disease but different type' or 'same type but different disease'. Therefore, not only is the 'umbrella trial' suitable for 'different treatment of the same disease' needed, but also the 'basket trial' suitable for 'different diseases with the same treatment'. This indicates that evidence-based TCM research must not only introduce an 'umbrella trial' suitable for 'differentiating diseases in Western medicine and typing in TCM', but also a 'basket trial' model that focuses on TCM syndromes and is not limited to a certain disease of modern medicine. The introduction of these new clinical trial methods is bound to subvert our current understanding of evidence-based medicine and fundamentally change the concept of evidence-based medicine based on 'randomized controlled trials' as the gold standard.

In recent years, with the development of precision medicine, EBM has paid more and more attention to clinical evidence based on individual differences, which is patient-centered or in the actual complex medical environment. The diagnosis and treatment based on the individualized clinical evidence of the patient is consistent with the TCM treatment based on syndrome in a certain sense. Single-case RCT reflects the treatment effect of individuals with specific clinical characteristics through its own before and after comparison, which is more suitable for reflecting the advantages of individualized treatment. Japanese Kampo medicine introduced the research method of single-case RCT earlier to better evaluate the individualized efficacy of Kampo medicine, which is worthy of reference for evidence-based research in TCM. On the other hand, in recent years, the method of real-world research (RWS) has attracted much attention in EBM research. Real-world research pays more attention to external effectiveness and aims to evaluate the actual efficacy of interventions in the real medical environment. It is more representative of the treatment results in the actual clinical environment, so it is widely used in disease treatment, diagnosis, prognosis, etiological analysis, etc. At present, how to design and implement RWS, how to make full use of multi-channel data, and how to combine RWS with medical big data are all in the exploratory stage around the world (Expert Committee on Implementation Specifications for Real World Research in Clinical Medicine in China 2017). Large-scale cohort studies based on the real world usually have detailed research designs, detailed implementation plans and memo documents before implementation. For routine special work in the implementation process, the research group usually formulates internal operating specifications (IOP) and standard operating procedures (SOP) to ensure the unity of operation. The research design and the formulation of the implementation plan need to absorb valuable references from the traditional RCT trial rules.

However, regardless of whether it is based on traditional RCT clinical trials, the 'basket trial' model that has emerged in recent years, or real-world research, as long as the research is conducted with TCM syndromes as a reference frame, will face problems such as the definition and diagnostic criteria of TCM syndromes. This brings us back to the problem of standardization of syndromes advocated and vigorously promoted by academician Wang Yongyan in the 1980s. In the 1980s, there was no basis for methodological research in the research of syndrome standardization. The structure of syndromes distribution and the quantification of syndrome description were not included in the research scope, and the identification problems that require fuzzy logic and corresponding intelligent identification methods did not have a scientific and technological basis. Since then, 40 years have passed. The methodological thinking of the standardization of syndromes in the TCM community has involved in-depth issues such as the structuring and quantification of syndromes, and has put forward feasible solutions. With the advancement of information processing and artificial intelligence technology, fingerprint recognition, face recognition, and voice recognition based on fuzzy recognition have been technically mature, and the corresponding intelligent recognition system has been successfully applied to people's daily lives. In other words, the realization of the standardization of TCM syndromes currently has a sufficient foundation in terms of methodology research, algorithm logic, and technical means.

Obviously, as personalized medicine, the problems faced by precision medicine moving toward EBM, and the development of evidence-based TCM also have to face it. The evidence-based reference system for precision medicine is no longer the conventional DCS, but a brand-new DCS based on biomarkers. Similarly, the evidence-based TCM cannot be based on the DCS of modern medicine as a reference, but must be based on the TCM syndrome system. As mentioned earlier, whether it is based on the conventional DCS or the personalized DCS being constructed by precision medicine, the current methods of evidence-based medicine cannot solve the overall comprehensive problem of medicine. As for the TCM syndrome system that describes the overall state of the human body from the holistic level, the EBM based on syndrome does not have the integration problem of 'reconstruction on the basis of analysis'. However, if the current situation of evidence-based research in TCM based on the modern medical DCS as the frame of reference is not fundamentally changed, it will inevitably further split the integrity of TCM syndrome system, and degrade TCM to the status of 'abandoning medicine and keeping herbal medicine' of Kampo medicine in Japan. Evidence-based research based on the TCM syndrome system is difficult to continue with traditional 'randomized controlled trials', it should rely more on single-case RCT research methods and statistical analysis based on large population cohorts in the real world. We will discuss this issue further in later chapters.

Like Kampo medicine in Japan, based on evidence-based research using modern medical disease classification as the frame of reference, clinical guidelines and clinical pathways for each disease are formulated, and effective proprietary Chinese medicines are screened for inclusion in the medical insurance list. This method is of positive significance for the use of proprietary Chinese medicines as complementary and alternative treatments in modern medicine, and it is undoubtedly a huge advancement in the field of clinical treatment of modern medicine, which will improve the level of clinical treatment of modern medicine. However, if, with the support of the national level, this evidence-based medicine model is used as the main model for TCM medical institutions to diagnose and treat diseases with TCM methods, the TCM holistic concept and the people-oriented treatment based on syndrome will disappear. In this way, TCM, which is dying after 50 years of 'baptism' of the wave of 'integration of Chinese and Western medicine (ITCMWM)', will face a more severe survival crisis in this new round of 'development' based on the concept of EBM.

Following the evidence is needed for the development of TCM. However, if the core concepts that are more important to TCM are discarded in order to be able to be more rigorous evidence-based: the holistic concept and the treatment based on syndrome, hasn't this turned national support to TCM into a new round of 'strangulation' for TCM?

Twenty years after the beginning of evidence-based TCM research, some visionary scholars in TCM are clearly aware of this problem and have put forward a call for 'creating an individualized evaluation method of EBM in accordance with TCM treatment based on syndrome'. The problem is that the current framework of evidence-based medicine is far from enough to accommodate the overall concept of transcending disease medicine and the individualized principles that are difficult to cover in conventional disease classification. Following the evidence is just one of the core concepts that TCM needs to follow in its future. Should TCM be forced into the framework of EBM to seek development, or should it promote the development itself based on the holistic concept and the principle of individualization while introducing the concept of following evidence? How will TCM, which is moving towards evidence-based, grasp the future direction? Obviously, after the wave of ITCMWM, the development of TCM is facing a new round of major choices.

6.6 Methodological exploration of medical development triggered by the collision of TCM and Western medicine

Research on the methodology of TCM began after Western medicine entered China. Tang Zonghai, one of the representatives of ITCMWM in the late Qing Dynasty, said that Western medicine was 'detailed in the structure and function of the entity but ignored the effect of gasification' in his 'Essence of integrating traditional Chinese and Western Medicine'. Zhu Peiwen, another representative of the integration of TCM and Western medicine in the early days, said that TCM is "excellent at reasoning and speculation, while the observation and description of the structure and function of the entity is poor." In the cognition of the methodology of TCM, some people think that Western medicine emphasizes local anatomy, while TCM emphasizes holistic concepts. Some people think that Western medicine is based on disease differentiation, while TCM is based on syndrome differentiation. Some people think that Western medicine focuses more on external factors, while TCM often focuses on internal factors, and so on.

6.6.1 The profound revelation of the mysteries of TCM from the perspective of cybernetics and system theory

In 1979, Hua Guofan and Jin Guantao (1947) published 'TCM: A Miracle in the History of Science' in the Journal of Dialectics of Nature, which profoundly revealed the most important methodological features of TCM from the perspective of cybernetic black box theory: The basic theory of TCM was established by the method of establishing a black box behavior model by investigating input and output similar to cybernetics. The identification and regulation-control of the human body state in TCM are based on the description of the state of the human body by this model (Hua Guofan and Jin Guantao 1979). In the early 1980s, I was studying TCM at Beijing University of TCM. Inspired by this point of view, I cooperated with Mr. Zhao Xiaojun to conduct systematic methodological research on traditional Chinese and Western medicine using the thinking methods of cybernetics and systems theory. Mr. Zhao Xiaojun (1959) was majoring in automatic control at the University of Science and Technology Beijing at that time. Before entering the university, he had three years of experience in learning TCM in mentoring way and had a certain understanding of TCM. The first draft of our collaborative work 'Modern Methods of TCM' (Yuan Bing and Zhao Zheng 1986) was completed in 1982, which was published by Hubei Science and Technology Press in February 1986. The book was appraised by Mr. Qian Xuesen who vigorously advocated 'human body science' at that time, as "the first work to systematically study TCM using the method of cybernetics" (Yuan Bing 2010).

Based on the consistency of the methods used to build human body models in TCM and systems science to build complex system models, based on the effective methods and successful experience gradually formed in the black box research of cybernetics, in the 'Modern Methods of TCM', we deeply analyzed some problems in the establishment of TCM models, and proposed the modern TCM method that improves the TCM model system based on scientific norms and makes it structured, standardized, and quantitative. The model established based on this method not only conforms to the methods and core characteristics of TCM, but also conforms to modern scientific norms. Based on such a model, system simulation technology can be used to realize the computer simulation of the model, and modern detection methods and treatment technologies can also be easily integrated into TCM and become the diagnosis and treatment methods of TCM.

The method proposed in this book runs through the empirical concept from beginning to end, that is, the theoretical model and SDS of TCM should be completely laid on the basis of empirical evidence. Use the evidence provided in practice to verify, adjust and improve the model, so that it can more accurately and fully reflect the physiological and pathological laws of the human body; through the evidence provided in practice, continuously improve the SDS extended from the model to enable it to identify the pathological state of the human body more accurately, and at the same time make the state-based regulation/control more precise. As a result, TCM will break through the limitations of tradition and be sublimated into a strictly scientific theoretical medicine that conforms to the emerging scientific norms, can directly face the complexity of the human body, and integrates the technology and experience of traditional Chinese and Western medicine. In such a medical system, the holistic concept and the principles of the treatment based on syndrome as the core of TCM will be more perfectly reflected.

'Modern Methods of TCM' is a booklet of less than 100,000 words, divided into two parts. Part I gives the method to realize the scientificization of the theoretical system on the basis of maintaining the core concept of TCM and make it lay on the empirical basis. It includes the establishment and improvement of human body models, the classification of diseases and the art of regulation and control, the standardized methods of syndromes and appearances (including symptoms, signs and detection indicators), the introduction of modern scientific and technological detection methods, and the establishment and improvement of models of treatment methods (CHM, herbal formulas and other treatment methods) that are compatible with human models. It reveals the treatment mechanism and advantages of TCM treatment methods from a methodological point of view. Part II is called 'Computer diagnosis and treatment system for all syndrome domains and all disease domains'. The system abandons the concept of single disease domain based on 'TCM classification within a single Western medicine disease area', and juxtaposes the disease system with the syndrome system to ensure that the holistic concept of TCM and the treatment based on syndrome are perfectly embodied. As a result, a set of 'patient-centered' thinking methods of TCM wisdom diagnosis and treatment that can be used for comprehensive analysis and regulation-control of complex conditions with multiple syndromes and multiple diseases has been formed.

As mentioned earlier, diseases in modern medicine are usually classified and defined based on specific causes, disease locations, pathological changes, and clinical manifestations. Therefore, the detected pathogenic factors, specific pathological changes in specific disease locations, and specific clinical manifestations usually have significant specificity in the diagnosis of the disease. The TCM syndromes are usually introduced based on describing the functional state of the whole body or a certain part. They have a good correspondence with the treatment method, and the correspondence with the symptoms, signs, and detection indicators that can be collected from the patient is usually not specific. This is also an important reason for the difference between accuracy and vagueness (chaos) in the evaluation of curative effects of traditional Chinese and Western medicine mentioned by Wang Yongyan and Huang Luqi. The ambiguity of TCM syndromes is usually manifested in that syndromes are defined by a set of unspecific symptoms and signs, and these symptoms and signs often appear in the set of symptoms/signs that define other syndromes. Therefore, whether a certain syndrome is established or not usually depends on the tendency of multiple symptoms/signs to be superimposed. For this reason, the formulation of TCM syndrome diagnostic criteria is obviously different from the standardization of modern medical disease diagnosis. Not only must qualitative conditions be considered, but the tendency reflected in quantity must also be considered. The boundaries between syndromes are also often blurred. In fact, the exploration of identification and diagnosis methods for syndromes with vague characteristics began in China even before the standardization of syndromes in the early 1980s.

6.6.2 Research on Methodology of standardization of TCM syndrome differentiation and treatment system and on TCM artificial intelligence

In China, the application of artificial intelligence to the field of TCM developed at the same time as the personal computer entered China at the end of the 1970s. In 1976, Shortliffe (Edward Hance Shortliffe 1947) and others of Stanford University successfully developed MYCIN, a medical expert system for identification and treatment of bacterial infections, which opened the way for the application of artificial intelligence expert system technology in the medical field. Inspired by it, China started the research of the TCM expert system in 1978 in order to inherit the experience of famous and old TCM practitioners.

In January 1979, the 'Guan Youbo Computer Program for Diagnosis and Treatment of Liver Diseases' of Beijing Hospital of TCM was successfully developed and put into clinical use. Subsequently, under the vigorous promotion of the national TCM management department, TCM expert systems of various names sprang up like mushrooms, reaching the peak between 1983 and 1989. Nanjing University of TCM developed Zou Yunxiang's computer diagnosis, treatment, teaching, nursing and consultation system for TCM kidney diseases; Xiyuan Hospital of China Academy of TCM Sciences developed Qian Boxuan's menstrual disease diagnosis and treatment expert system; Guangzhou University of TCM developed Luo Yuankai's dysmenorrhea diagnosis and treatment system; the Capital Medical University developed a diagnosis and treatment system for Liang Zonghan's pediatric spleen and stomach diseases (Bai Chunqing 2011). Most of the expert systems developed during this period were single-disease-domain systems confined to a certain disease field. The 'Professor Dong Jianhua Fever Diseases Expert System' of Beijing University of TCM was successfully developed in 1986. The project was presided over by Professor Du Huaitang (1935) from Dongzhimen Hospital Affiliated to Beijing University of TCM, and was completed by me in cooperation with Professor Zhou Pingan (1939–2017), Dr. Jiang Liangduo (1948), and Dr. Hao Ruifu (1962). I am responsible for the medical logic design and software implementation of the project during the research process. In the design of the system, we adopted the thinking method of 'diagnosis and treatment of all syndrome domains and all disease domains', juxtaposed syndromes and diseases, and strove to make the system truly reflect the clinical thinking logic of TCM with syndrome differentiation and treatment as the core. The scope of diagnosis and treatment of this system covers all the syndromes involved in externally induced diseases in TCM. Because it is not restricted by the field of disease, it can even be used for the differentiation and treatment of certain diseases of internal injuries. The 'Computer System for TCM treatment based on Syndrome' developed by the team of Zhu Wenfeng (1941–2009) and Kong Lingren, is similar to our research ideas, and incorporates diseases, syndromes and herbal formulas data, including internal medicine, gynecology, and pediatrics, covering a wider range (Qin Dulie and Bao Yiwan 1989).

At that time, the development of the TCM computer diagnosis and treatment system seemed to have nothing to do with the standardization of syndromes, and the TCM community did not yet have sufficient understanding of the need to introduce scientific methods and mathematical algorithms for standardization. It wasn't until the TCM community encountered problems in the process of introducing evidence-based medicine into TCM in the past ten years that methodological research and the introduction of mathematical algorithms gradually attracted attention. However, the realization of computer diagnosis and treatment, including machine identification of syndromes and diseases, selection of treatment methods, prescriptions, and CHMs, will definitely involve syndrome identification standards, diagnosis algorithms, and optimization rules/algorithms for screening prescriptions and CHMs. In the 1980s, the development of computer diagnosis and treatment intelligent systems basically used the engineering methods commonly used in early artificial intelligence (AI) research: directly imitated the clinical thinking of TCM practitioners for reasoning and algorithm design, and then compiled programs to make the system show intelligent effect.

Objectively speaking, the algorithm designed in the research of artificial intelligence in TCM in the 1980s was quite successful in solving the fuzzy logic problems encountered in the standardization of TCM syndromes. I remember that in 1986, during the process of debugging the syndrome diagnosis rules and parameters of the 'Professor Dong Jianhua fever diseases expert system', seeing that the machine issued a beautiful prescription based on the input medical records, my collaborator Dr. Jiang Liangduo excitedly took it to Professor Dong Jianhua (1918–2001). The old professor was very happy: "Judging from the syndrome differentiation results and the medication habits of the prescription, this prescription must be mine, but the compatibility is so meticulous that even I myself would not be able to reach this level without taking the time to carefully consider it." In fact, due to the improper parameters and rules, many diagnoses and prescriptions made by the machine can be described as nondescript, and sometimes even make people laugh. Obviously, based on the results of medical record iteration and clinical verification, it is indispensable to adjust and modify the parameters and logic rules. However, in the process of adjusting the parameters and rules, it is often found that the adjusted results have increased the coincidence rate for the original inaccurate medical records, and the wrong results have appeared for the originally conforming medical records. This is called 'overfitting' in the field of machine learning modeling. This is a fatal problem that prevailed in the development of TCM artificial intelligence systems in the 1980s.

How to ensure that the modified parameters and logic rules are still suitable for all verified cases, or are statistically the best overall? This is not only a technical problem of how to design a combination of parameters and rules, but more importantly, it is an intelligent algorithm problem of finding the best combination of parameters and rules based on iterative verification and data analysis of a large number of clinical cases. 'Deep learning' is a concept that has only emerged in recent years. It is a technical field that may only emerge after computer computing capabilities have developed to a certain level and artificial intelligence algorithm research has made decisive breakthroughs. This is a problem that was impossible to solve in previous studies, and was not even recognized by people. The TCM expert system and computerized system of the treatment based on syndrome in the 1980s were systems developed based on diagnosis and treatment rules summarized by TCM experts. Since there was no self-learning ability to flexibly adjust these rules and parameters and optimize system performance on the basis of a large number of practices at that time, these systems have always had problems in the accuracy of decision-making and judgment and the adaptability of clinical applications. After a period of clinical trials, they all disappeared.

Today, intelligent recognition technologies such as fingerprint recognition, face recognition, and voice recognition, which have the same ambiguity as TCM syndromes identification, have become mature technologies and are widely used in various fields of society and people's daily lives. With the advancement of artificial intelligence deep learning technology, humans can already adjust and optimize the rules and parameters of intelligent decision-making based on machine learning of large amounts of data. However, without the design of engineering methods, and without unified and standardized diagnostic decision-making standards, it is impossible to generate data suitable for machine learning. Therefore, the organic combination of the engineering algorithm of early artificial intelligence application and the latest machine learning model based on large amounts of data, provides a practical way of thinking and method for the formulation and continuous improvement of the diagnostic criteria of TCM syndromes with fuzzy characteristics (Yuan Bing and Fan Gang 2018).

However, the standardization of TCM syndromes must not only solve the precise judgment of vague things, but also the structure of TCM syndrome distribution and syndrome system. As long as we regard TCM as a science that can be falsified, it means that we can design the syndrome system from the top level based on the structured human model and in accordance with the principles of completeness and independence, and adjust and improve it on the basis of practical verification. As a result, it can be expected to establish a structured syndrome system that conforms to scientific norms.

6.6.3 Re-understanding of TCM from the perspective of complexity science has opened up a new path for the development and scientization of TCM

Methodological research is important, which enables us to see the essence of things through intricate phenomena, and see the path of development that can pass in the future through the layers of fog that are shrouded in front of us. Due to the universal similarities in the concepts and research methods in different fields of natural sciences, the successful methods and experiences applied in the development of relatively mature fields of natural sciences can be easily transplanted to other similar fields to make the development of these fields reduce the exploration process of 'crossing the river by feeling the stones'. In particular, cross-sectional disciplines such as cybernetics, systems theory, and information theory that put aside the individual characteristics of various systems and specialize in the state identification, regulation, and information transmission laws of general systems. Their research results have more universal applicability in various scientific fields.

The method of building models has become a common method to research various objective objects in natural science today, which is commonly used in all fields of natural science, and has even appeared in textbooks about science in elementary and middle schools. Establishing a system model and then introducing a batch of state variables to describe and regulate/control the state of the system is currently a common method used in natural science and its application fields to deal with various 'black box' systems. With the development of science today, scientists have generally realized that almost all objective objects that exist in nature are 'black boxes' that cannot be completely opened. Such a 'black box', even if decomposed to the most 'tiny' and 'simple' elementary particle level, cannot be fully opened due to the limitations of technical means and the principle of 'unpredictable'. Scientists can only infer its internal structure through the investigation of its input and output correspondence, introduce state variables to describe it, and realize adjustment and control to its state by detecting and controlling the state variables.

However, although biology and medicine closely follow the advancement of modern science and technology at the technical level, at the theoretical level, especially the methodological level of studying life, they seem to be far away from natural science, just like the speed difference between wind, horse and cow. It's no wonder that even today we still hear from time to time the remarks from medical people that 'medicine is not science'. This cannot be said to be unreasonable. Perhaps the founders of modern medicine, such as Galen, Harvey, and Bacon, introduced medicine into a wrong path from the beginning. Perhaps in the past few hundred years, biology and medicine have established magnificent 'high-rise buildings' along the path pioneered by these ancestors, and it's hard to return. In hundreds of years of research and development, medical scientists have become accustomed to exploring what can be seen and touched in the living body. Use a microscope if it can't be seen with the naked eye, and use an electronic microscope if it can't be seen with an optical microscope. From organs to tissues, from cells to molecules, biology and medicine have become accustomed to building their own model systems based on visible and tangible entity structures. As everyone knows, through dissection and analysis, what we can see, and touch is only an isolated, static entity that is disconnected from the whole. Even at the micro level, the dynamic relationship between one part and other parts is either invisible to us or too complicated to grasp, which means it is still a 'black box' that cannot be opened. The usual research methods and modeling methods of natural sciences in other fields still have meaning in the face of complex living bodies. It is precisely because medicine and even life sciences have not kept up with the development of natural sciences in methodology that has caused it to fall into today's crisis. This issue will be further explained in the following discussion.

The development of modern medicine with a history of only 400 years is severely disconnected from natural science, and it seems that TCM with a history of two thousand years and contemporary natural science should be further apart. However, as mentioned above, the comparative study of TCM methods and modern scientific methods has profoundly revealed that the method of establishing a theoretical system in TCM is essentially the current method of establishing theoretical models commonly used in natural science; while the essence of TCM treatment based on syndrome is the state description and state regulation-control method based on state variables widely used in modern science. In other words, the development of modern medicine lags far behind the natural sciences in terms of methodology. And under the influence of ancient holistic philosophy, the methods that TCM has unconsciously applied for thousands of years are precisely the methods that are generally recognized in various fields of modern natural science.

The development of cybernetics and system theory, which are cross-disciplines with methodological characteristics, as well as their application practice of various research objects in different fields, have accumulated a lot of successful experience and formed general principles of application. When we investigate and compare the status quo of the application of these methods in TCM based on these experiences and general rules, we will naturally find some problems in TCM. The establishment and improvement of the human body model, etiology model, medication and formula model, the establishment and improvement of the SDS, and the standardization of syndromes, symptoms, and signs proposed in 'Modern Methods of TCM' are the results of exploration based on such an idea. Today, the new biomarker-based disease system to be established in precision medicine and the TCM syndrome system show profound structural similarities; the association between new personalized diseases and traditional diseases in modern medicine, and the association between syndromes and diseases in TCM shows a high degree of consistency. From this, it can be imagined that the combination and association of syndromes, the structural relationship among syndromes, and syndrome-based comprehensive treatment and other similar problems are also faced by the SDS based on biomarkers in the development process. However, dealing with these problems at the micro level will face an unimaginable explosion of knowledge, which will be unbearable for future science and mankind. This determines that the path pioneered by precision medicine will not become the mainstream of medicine integration and individualization, and it also determines that precision medicine and systems biology must be a 'short-lived flower-like epiphany' in the long history of scientific development.

In 2016, Professor Zhao Zhongzhen (1957), a well-known appraisal expert of CHM in Hong Kong in his 20-episode CHM knowledge popularization program 'Zhongzhen says Materia Medica', through the image metaphor of the relationship between ingredients and Chinese and Western food, explained the problem of how to define CHM and Western medication in a simple way. He said: "Whether a medication is a CHM or a Western medication is not determined by the medication itself. A medication that can be guided by the theories of TCM is a CHM." The implication is that a medication that can be guided for application by modern medical theories is Western medication. And if a medication can be applied under the guidance of these two theories respectively, it is both a CHM and a Western medication. Regarding this issue, the book 'Modern Methods of TCM' stated this way: "The two different treatment methods of traditional Chinese and Western medicine are formed on the basis of adapting to the human body theory of the two major medical systems of traditional Chinese and Western medicine." The reason why they are divided into two major categories, and each is only dominated by the corresponding system, is because they are developed for the purpose of disease treatment in different systems. Each system always focuses on seeking effective treatment methods for the state variables of its own system, and in the process, the therapeutic actions of these methods relative to the human body model of its own system were learned. "In this sense, any physical, chemical, biological means or process (including those belonging to modern medicine), in principle, can be used as therapeutic means of TCM as long as it can improve the state deviation (syndrome or disease) relative to the human body model of TCM" (Yuan Bing and Zhao Zheng 1986).

In November 2000, the article 'Establishing a Research Method of CHM in Accordance with the Characteristics of TCM' made a clearer statement: "Whether a medication is a CHM or a Western medicine is not determined by the characteristics of the medication itself. The important thing is whether its properties and effects are expressed in terms of TCM theory or Western medicine theory. If its properties and therapeutic effects are expressed in the theory of TCM, it can be used in TCM, so it is a CHM; otherwise, it is a Western medication. And if we can use two theoretical terms to describe its properties and therapeutic effects, it is both a CHM and a Western medication. That is to say, the existing Western medications can be incorporated into the theoretical system of TCM as long as they have completed the description of their characteristics and therapeutic actions in the theoretical terms of TCM" (Yuan Bing 2000). An example is the gypsum aspirin soup created by the famous Hebei doctor Zhang Xichun in the period of the Republic of China, which uses the action of aspirin in inducing sweat and dispelling exogenous evils to treat exogenous syndromes. Here, aspirin had been actually a CHM.

Regarding the principles of defining CHM and Western medication, Professor Zhao Zhongzhen summarized the conclusions based on his in-depth understanding of the CHM and Western medication system. And our understanding of CHMs and Western medications 30 years ago is far from that of Professor Zhao Zhongzhen, who is very knowledgeable today. Our conclusions can only be derived from methodological reasoning based on the prevailing rules of model methods. The coincidence of the conclusions drawn by the two channels has profoundly explained the forward-looking significance of methodological research. There is an old Chinese saying: 'Stones from other hills may serve to polish the jade of this one'. Due to the various similarities among the objective object structures and functions in different research fields in nature, the methods and principles used in different research fields of natural sciences can be referred to each other. If we regard TCM as a science, then the establishment of models in other fields of natural sciences, the methods used to standardize the knowledge system, and experience will undoubtedly provide useful enlightenment and reference for the scientificization of TCM theoretical models, and standardization of the TCM syndrome/symptom system.

Development means change. When the development of TCM faces the need to make some changes to this legacy left by our ancestors, the first question we face is: as a legacy left by our ancestors, can TCM change? If certain changes cannot be made based on scientific principles and practical tests, how can TCM be called science? If it can be changed, then is the new body of knowledge after such changes still TCM or not?

The earliest surviving medical work 'Huangdi Neijing, The Yellow Emperor's Inner Canon' in China is regarded as the 'classic' of TCM. Because it is not clear what TCM is in the long historical process, every step in the development of TCM, medical scholars always find the 'source' from this classic, and then expand on the basis of the 'classics'. If there is no classical basis, the development and changes made will hardly be recognized by other scholars. The development on the basis of 'classics' had been the habit of this discipline, which continues to this day. This view of development is simply unacceptable from the perspective of the history of science. Just imagine, after the development of physics entered the scientific age, whether Newton, Einstein or Planck (Max Planck 1858–1947), when establishing their own new theoretical system, who would go to Aristotle's era to 'cite scriptures' to find the source?

Modern medicine, in its development of recent decades, has continuously gained inspiration and reference from the knowledge and treatment experience of TCM, and then used its own methods and techniques to explore it, and incorporate the research results into its own theories and application system. The learning/borrowing from TCM did not change the reality that modern medicine is still modern medicine, on the contrary, modern medicine has been continuously developed and strengthened. Today, methodological research profoundly reveals the scientific nature of TCM's method of understanding the human body and diseases and the theoretical framework, and reveals the effectiveness of TCM's thinking and methods of regulating/controlling the human body based on the state of the body. When we regard the theory of TCM as the same theoretical model as today's scientific theories, and after we understand these essential characteristics of TCM, we have to ask why TCM can't introduce some of the achievements of modern science and modern medicine to develop and improve itself? Could it be that by introducing some modern science and modern medical technology and methods, TCM is no longer TCM?

When we regard TCM as a science, and grasp the core of it as a subject, jump out of the fetters of ideas, develop and improve the TCM theory from a new perspective based on scientific concepts and empirical principles, the previous obstacles disappear and everything becomes simpler: The standardization and structuring of TCM theoretical models, syndrome systems, symptoms, and signs have become technical issues of how to carry out structured design from top to bottom.

Chapter 7

The Medical Revolution Triggered by Precision Medicine and Its Scientific Limitations

In recent years, with the progress of the Human Genome Project, as the relevance of gene mutations and human diseases is increasingly revealed, scientists have discovered that by introducing biomarkers related to genetic mutations, the selectivity of the scope of application of certain medications can be more accurately defined. In 2011, the National Academy of Sciences, the National Academy of Engineering, the National Institutes of Health, and the National Council of Science jointly launched the 'Towards Precision Medicine' initiative, and the National Research Council report, 'Towards Precision Medicine', which was drafted by Dr. Maynard V. Olson (1943), a well-known genomics scientist, was officially published at the same time. This report puts forward an epoch-making recommendation: through genetic association research and close integration with clinical medicine, precise treatment and effective early warning of human diseases can be achieved.

On January 20, 2015, President Obama (Barack Hussein Obama, 1961) proposed the 'Precision Medicine Initiative' (PMI) in his State of the Union address, and called on the United States to increase medical research funding, promote individualized genomics research, and lead medicine into a new era. The realization of precision medicine will enable human beings to formulate individual medical plans for patients of cancer and other diseases based on personal information, thereby standardizing gene-matched cancer therapy like blood transfusion matching blood type. This would make finding the correct medication and dose as simple as taking a temperature, giving the right person the right treatment at the right time. On January 30, President Obama formally approved the 'Precision Medicine Initiative' and proposed that Congress invest $215 million in the plan in fiscal year 2016 to promote the development of personalized medicine. As a result, precision medicine entered the public's field of vision and quickly became the focus of worldwide attention (Terry 2015).

7.1 The core characteristics and practical significance of precision medicine

Personalized medicine is a medical model that formulates treatment plans based on the individual characteristics of patients. Obviously, for a medicine to be called personalized medicine, it must have a description system that can personally describe the characteristics of patients' diseases. Traditional Chinese Medicine (TCM) is a personalized medicine, because its treatment system based on syndrome is a state description and regulation-control system that can individually describe and regulate human diseases.

The definition of precision medicine in the US 'Precision Medicine' plan is to formulate a personalized treatment plan based on the individual characteristics of each patient. Obviously, the 'precision' here is not limited to our conventional understanding: the precision of disease diagnosis, the precision of understanding and grasping the actions of medications and treatment methods, and the precision of selecting treatment options for diseases. The precision of precision medicine includes the 'precise' grasp of the individual characteristics of the human body and diseases. In other words, the reference frame of the precision is different from the past, and it is no longer the DCS that we are familiar with. It is necessary to introduce a description system that is independent of the conventional DCS and can personally describe the characteristics of human diseases, the biomarker system. As a result, personalized medicine, a medical model that was only a concept before, began to have practical significance in modern medicine (Yuan Bing 2016a).

7.1.1 Proposition of the concept of precision medicine and its original purpose

The full name of the strategic research report proposed by the National Research Council in 2011 is 'Towards Precision Medicine: Building a Knowledge Network for Biomedical Research and New Method of Disease Classifications'. The expert committee responsible for writing this report is called the 'Committee on a Framework for Developing a New Taxonomy of Disease'. The summary of the report clearly states: "The committee's mission is to explore the feasibility and use of establishing a new standard for the classification of human diseases based on molecular biology, and to construct a possible framework for this" (National Research Council 2011). This more than one-hundred-page report uses 6 sub-headings to summarize in the abstract:

1. The new classification standard will lead to better health care;
2. The time is right to modernize the disease classification standards;
3. New disease classification standards should keep pace with the times;
4. A knowledge network about diseases will enable the realization of new classification standards;
5. A new model based on group research will promote the development of disease knowledge networks and new classification standards;
6. The relocation of resources will help the development of disease knowledge networks.

From the title to the content of this report, it always revolves around such a strategic goal: to build a new standard for the international classification of diseases based on molecular biology. The report pointed out: Once this new standard for disease classification is established, it will be possible to achieve precision medicine that can make treatment plans based on the individual characteristics of patients (National Research Council 2011).

In the view of the author of the 'Towards Precision Medicine' report, the establishment of this new classification standard, whether it is used to refine or improve the ICD classification standard, or to develop into a new classification standard that coexists with ICD, will "significantly improve the quality and quantity of various information used in biomedicine, so as to better discover pathogenic mechanisms, improve disease classification and improve medical care" (National Research Council 2011).

The report not only defines the holistic idea of 'building a disease knowledge network' to achieve its strategic goal of 'new standards for disease classification', but also proposes corresponding strategic measures: that is, using the current emerging data-intensive biology such as omics to implement its strategic thinking through the integration and sharing strategy of big data. "The establishment of a knowledge network and its application in research and clinical practice depend on the availability of large-scale databases; these databases fully integrate various knowledge of human diseases and are organized in a hierarchical form. These databases will lay the foundation for new classification standards" (National Research Council 2011).

In the view of the author of the report, the core of this biological database is to build an 'individual-centric' data sharing platform that integrates various physiological and pathological data of individuals from molecules to phenotypes to construct individual disease knowledge networks.

This 'individual-centric' biological database is different from traditional biological databases, which usually collect data of the same type from tens of thousands of individuals according to their specific data types; and different data types from the same person are usually allocated to the corresponding database. In a sense, traditional biological databases and evidence-based medicine are based on population samples and statistics, and are typical non-individualized research models. The strategic goal of precision medicine is to achieve individualized health maintenance and individual precise diagnosis and treatment. Obviously, it cannot rely on traditional databases. Therefore, the development of an 'individual-centered' biological database will inevitably become the core task of achieving the strategic goals of precision medicine (National Research Council 2011).

What the Towards Precision Medicine report plans is not a short-term behavior, but a long-term task that lasts for decades. "The development of an information sharing platform, disease knowledge network, and a new classification standard requires long-term consideration. In a sense, this challenge is the same as building a European cathedral: One generation started this work, and the next generation can complete it." More importantly, once this strategic goal is achieved, the impact on health medicine will be long-term. It will completely change the current classification of diseases and make it possible to realize precision medicine characterized by individualized diagnosis and treatment. As a result, "not only can the current biomedical research capabilities be raised to a new level, but in the future, it will bring incalculable improvements to the level of clinical medicine" (National Research Council 2011).

The strategic effect focused on by the Towards Precision Medicine report is not to cure specific diseases or develop certain technologies, and is not even limited to the medical and health field. The author of the report believes that this new classification standard based on the disease knowledge network can not only promote evidence-based medicine based on large-sample statistical research to the direction of individualized precision medicine based on biomedical big data, but can also significantly improve the biomedical research on the physiological and pathological regulation-control mechanisms of the human body. The author of the report further emphasized: "The meaning of these views and recommendations put forward by this committee has gone far beyond the scope of disease classification, and has a great impact on almost all companies and their stakeholders engaged in biomedical research and healthcare" (National Research Council 2011).

The precision medicine plan based on this report takes tumor treatment as a short-term goal, and its long-term goal is to expand its application to many fields such as health and health care. In September 2015, the US NIH officially released the implementation plan of the 'Precision Medicine Initiative Cohort Program' (PMI-CP) based on this plan. PMI-CP "will establish one million or more American volunteer research cohorts to constitute a research platform to expand our knowledge of precision medicine and benefit the American people many years later" (PMI Working Group 2015). Francis Collins, director of the NIH, said when interpreting the precision medicine plan that he envisioned a large database of one million or more volunteers participating in this research. "Participants will be required to collect the following biological specimen data, such as cell numbers, proteins, metabolites, RNA and DNA, as well as whole-genome sequencing, behavioral data, and their electronic health records." One year after this implementation plan was released, NIH changed the name of the 'Precision Medicine Pilot Cohort Project' to the 'All of Us Research Program'.

Whether it is from the Towards Precision Medicine report, or the Precision Medicine Plan formulated based on this report, it can be seen that American scientists want to count as widely as possible the correlation between various measurable indicators of the human body and the occurrence of various diseases through the 'big data' analysis of a large number of samples. As a result, a description system that can reflect the individualized disease characteristics of the human body will be established, which will lay the foundation for a more in-depth, accurate, and comprehensive reflection of the individualized characteristics of the disease process (Yuan Bing 2018b).

7.1.2 Changes in clinical trials and efficacy evaluation systems brought about by precision medicine

The rise of precision medicine has initiated the transformation of mainstream medicine from disease medicine to personalized medicine, and it is also difficult to avoid collisions and conflicts with the clinical evaluation system based on disease medicine. In 2014, the American Association for Cancer Research (AACR) proposed that innovative clinical trials for precision cancer medicine can be divided into two categories. One type is called 'Basket Trial'. Specifically, a certain medication with a clear target is a basket. The basket trial is to put different cancers with the same target gene into a basket for research, so as to promote research and development of medications against tumors of different origins with the same target gene. The second type of clinical trial is called 'Umbrella Trial', which is to gather the same type of cancer with different driver genes under the same umbrella to promote medication development for tumors of the same origin with different target genes. For example, put NSCLC patients with different driving genes *KRAS*, *EGFR*, *ALK* together, and then give different targeted medications according to different target genes. The biggest advantage of the Umbrella trial is that it concentrates very rare mutation events and turns rare events into 'common' events. This is of special significance for accelerating clinical trials for rare diseases or for an individual's chance of obtaining precise treatment (Wu Yilong 2015). The opening of these two types of clinical trials means that cancer patients in the future will be able to use effective treatment medications faster, instead of the long waiting time of 7 to 10 years as in the past.

In May 2017, at a milestone, the FDA approved for the first time the indications of anti-tumor therapies that are not based on tumor source, but based on biomarkers. KEYTRUDA (pembrolizumab) developed by Merck & Co. (MSD) was approved to treat patients with solid tumors with high microsatellite instability-high (MSI-H) or mismatch repair deficient (dMMR). Previously, the FDA had been approving cancer therapies based on where the cancer originated in the body (such as lung cancer or breast cancer) (U.S. Food and Drug Administration 2017). This time the medication was approved for more types of solid tumors, and its indications were based on two 'biomarkers'. In other words, as long as the patient's tumor carries one of these two biomarkers, the medication can be used for treatment no matter which part of the body is suffering from a solid tumor.

MSI-H and dMMR are two common genetic abnormalities. In tumors with these two mutations, the DNA repair mechanism in the cell is often affected and cannot function normally. Tumors with these abnormalities are widely distributed and can appear in multiple locations such as the colorectal, endometrium, gastrointestinal tract, breast, prostate, bladder, and thyroid, etc. Therefore, distinguishing these cancers by the characteristics of genetic variation rather than the location of the disease has a better guiding significance for the treatment (U.S. Food and Drug Administration 2017). This achievement was approved precisely based on the 'basket trial' for the medication.

In November 2018, the FDA approved the results of the first 'basket trial': involving the small molecule inhibitor Larotrectinib carrying tropomyosin receptor kinase (*TRK*) gene fusion mutations for patients of adults and children with solid tumors who suffered from with *NTRK* gene fusion and had previously failed treatments (U.S. Food and Drug Administration 2018). According to Dr. David Hyman, the global research leader of a clinical trial of Larotrectinib, "*NTRK* gene fusion is a rare cancer driver, and the FDA's approval of Larotrectinib for the treatment of such tumors is an important milestone." It is "the first approved therapeutic medication which targets this genetic change and has nothing to do with the type of tumor" (Bayer 2018).

Larotrectinib had an overall effective rate of 75% in all 55 patients included in clinical trials, a complete remission rate of 22%, and a partial remission rate of 53%. These patients had 13 different types of solid tumors, including lung cancer, thyroid cancer, and melanoma, GIST, colon cancer, soft tissue sarcoma, salivary gland tumors and infant fibrosarcoma, but all had a common target molecule-TRK gene fusion (U.S. Food and Drug Administration 2018).

The concept of precision medicine has greatly promoted the progress of tumor treatment, discovered more effective treatment methods, reduced unnecessary drug side effects, and significantly improved patients' survival rates and quality of life. Some success stories are:

1. *Treatment of chronic myeloid leukemia (CML)*: Chronic myeloid leukemia is a blood cancer for which previous treatments have had limited effectiveness. However, through precision medicine methods, researchers have discovered that the *BCR-ABL* fusion gene is the main causative factor that causes abnormal leukemia cells to proliferate. Subsequently, scientists developed a class of drugs called tyrosine kinase inhibitors, such as Imatinib. It can precisely interfere with the activity of the BCR-ABL gene and significantly reduce the number of leukemia cells in CML patients, thereby effectively controlling the disease (Druker et al. 2001, Hochhaus et al. 2017).
2. *Treatment of HER2-positive breast cancer*: HER2 (human epidermal growth factor receptor 2)-positive breast cancer is a type of cancer that depends on the HER2 protein (Slamon et al. 1987). By testing whether a patient is HER2-positive, doctors can develop a more precise treatment plan for the patient, such as using specific HER2-targeted therapies such as Herceptin (Baselga et al. 1999), Pembrolizumab (Perjeta), and Tenipex enzyme (T-DM1). They can target the HER2 receptor and inhibit its overactivity, slowing or stopping the growth of breast cancer. These medications are now often used in combination with chemotherapy or other medications, thereby significantly improving the effectiveness of treatment.
3. *Immunotherapy for lung cancer*: In immunotherapy, by analyzing the tumor tissue of lung cancer patients, doctors can determine whether there are targets that can be attacked by the immune system, such as the PD-L1 protein. They could then choose to use immune checkpoint inhibitors, such as PD-1 inhibitors, to boost the patient's immune response and thereby improve treatment effectiveness. This method has shown good results in improving the survival time and quality of life of patients with advanced lung cancer, especially those who have progressed after chemotherapy (Brahmer et al. 2012, Topalian et al. 2012).

On July 15, 2020, 'Nature' magazine announced the results of the largest lung cancer 'Umbrella Test' to date, which is the result of a recent study by the National Lung Matrix Trial (NLMT) in the United Kingdom. In this trial, researchers screened 302 patients matching the genetic variant with the experimental therapy from 5467 lung cancer patients, divided them into 22 different treatment groups, and used medications that specifically block the gene or other important proteins in the pathway. The result was that with the exception of EGFR, PD-L1, ALK, and RET medications that were already on the market before the publication of this trial, and MET inhibitors produced a 75% response rate, most medications only had a low response rate (< 30%) and a short response time (< 1 year). Some subtypes were terminated early because the number of volunteers recruited was too small (Middleton et al. 2020).

The total number of therapeutically valuable targets discovered so far does not exceed 1,000, and the success rate of medication development for new tumor targets is still in the single digits. In light of this, it is not surprising that only four of the 22 treatment groups had a higher response rate. After all, finding a gene variant that carries cancer is a completely different concept from finding a medication that responds to that gene. Moreover, studies have shown that tumors are usually not caused by a single gene mutation, and there are also multiple studies showing that normal cells within tumor tissues also carry a large number of gene mutations. This shows that mutations must accumulate to a certain extent before they become tumors, and finding a targeted medication for one of the targets may not necessarily destroy the tumor. The overall response rate of the basket trials that have been conducted so far is also low (Middleton et al. 2020).

It was not easy to select 302 patients who matched the genetic variant with the experimental therapy from the 5467 lung cancer patients. However, 302 patients were divided into 22 different treatment groups to conduct randomized controlled trials, with an average of less than 14 people in each treatment group. For the treatment group with a small number of people, what can the results of the trial explain? The reform direction of clinical trials and efficacy evaluations such as basket trials and umbrella trials are unquestionable. However, due to the large number of treatment groups that need to be set up due to the excessive number of personalized differential typing, the organization of randomized controlled trials has become more and more difficult, and the cost is increasing. As a result, the National Health Research Project uses real-world cohort studies to replace randomized controlled trials, demonstrating the trend of changing the mainstream clinical research model of personalized medicine.

The introduction of new biomarker indicators can undoubtedly differentiate diseases further based on the individual characteristics of patients. In other words, according to the differences in the biomarkers of patients suffering from the same disease, the disease can be divided into different personalized types. However, clinical studies based on the experimental model of the basket test and the umbrella test show that biomarker abnormalities discovered to refine one disease classification often appear in other diseases as well. As mentioned earlier, abnormalities of MSI-H and dMMR can occur in tumors in multiple locations such as the colorectal, endometrium, gastrointestinal tract, breast, prostate, bladder, and thyroid. In other words, the biomarker system developed by precision medicine has a many-to-many network relationship with the conventional disease classification system (DCS), which is similar to the relationship between syndromes and diseases in TCM. Therefore, the significance of the biomarkers which used in precision medicine to calibrate the target is not limited to the more detailed classification of modern medical diseases. As conceived in the 'Towards Precision Medicine' report, with the advancement of precision medicine research, and new targets constantly being discovered, a new DCS that is different from traditional one will appear under the modern medical framework. This is a target system defined by biomarkers, and it will be an independent system alongside conventional DCS. On this basis, a targeted medication system that specifically targets these biomarker abnormalities rather than diseases will also be formed. As a result, modern medicine will enter the era of combining disease treatment and 'target' treatment.

The new DCS to be established by precision medicine emphasizes the difference from the conventional DCS, and focuses more on describing the patient's personalized disease information. This system distinguishes diseases more accurately from a different perspective from conventional disease classification, and will naturally lead to the discovery of targeted medications with abnormal personalized indicators. As a result, modern medicine will enter the era of 'same disease with different treatment' and 'different disease at the same time' like TCM. Does this mean that modern medicine began to establish its SDS, and began to establish state medicine based on state description and state regulation-control under the modern medical framework?

7.2 State medicine: A medical model based on state description and state regulation-control

Introducing a group of state variables to describe the state of the system, and changing the state of the system by changing the state variables, is a commonly used method for regulation-controlling application systems in various fields related to science today. Let's take the description of the state in a closed container as an example: introducing three variables of temperature (T), humidity (W) and pressure (P), these three state variables constitute the SDS of this closed container. Assuming that the temperature can take an integer value in the range between –99°C to 100°C, the humidity is expressed by relative humidity and can take an integer value in the range of 1% to 100%, the pressure can take an integer value in the range of 1 mPa to 100 mPa. Therefore, a set of values composed of Ti, Wi, Pi represents the state of the system Si at time i.

Si = (Ti, Wi, Pi) [Among them: Ti = –99 ~ 100; Wi = 1 ~ 100; Pi = 1 ~ 100]

How to control the state of this system? We have methods for changing the pressure: blowing and exhausting; there are methods for changing the temperature: heating and cooling; there are methods for changing the humidity: humidification and dehumidification. Obviously, the use of these methods changed the values of state variables, which also changes the state of the system accordingly. The regulation-control of application systems based on natural science, such as the regulation-control of common greenhouses, the regulation-control of soil composition, the regulation-control of steelmaking blast furnaces, etc., all adopt this method. The regulation-control of social systems and economic systems also usually adopts this method.

Therefore, we can define state medicine as follows: describe the health state of the human body by introducing a set of state variables, and realize the regulation-control of the health state of the human body through the regulation-control of state variables.

Let's look at the TCM treatment system based on syndrome:

If we regard TCM syndromes, such as qi deficiency, blood deficiency, yin deficiency, yang deficiency, qi stagnation, blood stasis, and damp-heat as state variables, then the state of the human body at a certain moment can be represented by this set of state variables:

deficiency of qi (2), deficiency of blood (0), deficiency of yin (0), deficiency of yang (0), stagnation of qi (2), blood stasis (1) ,...

[Among them, the number in brackets after the syndrome represents the degree of the syndrome.]

If we introduce 100 state variables (syndromes) to describe the health state of the human body completely, the state of the human body can be represented by 1 point in the 100-dimensional state space:

$S = x_1, x_2, x_3, \ldots\ldots, x_n$ *Among them,* $n = 1,2,3,\ldots\ldots 100$

Regulation-controlling the pathological state of patients, TCM is also achieved by influencing abnormal state variables (syndromes). For example, qi deficiency needs to replenish qi, blood deficiency will replenish blood, qi stagnation needs to promote qi, and blood stasis will promote blood circulation. Obviously, the TCM treatment system based on syndrome is a state medicine system. From a methodological point of view, the regulation of human diseases by TCM is consistent with the methods currently generally adopted by natural sciences/social sciences on the regulation-control of applied systems.

With the SDS, it is possible to uniquely determine the individualized state of the body in the course of the disease. For a disease with a single syndrome, we can approximate it as a state where the corresponding state variable is abnormal, and other state variables are normal (or the degree of abnormality can be ignored). For complex diseases with multiple abnormalities in multiple parts and syndromes, the individualized state can be uniquely combined by abnormal state variables, and the regulation-control scheme for individualized states can also be achieved by combining regulation-control methods for abnormal state variables.

In modern science, there are principles to follow for introducing state variables to establish a SDS. Generally, the choice of state variables varies from system to system, but the same system can also be described by different state variable groups. That is to say, the choice of state variables has a certain degree of freedom. To describe the patient's personalized state comprehensively and accurately, there are also requirements for the introduced state variables and their mutual structural relationships. To build an SDS, the selection of state variables should generally meet the following requirements:

1. The selected state variables must have specific meanings, which can characterize some basic characteristics and behaviors of the system, and therefore have practical significance in guiding the regulation-control of system states. The introduction of meaningless state variables will only increase the complexity of the system in vain and increase the difficulty of identification.

2. The set of introduced state variables should be complete, that is, the number of state variables must be large enough to be able to completely describe the state of the system. If the characteristic or functional state of a certain aspect of the system does not introduce corresponding state variables to characterize it, the corresponding state cannot be identified, let alone its regulation and control.
3. The introduced state variables should be independent, which means that any state variable cannot be expressed as a function of other state variables. In other words, the introduced state variables cannot have redundancy such as repetition and partial inclusion. Otherwise, the structure of the state variable system will be confused.
4. The most simple and applicable, that is, under the premise of meeting the accuracy required for system regulation-control, the number of state variables should be as few as possible, and the correlation between them should be as simple as possible. In this sense, not all symptoms, signs, and test indicators that we can find must be included in this system. On the premise that the values (syndrome) of the system state variables can be accurately distinguished, the fewer symptoms/signs/detection indicators introduced, and the simpler the collection and detection, the better.

The state description system (SDS) requires 'enough' state variables to ensure the integrity of the SDS, but there is no absolute correlation between the integrity and the number of state variables. That is to say, the large number of state variables introduced does not mean that the state variable system must be complete. On the contrary, the small number of introduced state variables does not mean that the state variable system must be incomplete. For describing the state of the human body, completeness means that the functions and behaviors of any aspect of any part of the human body are characterized by corresponding state variables. The treatment system based on syndrome used for state description and regulation-control in TCM is constructed by introducing state variables on the basis of functional models that fully describe human physiological and pathological activities in ancient China. The independence of state variables means that none of the state variables can be the result of the combination of other state variables. A system whose state variables are not independent will inevitably have redundancy, overlap and structural confusion, and will not have rigorous logic. Based on such a system, it is impossible to achieve a unique expression of personalized status (Yuan Bing 2017b).

To ensure the integrity and independence of the SDS, in the process of establishing the SDS, people usually introduce state variables based on a model that can completely describe the behavior and function of the system. Only when the selected state variables are independent can a state description conforming to scientific norms be established, and the unique expression of the personalized state of the system can be realized. The rules and methods for establishing the SDS belong to methods of a technical layer in modern science. When the application of this method is raised to the methodological level involving the development of a discipline, and the concepts of completeness and independence are introduced into the screening of state variables and the construction of state description systems, its impact on the development of medicine, a discipline not considered strictly a science, would be extremely huge.

While modern medicine with disease as its core is expanding into new disease categories, it is also using disease as a reference system to continuously explore new detection methods and more effective treatment technologies, establish more accurate disease classification, diagnosis, and treatment rules, and conducts more in-depth research on the occurrence and development of diseases. After more than 400 years of development, it has established its huge clinical medical system today. State medicine describes diseases based on the state of the human body as a frame of reference. Similarly, the detection methods, diagnosis rules and treatment methods for the state, the correlation among the various state variables and the evolution law of the human state in the disease process, constitute indispensable parts of state medicine. State medicine uses state variables to personally describe the health of the human body, and this is precisely the way that modern science usually uses to personalize the description of research objects.

To realize regulation and control based on states in state medicine, it is indispensable for the study of medications and treatment methods for various pathological states. Schwaederle et al.'s genetic study on cancer pointed out that 90% of patients carry at least one genetic mutation which is the target of an approved medication or those still in clinical trial. This finding is not an isolated case, and it exists in many different malignant tumors. Therefore, for most patients, genetic testing of cancer patients can provide medical practitioners and patients with at least one treatment plan (Schwaederle et al. 2015). Obviously, since disease medicine cannot further distinguish the selectivity of medications for the same disease, a clear disease diagnosis cannot uniquely determine a medication treatment plan. However, the indiscriminate application of medications only based on disease diagnosis that cannot reflect the individual characteristics of patients, will greatly reduce the clinical effectiveness of these medications. With the advancement of genetic medicine research, people gradually realized that the unsatisfactory clinical effects of these medications cannot be completely attributed to the medications themselves, to a large extent due to insufficient identification and differentiation of the individualized status applicable to the medication.

Obviously, the research of state medicine on the actions of medications and treatment methods must be based on the corresponding SDS, and the clinical selection of medications (treatment methods) must also be based on their therapeutic actions for the human personalized state. That is to say, with the emergence of state medicine, the clinical pharmacology system not only has a medication system for diseases, but also has a medication system for various personalized states of the body.

Due to the gap between disciplines, the medical profession has been emphasizing the particularity of medicine. In the eyes of many medical scientists, it seems that the methods and concepts that are universal in other fields of natural science are not applicable to medicine. In the 400 years since the rise of modern medicine, the disciplinary framework and applied methodology of medicine have been following the traditions of the Renaissance era. Today, with the emergence of the need for personalized state description of the human body brought about by the rise of precision medicine, the limitations of the traditional single-factor analysis and disease description as the core methods have become increasingly clear, which has triggered a disruptive change in the modern medicine system. Whether the process of establishing a personalized DCS initiated by precision medicine will eventually lead to the birth of modern medicine's state description and regulation-control system is still uncertain. However, the implementation of the concept of individualization indicates that modern medicine is on the right path to finally establish an individualized state description.

7.3 Problems that the personalized disease classification system (DCS) will face when it moves toward the state medicine system

Although the disease medicine of modern medicine keeps pace with the development of modern science and technology at the technical level, its ideas and methods of disease treatment are far behind modern science. Its description of the condition of the diseased body is usually on the basis of classification, centered on the disease. The diagnosis and treatment of many different diseases of a patient as a whole is usually completed by clinicians in different departments separately. So far, modern medicine does not have the concept of state description, nor has it formed a regulation-control method system that can effectively deal with various abnormal state variables. From this point of view, it is not unreasonable that many natural scientists believe that medicine is not a science.

7.3.1 The introduced biomarkers need to form an organized structure according to certain rules

Precision medicine introduces biomarkers that can describe the personalized characteristics of patients, and begins to build a new DCS, which is undoubtedly a solid step towards state medicine that can individually describe and regulate/control the state of the body. However, can this new DCS eventually lead to the state medicine that achieves a comprehensive description and regulation-control of human health state? At present, the research of precision medicine has just started, and the characteristics of the state description system (SDS) such as the completeness and independence of state variables and the principle of 'the simplest and applicable' have not yet entered the view of medical scientists. With the increase in the number of biomarkers and target medications discovered and the relationship among different biomarkers revealed, they will be unavoidable problems that include the independence of the introduced biomarkers, the completeness of the biomarker system in describing the personalized state of the human body, and the structuring of this new DCS. What does it mean for modern medicine to solve these problems?

The progress of biology and medicine includes the deepening of understanding along the whole→system→organ and tissue→cell→molecular level, and there is also the expansion of knowledge in the same level along the breadth direction. The latter includes the discovery of previously unknown functional characteristics at a certain level of the human body and the understanding of the law of disease development and transmission in the human body; it also includes, for example, the research expansion in the field of the proteome, metabolome, etc., at the molecular level, which was not well understood before this. This is especially true when the research goes deep into the internal organs, tissues, cells, and molecular levels of the human body that our naked eyes cannot see completely and our senses cannot fully perceive, as well as the dynamic processes involving the interrelation of various parts and elements of the human body. For this reason, we can only introduce a set of state variables that are as complete as possible based on the existing knowledge of the structure and function of the human body at a certain level. And with the discovery of new functional characteristics of the human body at this level, it is perfected and supplemented by introducing new state variables.

With the deepening of research, it is possible to gradually improve the completeness of the SDS by continuously introducing new state variables. To ensure the independence of state variables requires the introduction of relatively independent state variables based on their relevance when constructing the SDS. And in the process of adding new state variables to expand the SDS, based on the correlation of state variables, it is necessary to arrange the structure of SDS and remove its existing redundancy. Using a real-world research method, when the scope of the research includes the existing state variables of the SDS and the new state variables to be added, the correlation data among the state variables can be obtained through statistical-based correlation analysis. On this basis, it is technically possible to remove redundancy and arrange the structure of the SDS. However, these works are far beyond the capabilities of the methods of randomized controlled clinical trials, and must be achieved after real-world research have become the mainstream method of clinical trials. On the basis of ensuring the independence of state variables, with the introduction of new state variables, the completeness of the SDS will be continuously improved, so that it increasingly tends to satisfy the principle of 'the simplest and applicable'.

TCM introduces state description from the holistic level, and at present, only about 100 state variables (syndrome) can already describe the basic physiological and pathological activities of the human body and guide the effective overall regulation-control of various human diseases. Modern medicine introduces state variables to establish state description from the molecular level, which will inevitably greatly increase the complexity of the SDS and the complexity of the corresponding regulation-control. And, due to the non-single nature of the actions of modern medical regulation-control methods (such as medications), it will inevitably produce a variety of side effects that are unbearable for humans. These are all important issues to be considered to determine whether this medical model can ultimately be accepted by science and mankind.

7.3.2 The complexity problem posed by huge number of biomarkers and their combinations

As mentioned earlier, human genome research has revealed that there are about 20,000 protein-coding genes in the human body. If we use these coding genes as state variables, and the standard state and mutation state of each coding gene as different values of the state variables, hundreds of millions of gene mutations and their astronomical combinations can be uniquely expressed by the SDS containing about 20,000 state variables. And with further introduction of the proteome, metabolome and other omics information related to human diseases and human behavior data to describe the individual characteristics of the human body completely, we will face a much more complicated SDS.

The problem is that different variants of the same coding gene often do not have a linear relationship in expression, and the target medications for them are also different in many cases. This means that these relatively independent gene variants can only be expressed by different state variables, which means that each relatively independent variant of each coding gene needs to introduce a state variable. In this way, more than 20,000 protein-coding genes constitute more than 20,000 subsystems containing several state variables, and the totality of all state variables belonging to different subsystems constitutes the state space that expresses personalized genetic variations of human beings. Assuming that only 1% of the hundreds of millions of state variables are independent, the state that can be described by the pairwise combination of millions of independent state variables will also exceed the trillion-level scale. With the expansion of biomarkers to the fields of proteins and metabolites, as well as the introduction of omics information and human behavior data related to human diseases, the scale of the SDS will be even more unimaginable.

Such a large and complex medical system, whether it is learning, understanding or grasping, is difficult for human intelligence to deal with. Therefore, as state variables increase, it is imperative to simplify them. The elimination of redundancy based on the correlation analysis of state variables can undoubtedly greatly simplify the SDS, but the simplification based on this method is limited. If the independence of the state variables included in the SDS meets the requirements of the system, further simplification may damage the completeness of the system, which means that based on this method, the simplification has reached its limit. In this case, is there room for further simplification of the system?

The reality system is usually hierarchical and the complexity of the SDS is usually closely related to the initial level of the introduced state variables. The lower the initial level of the introduced state variables, the larger the scale of state variables required to describe the system completely. For example, a school has 6 grades, each grade has 10 classes, and each class has 50 students. From school to grade, to class, and then to specific students, there are different levels of the school. Starting from different levels such as grade, class, or student/teacher to establish a state description, the complexity of the system is different. Obviously, the SDS based on the class as the basic unit increases the number of elements by 10 times compared with the system based on the grade as the unit, and the number of elements increases by 50 times again when the individual student is the unit. For a class composed of 50 people, if the number of students included in the system is less than 50, then the system is incomplete. The most basic element of the system is upgraded to the class, and the scale of the SDS is 50 times smaller. As long as 10 classes are included in the system, the completeness of the system will not be affected. However, the SDS established with the class as the basic element cannot reflect individual attributes and characteristics of the students in the class.

Therefore, it is not difficult to understand that there are two different levels of syndromes at the holistic level and at the subsystem level in the SDS of TCM, and it is also easy to understand the difference and structural relationship between the SDS of TCM and the state description expected to be established in precision medicine. Both are SDSs based on the real human body system. The difference in complexity and system scale is mainly due to the difference in the initial level of the introduction of state variables. This point will be further discussed in later chapters.

7.3.3 The integration issues with combination medication targeting multiple abnormal biomarkers

It takes a long process for precision medicine to establish a personalized DCS, and it will also require a long process for this personalized DCS to achieve a scientific SDS, which requires a large number of experimental analyses and clinical trials. And does the establishment of such a system mean that the current problems of modern medicine could be solved more perfectly?

In the treatment of cancer, clinical oncologists have found that a single medication is often ineffective in controlling tumor growth because tumor cells can evade a single treatment in a variety of ways. Therefore, medical scientists began to explore the development of medication combinations that can target multiple key targets of tumors at the same time. This strategy is called "multi-target therapy." The advantages of medication combinations are:

1. *Increase efficacy*: Multi-target medication combinations can attack multiple vulnerabilities of tumors at the same time, thereby more effectively inhibiting tumor growth.
2. *Reduce drug resistance*: Tumor cells are more likely to develop drug resistance when faced with a single drug. Combinations of medications can reduce this risk because cells resistant to one medication may still be sensitive to other ingredients.
3. *Better reflect the principles of personalized treatment*: The principles of precision medicine are applicable to the design of medication combinations. Physicians can choose the most suitable combination treatment plan based on the patient's tumor genotype and phenotype.

Currently, several cancer-specific medication combinations have been approved and used clinically. For example, for HER2-positive breast cancer, a common treatment regimen involves using a combination of Herceptin and Perjeta, two medications that target different parts of the HER2 receptor. Developing medication combinations requires a deep understanding of tumor biology and drug-drug interactions. Additionally, medication combinations may cause additional side effects and therefore require careful monitoring and management.

The introduction of targeted medications in precision medicine is usually obtained through experimental/clinical studies with the target as the frame of reference (Swain et al. 2015, Schneeweiss et al. 2013). When a patient has more than one abnormal target, it usually requires more than one target medication to be used in conjunction. For situations that are associated with multiple abnormal targets, or the patient suffers from multiple diseases at the same time, in order to more comprehensively control the disease, it may be necessary to take several or even dozens of target medications at the same time. At present, precision medicine research has revealed that it is quite common for a disease to be associated with several abnormal targets, and it is common for a patient to have multiple diseases at the same time. If this combination of medications does not produce good therapeutic effects and will cause a large number of medication-induced diseases, even if such a SDS is constructed by humans with a lot of manpower and material resources, its practical value will be greatly reduced.

Precision medicine's method of regulation-controlling diseases ignores the self-organizing characteristics of complex systems, and ignores the constraints of the upper level on the next level determined by the own nature of the organism, just as the command of a general involves controlling every action of every soldier under his jurisdiction. In fact, there are much simpler ways to regulate/control self-organizing and adaptive systems. In most cases, the realization of overall comprehensive regulation-control does not need to be controlled to the micro level of the human body, and therefore there is no need to establish such a complicated state description at the micro level.

If the personalized DCS of precision medicine does not move toward the SDS, its role will be limited to distinguishing patient-specific personalized characteristics, and looking for targeted regulation-control methods for a single personalized characteristic. Like traditional disease medicine, this kind of medicine still cannot get rid of the color of reductionism, and cannot achieve comprehensive treatment of diseases. In other words, the rise of precision medicine has started the process of modern medicine from disease medicine towards personalized medicine, but this trend may not lead to the establishment of state medicine that conforms to scientific norms under the framework of modern medicine. In the post-precision medicine era, modern medicine cannot still reach the height that TCM can achieve in terms of the overall integration of the human body.

7.4 The scientific limitations of precision medicine based on the concept of reductionism

The basis of precision medicine is personalized diagnosis, and the core is personalized medication. The precision medicine pioneered in the United States is based on the maturity of detection and diagnosis technologies such as gene sequencing, as well as the rapid development of artificial intelligence and big data analysis technologies. Using gene sequencing to find the targets of cancer patients' gene mutations, using targeted medications for 'precise strikes' and then monitoring the markers to accurately track the treatment effect and adjusting the treatment plan at any time, is now the typical process of precision medical treatment of tumors. Such precision treatment can replace the current carpet-bombing methods such as radiotherapy, chemotherapy, and surgery in tumor treatment, which can not only improve the efficiency of treatment, and reduce the suffering of patients, but also reduce the economic burden of patients.

With the completion of the Human Genome Project, scientists have discovered that genetic factors are intrinsically linked to the occurrence of many human diseases in the process of deciphering the correlation between genetic variation and human diseases. The medical community is looking forward with confidence that, based on personal genome information, combined with proteome, metabolome and other relevant internal environment information, humans will take a big step forward in the accurate diagnosis of diseases. 'Incurable diseases' such as tumors will gradually be overcome and become chronic diseases such as hypertension and diabetes. Patients can survive with tumors for a long time and have a better quality of life.

The progress of genetic medicine research also shows that the influence of genes on an individual's future diseases and health conditions is not decisive in most cases. The occurrence and development of disease is a complex process affected by multiple factors. The appearance of a certain result may be caused by many different links in the entire causal relationship chain. Therefore, it is impossible to achieve accurate diagnosis by relying only on gene sequencing, and comprehensive analysis should be performed in conjunction with relevant internal environmental information such as proteome and metabolome, as well as clinical manifestations, behaviors, physiological indicators, and environmental parameters. Moreover, precision is relative, which is relative to the accuracy that can be achieved in previous disease diagnosis, and it is unrealistic to require accuracy in an absolute sense.

With accurate diagnostic indicators as a frame of reference, it is bound to advance the research and discovery of personalized precision medications. The progress of the treatment of non-small cell lung cancer showed the typical process of the development of precision medication. Non-small cell lung cancer accounts for about 80% of all lung cancers. About 75% of patients are already in the advanced stage when their illness discovered, and the 5-year survival rate is very low. There are nearly 20 pathogenic genes in non-small cell carcinoma, and different oncogenes require different medications. To treat this type of cancer, in the 1960s, the medical community used cytotoxic medications with an effective rate of less than 5%. In 2003, it was discovered that EGFR is an important pathogenic factor, and the targeted medication gefitinib was used instead, and the effective rate increased to 10%. In 2005, it was discovered that mutations in EGFR were sensitive markers, and the effective rate of new medications was increased to 70% to 80%, which could extend life by 30 months (She Huimin 2015).

However, whether through oral, intramuscular injection or intravenous drip and other different routes of administration, the medications that enter the human body will eventually flow through various organs and tissues through the blood and spread throughout the body. Due to the extensive similarities among tissues/cells of different parts of the human body, we cannot ensure that the medication does not affect other tissues except the target wherever it goes, and these actions are not always positive, that is to say, the side effects of the medication are usually difficult to avoid. The so-called precision can only be as far as possible to find medications that have better actions on the target but fewer side effects on other parts (or elements). This situation will not change because of the emergence of the concept of 'precision medicine'. Of course, chemotherapy medications for tumors are an exception. At present, even when compared with medications for other diseases in modern medicine, the daunting side effects of chemotherapeutics show that they are too far away from the requirements of precision.

Advances in precision diagnosis technology will undoubtedly make surgical operations and non-invasive or minimally invasive physical therapies such as helium-neon knives, lasers, and infrared hyperthermia more precise in the positioning of the treatment site, and minimize the harm of these treatments on the human body. However, since the human body is a whole in which the functions of various parts are closely related, precise intervention of a part will inevitably directly or indirectly affect the related parts or elements, and even the whole human body. Precision medicine can only reduce this impact and damage as much as possible, and it is impossible to completely avoid it.

Generally, the medications used include biological products, chemically synthesized compounds and extracted plant extracts. The composition and structure of chemical synthetic medications arc relatively simple, while biological products and plant extracts are often a mixture of multiple ingredients. But even with chemical synthetic medications, in most cases, the action on the human body is not single, there are positive therapeutic actions, but also negative side effects. Its action on the target as the frame of reference during research is 'precise', but the part (or element) affected by its side effects cannot be said to be 'precise'.

Due to the complexity brought about by the close correlation between various parts and elements of the human body, in many cases, the diseases of the patients involve abnormalities in more than one part or element. On the premise of 'precise' diagnosis of more than one disease (A, B, C, D) involved in the patient, how to choose "precise" medications developed on the concept of "precise" for "precise" treatment?

Since the medications we use have multi-faceted effects, including therapeutic effects and side effects, medications that are 'precise' for A may not be 'precise' for B, and those that are 'precise' for B may not be 'precise' for C or D. The 'imprecise' mentioned here is probably a side effect. When we have to consider multiple parts and elements of the disease as a whole, the concept of 'precision' is already meaningless. A more useful concept is that it should be 'optimal', that is, to choose a relatively better comprehensive treatment plan for the disease or the patient as a whole. The research of 'precision' medications in precision medicine requires 'precision' and effectiveness on the target, and at the same time, as few side effects as possible. In fact, this is the goal that pharmaceutical researchers have been working on before the emergence of the concept of precision medicine. It is only after medicine has entered the era of 'precision' that there are higher requirements for finding precise medications with highly specific and fewer side effects.

To deal with diseases caused by specific parts or specific factors, precision medicine undoubtedly has great potential and is easy to achieve results. But the reality is that the diseases that can be attributed to specific parts or specific factors are very limited, and most diseases (or patients) involve abnormalities in the structure and function of more than one part (or element). In this case, even if we have 'precision' diagnosis, 'precision' medications and treatment methods, 'precision' selection of medications and formulation of 'precision' comprehensive treatment plans will still be technical difficulties that precision medicine has to face. Here, on the one hand, it is necessary to consider the harmfulness of various diseases that have been diagnosed with 'precision' diagnosis, and the importance of the parts (or elements) involved in the disease to the human body's life-sustaining activities, thereby determining the priority and order of treatment. On the other hand, it is necessary to consider the correlation between the various parts (or elements) of the disease, as well as the correlation between the structure and function involved in each part of the disease, so that the different measures of the treatment plan have better synergy. At the same time, it is necessary to balance the different side effects of a variety of precision medications as a whole, or add measures to offset these side effects in a comprehensive plan. Obviously, such precision medicine has been no longer limited to the precise treatment of diseases caused by specific parts or specific factors, but focused on the 'precision' of the overall state of the human body at the holistic level. The target positioning of precision medicine is upgraded from the establishment of a personalized DCS to the establishment of a personalized SDS, and the 'precision' of disease diagnosis is upgraded from a disease defined by a single specific biomarker to a personalized status defined by multiple biomarkers. This will undoubtedly increase the coverage of the actions of the developed medications and enhance the comprehensiveness and integrity of the treatment plan. In short, 'precision' for the overall state of the human body is the highest state of precision medicine.

As mentioned earlier, TCM treatment is mainly based on syndrome, but it also has a DCS. Clinical treatment of TCM is usually a combination of syndrome identification and disease identification. The relationship between the biomarker system established by precision medicine and the DCS of modern medicine is the same as the relationship between syndromes system and diseases classification system of TCM, which is a many-to-many network relationship. One disease can be divided into different syndrome types, and one syndrome also appears in many different diseases. Obviously, modern medicine, which previously only had a treatment method based on disease differentiation, has begun the process of constructing its unique and personalized 'syndrome' system with the development of precision medicine. As a result, modern medicine will also realize 'different diseases with the same treatment' and 'same disease with different treatments', opening the era of combining treatment methods with disease differentiation and treatment with 'syndrome differentiation'.

However, the personalized medical system constructed by precision medicine based on biomarkers such as genes, proteins, and metabolites is still based on the concept of reductionism characterized by analysis. Whether it is abnormal biomarkers in the disease process or targeted medications for biomarkers, research is conducted in different categories. In reality, the occurrence of human diseases is often accompanied by abnormalities in the structure and function of multiple tissues and organs, which are manifested as abnormal changes in numerous biomarkers. In order to achieve effective treatment as much as possible in clinical practice, clinicians often have to consider comprehensive medication for multiple diseases and multiple abnormal biomarkers at the same time. So, even if people find a fairly effective medication for each target, if multiple such targeted medications are used simultaneously for patients with multiple abnormal biomarkers, can we think of what will happen?

Due to the complexity of the interrelationship between the various parts of the human body, various parts and links of the disease usually affect each other. When a medication enters the human body, it usually does not only affect the target site or link. Taking different target medications at the same time may not only cause changes in medication structure and efficacy due to chemical reactions among medications, but also produce different synergistic or antagonistic effects in the human body depending on the combination of medications and the ratio of the combination. As a result, the regulation-control of an entire body containing multiple biomarker abnormalities cannot simply be attributed to the sum of precise regulation-control of each abnormal biomarker. Obviously, the precision medicine that people are looking forward to today, and the personalized medical system that will gradually develop under the banner of precision medicine, still cannot solve a series of problems caused by the complexity of the human body, the complexity of the disease and the treatment process.

For precision medicine to achieve 'precision' in treatment, precision in three links is indispensable: precision in disease diagnosis and recognition of human body state, precision in grasping the effects of medications and treatment methods, and precision in selecting medications and treatment methods in clinical applications. In recent years, with the maturity of gene sequencing technology, and the gradual practical application of 'big data' and artificial intelligence technology, the 'precision' of medicine, a previously long-distance goal seems to be 'desirable' and 'available'. Therefore, as soon as the concept of precision medicine was proposed, it immediately became a hot topic of worldwide attention.

Strictly speaking, although 'precision medicine' has the meaning of scientific method level, it is mainly a technical level concept, and cannot be expected to solve all the problems of medical methodology and medical architecture level. In other words, the problems faced by modern medicine based on static structural research and single factor analysis will not disappear as medicine enters the era of 'precision medicine'. These include:

1. The subdivision of disciplines, the rapid expansion of the total amount of medical knowledge, and the proportion of knowledge that a person can master throughout career life relative to the total knowledge in this field is becoming less and less. However, the precise 'personalized treatment' of the human body requires a comprehensive understanding and overall grasp of the state of the human body.
2. Medications developed with the element of a certain lesion site or link as the controlled amount, its side effects are always unavoidable. The difficulty of studying 'precision' medications will not be reduced by the emergence of the concept of 'precision'.
3. For complex disease processes that are related to each other, medications with their own side effects are used to respectively treat by category. In this way, even if 'precision' is achieved from the point of view of the diseased location or elements, how can the unavoidable side effects reflect 'precision' for other parts or the holistic level of human body?

Obviously, the impact of precision medicine has significant limitations. It is not the whole of the development of modern medicine, or even a decisive factor in promoting the development of modern medicine, but it may become a powerful booster and catalyst for the development of modern medicine and even traditional medicine.

7.5 Precision medicine and TCM: Both personalized medicine system in different levels

The treatment system based on syndrome of traditional Chinese medicines (TCM) is formed naturally during TCM development. Since ancient times, relying on this SDS similar to modern science, TCM has always used methods that adapt to people, places, and time to diagnose and treat diseases. As mentioned above, we can see that the goal of the American Precision Medicine Program is also a personalized medical system that can accurately identify the individual characteristics of patients and formulate treatment plans from person to person.

Therefore, in the near future, facing the same human body, in the medical field, we will see two different personalized medical systems, and even two different SDSs. So, what is the structural relationship between these two systems? What are their respective characteristics in the prevention/treatment of human illness? What kind of pattern will they evolve into in the future under the large framework of Eastern and Western medicine? Which will better represent the future direction of medicine?

7.5.1 Two-level state description systems (SDS): The difference in the integrity of state description and the structural relationship between them

People understanding of the human body itself starts from the observation of the external behavioral characteristics of human beings as a whole. In ancient times when the holistic view of nature was dominant, analogies and speculations of similar things were the main way for people to realize unknown things. The reasons why this method was used were, on the one hand, people's belief in the integrity of the natural world and the similarity of the way of organization of things in nature; and on the other hand, it was the last resort due to the limitation of the development level of science and technology at that time. Today, with the development of complexity science, people realize that the establishment of a behavior model of a system through metaphor and analogy, and continuous improvement and revision of the model in empirical testing, is the only feasible way for people to understand complex systems and grasp their overall laws.

In the course of more than 400 years of scientific development, modern science has gone through a process from analyzing and decomposing research objects to integrating research objects based on partial and static understanding obtained from analysis and decomposition. To grasp complex systems, one must directly confront the complexity and adopt a holistic approach to study. This is a profound understanding gained after modern science has experienced the process from analysis to integration and gradually realized the limitations of the "analysis-integration" cognitive method. Today, with the highly developed modern science and technology, the reason why TCM with a history of more than two thousand years can still create medical miracles that amaze modern medicine is that it used this method to establish the human body model and the SDS of the human body. Moreover, in some aspects of the application of this method, it has even reached a level that today's complexity science can't reach.

As mentioned earlier, the study of the human body in modern medicine characterized by anatomy and analysis follows the order of organ-tissue-cell and has reached the molecular level. To establish the state description of the human body at the molecular level, it is not only necessary to understand the static information of which elements the human body is composed of at this level, but also to understand the structural relationship of these elements in the human body and the dynamic connections such as interaction and mutual influence. Genomics, proteomics, lipidomics, glycomics, transcriptomics and other omics research are carried out at this level. The description of the state of the human body to be established by the 'The All of Us Research Program' launched by the United States in 2016 is also based on this level.

As a result, two state description systems (SDS) are formed at different level: one is a relatively macroscopic SDS based on the external behavior characteristics of the human body as a whole, for example the syndrome system of TCM; the other is an SDS based on molecular elements such as human genes, proteins, and metabolites, which is the long-term goal of the American Precision Medicine Program. Of course, the state variables introduced in the state description of precision medicine will also contain some general health information and data related to physiological and pathological behaviors. Therefore, to be precise, its state description is based on the upward expansion of the molecular level.

Generally speaking, the lower the level of state variables, the greater the total number of variables needed to fully express the physiological and pathological activities of the human body. Conversely, the higher the level of state variables, the better the degree of abstraction, and the smaller the number of state variables required to fully describe the state of the system. From the perspective of precision medication research, medications that are studied with relatively macroscopic state variables as controlled quantities have relatively macroscopic actions and a wider coverage, usually covering multiple state variables at the microscopic level. The overall effect of the medication is usually the sum of direct or indirect actions on multiple state variables at the microscopic level. From the perspective of identification and control, the more macroscopic the state variables, the better the degree of abstraction, the simpler the SDS will be, and the easier it will be to understand and control.

For describing and identifying the state of the human body, the use of macroscopic state variables requires lower accuracy; for the treatment of diseases, the use of medications target to macroscopic state variables will reduce the accuracy requirements for understanding the actions of medications. Just like on a high-altitude airplane, if you want to observe the beards of soldiers on the ground, you need high-powered telescopes; if you want to observe mountains or rivers, it doesn't need to be so precise. Similarly, if the target of launching a missile is a vehicle, the strike volume uses the energy of ordinary artillery shells, of course, it needs to be accurately positioned and accurately controlled, but if it is a nuclear bomb and the target is a city, the accuracy of positioning and control will be greatly reduced.

Obviously, precision at the micro-level for individual variables does not imply precision for an individual's overall state. When the number of state variables is the same, the higher the level of state variables, the higher the accuracy of describing the overall state of the human body. The accuracy required for human body regulation-control is also closely related to the level of the state variables we want to regulation-control. The higher the level of state variables, the lower the accuracy of the control requirements, and the smaller the side effects of medications researched based on it. This is also our intuitive feeling in the process of studying Chinese herbal medicine (CHM) and Western medications. Chinese Materia Medica has a very vague description of the properties of CHMs, but it is basically sufficient for the clinical practice of TCM; the pharmacy of modern medicine describes the properties of medications in a relatively detailed manner, but from a macro perspective, there are more uncertainties and side effects.

The biological specimen data collected by the US Precision Medicine program, such as cell numbers, proteins, metabolites, RNA and DNA, and whole-genome sequencing, are all quite microscopic. In other words, for the SDS based on this level to achieve a complete description of the various states of the human body requires higher accuracy and a larger amount of data. Therefore, whether it is to achieve precision detection and diagnosis, precision medication research, or precision regulation-control of the human body, there is a long way to go. However, compared with the previous direction from the macro to the micro depth, the current medicine is on the right path from microanalysis to overall synthesis.

The theoretical system of TCM originates from the most basic functional division of the human body's physiological and pathological activities from the holistic level. In recent years, the progress in body constitution medicine research shows that the differences in the human body constitution can be described by syndromes that characterize the relatively stable state of the human body as a whole. About 10 kinds of syndromes describing the overall state of the human body in TCM basically cover the basic types of the human constitution (Yuan Bing 2013b). The human body SDS of TCM, that is, the treatment system based on the syndrome, uses less than 100 basic syndromes (state variables) and their combinations to cover all the basic pathological processes of the human body. As mentioned earlier, if the number of all pathological states in which no more than five state variables are abnormal is counted, these 100 state variables can be combined into more than 10 billion different pathological states. If the scope of calculation is not limited to five state variable abnormalities, this number will be much larger. Even so, this system is obviously much smaller than the SDS composed of hundreds of millions of genetic variants. But with the ability to describe more than 10 billion different states of personalization, the scale of this system is also considerable.

Obviously, the SDS of TCM is at the highest level of the human body and has a top level of abstraction, therefore, the demand for precision diagnosis and precision medication research will be lower. Therefore, it is not difficult to understand that CHM based on the TCM theory has better macroscopic and long-term efficacy.

Just imagine, if we use a SDS with 100 variables (such as syndromes in TCM) as a frame of reference to understand the state of the patient, when we have accurately identified 80 of these variables, we can say that the accuracy of understanding the patient's status has reached 80%; and if we use an SDS with 10,000 variables (such as the detection indicators of modern medicine) as a frame of reference to understand the patient's state, even if we have accurately identified

500 of these variables, we can say that we have only reached an accuracy of 5% of patient status understanding. This is why, although TCM is far less accurate in identifying symptoms and signs than modern medicine, its grasp of the overall state of the patient is far beyond the reach of modern medicine. Obviously, the accurate grasp of the human body's state is not only related to the technical problems of the measurement method and the precision of the measuring instrument, but also has a more important connection with the level of the SDS and the total number of state variables as the frame of reference. The lower the level of a SDS, the more the total number of variables introduced, the more difficult it is to achieve accurate identification using it (Yuan Bing 2017b).

In TCM, there is a mature SDS (TCM theoretical model and the treatment system based on syndrome) that can cover the basic physiological and pathological activities of the human body, and it has a higher level and fewer state variables. More importantly, in the thousands of years of clinical practice from ancient times to the present, a complete human body state regulation-control system has been established and perfected from theories, treatment methods, prescriptions, and medications. Obviously, TCM is much closer to the goal of personalized diagnosis and treatment of precision medicine. And to achieve the long-term goal of precision medicine, TCM has a higher starting point and initial conditions than modern medicine (Yuan Bing 2016a). In a sense, it can be said that what precision medicine wants to achieve as a long-term goal is exactly what TCM can do and is doing. It's just that the pioneers of precision medicine haven't realized it due to a lack of understanding of TCM.

7.5.2 How to ensure the independence of state variables and the completeness of the state description system (SDS)

Compared with the state description system (SDS), the structure of the DCS is relatively weak. When modern science establishes the state description of the research object, the independence of the introduced state variables must be considered. If an SDS has a large number of containment and linear correlation relationships, it is impossible to uniquely determine the state of the system, and it will greatly increase the complexity of the system and cause a lot of logical confusion. However, the classification system does not have strict restrictions on the independence of the introduced classification items. We often see the inclusion and partial redundancy between the classification items under the same category or the classification items of different categories under a classification system. The purpose of establishing a DCS in precision medicine is to achieve a personalized description and treatment of human diseases. However, if there are no corresponding restrictions on the classification items, the so-called personalization can only be rough, and it is impossible to achieve the degree of uniquely determining the state of personalization.

So, in the medical field, how to ensure the independence of the state variables introduced in the process of establishing the SDS? Let's take a look at how TCM is done and what are the problems. Then imagine what precision medicine can do to solve this problem in the process of transition from a DCS to an SDS.

As mentioned earlier, the introduction of syndromes (state variables) in TCM is based on the human model. The human model describes the functions and attributes of the human body and the relationship with the environment at the holistic level, and describes the functions and attributes of each part of the human body at the subsystem level. The state variables at the holistic level are introduced based on describing the functions and attributes of the holistic level. The independence of various functions and attributes makes the state variables describing them naturally have a certain degree of independence; the same is also true at the subsystem level. In the syndrome system of TCM, the non-independence of state variables is often caused by the duplication of functional provisions and structural confusion.

For example, in the human body model of TCM, based on the framework proposed in the 'Huangdi Neijing, The Yellow Emperor's Inner Canon · Suwen', the mental consciousness thinking activities are presided over by the 'heart', but there is another argument in 'Huangdi Neijing, The

Yellow Emperor's Inner Canon · Lingshu' that 'the brain is the sea of marrow'. Li Shizhen, a well-known medical scholar in the Ming Dynasty, even put forward the view that 'the brain is the organ of the soul'. Obviously, in TCM, the debate on whether the function of the human spiritual consciousness thinking activities belongs to the heart or the brain has existed since ancient times, and it has continued to this day. The syndromes introduced based on the functional provisions of 'the heart stores the spirit' include heart-qi deficiency, heart-blood deficiency, heart-yin deficiency, heart-yang deficiency and so on. The settings of these syndromes can well explain the problems that heart palpitations often occur at the same time as insomnia and forgetfulness. The heat disturbing the mind, the phlegm obsessing the heart orifice, and the 'heating into the pericardium' derived from the theory of febrile diseases in later generations are also syndrome provisions supported by the concept of 'heart accommodating the mind'. The introduction of syndromes supported by the concept of the 'brain governs mind' are mainly based on the clinical reality that the 'brain' that appears in stroke is closely related to the mind. In the earlier era of TCM, this part of the disease was attributed to the syndrome of liver-wind movement. However, the clinical characteristics such as sudden coma, unconsciousness, and hemiplegia that occur during stroke are significantly different from the "movement of wind" signs of spasms and convulsions caused by internal movement of the liver wind. In the Republic of China, Zhang Xichun introduced the syndrome of cerebral congestion to solve this problem in an intuitive concept, and correspondingly introduced the syndrome of cerebral anemia to describe pathological phenomena similar to dizziness and shock caused by insufficient blood supply to the brain in modern medicine. However, these syndromes have a certain conflict with the syndromes based on the concept of 'heart governing mind' (such as heart-qi deficiency, heart-yang deficiency). Obviously, the conflicts and structural confusion among these syndromes are entirely caused by the confusion of the function provisions of the relevant parts of the model (Yuan Bing and Zhao Zheng 1986).

In natural sciences, encountering such problems usually triggers a revolution in the theoretical structure, and finally replaces the previous theory with a new theory that can reasonably accommodate the above two aspects of facts. But since TCM is not strictly a science, it will naturally not follow the logic of scientific development. TCM scholars of all dynasties insisted on their own opinions, so that this chaos in the theoretical structure and the chaos in this part of the syndrome system caused by this is left unsolved to this day. Today's TCM even regards this kind of research that leads to the confusion of theoretical structure as the development of TCM. The structural confusion caused by repetitive definitions of the spleen and small bowel function are also similar problems.

Based on the understanding of the internal organs and structures of the human body in the era of 'Huangdi Neijing, The Yellow Emperor's Inner Canon', TCM introduced structures such as the five zang-organs, six fu-organs, and Qihengzhifu-organs (brain, marrow, bone, blood vessels, gall bladder, and uterus). Since modern times, with the development of modern anatomy, modern medicine has discovered more objects defined as organs in the human body, such as pancreas, thyroid, adrenal glands, and thymus. But TCM seems to no longer need these things. In TCM, even the introduced viscera have not all been incorporated into the syndrome differentiation system to guide clinical treatment. For example, most of the viscera belonging to the Qihengzhifu-organs have no syndromes introduced to describe their functional status. Based on the description of the position of the spleen in TCM, the spleen of the human body is actually a lymphatic organ, which has nothing to do with the function of 'the spleen governs absorption and transportation'.

Obviously, the introduction of internal organs of TCM does not require completeness in the anatomical sense. The completeness of the human body theory in TCM lies in the completeness required for the description of human physiological and pathological activities and disease analysis. In this sense, it is not necessary for TCM to retain organs that are meaningless in the process of disease analysis. Then, based on clinical practice, in order to meet the needs of a reasonable description of human physiological and pathological activities and disease analysis, what is wrong with adjusting the function provisions and relationships of the internal organs in the theoretical system of TCM?

In today's physics, how many traces of Newtonian physics still remain? The traces of physics in the Aristotelian era are even more nearly extinct. But no one doubts that physics today is still physics.

When we get rid of the shackles of thought, regard TCM as a science and examine the problems existing in the theoretical system of TCM according to scientific rules, constructing a new theoretical system is a much simpler matter than Newton's mechanical system, Einstein's relativity system, and Hawking's quantum gravity theory. Based on the understanding of the human body's physiological and pathological activities and the law of occurrence and development of diseases, the structure of the theory can be continuously adjusted and optimized, and the theoretical system can be continuously improved. Based on developing and improving the theoretical system, continuously adjust and optimize the state variables (syndromes) to gradually improve the completeness of the state variable system. Because the model that conforms to the scientific norms has a rigorous logical structure, and each state variable is usually introduced to describe a certain aspect of its function and characterize a certain aspect of its attributes, it is relatively easy to ensure their independence.

In TCM, the biggest issue involving the completeness and independence of the SDS is the unification of several major syndrome systems. visceral-qi-blood-body liquid syndrome system is suitable for the treatment based on syndrome of internal injuries diseases. Since the simple extension of the viscera-qi-blood-body fluid syndrome system cannot meet the needs of the description of the law of occurrence and development and treatment analysis of exogenous diseases, several other syndrome systems have emerged that are independent of the viscera-qi-blood-body fluid syndrome system. As far as the description and analysis of all human diseases covering exogenous and internal injuries are concerned, each of these systems is not complete, and there are many overlaps and redundancies in the syndromes of different systems. For example, the 'middle-jiao damp-heat' in the tri-jiao syndrome system is similar to the 'spleen and stomach damp-heat' in the viscera-qi-blood-body fluid syndrome system; the 'Yangming meridian syndrome' of the Six-meridians syndrome system, the 'qi-zone syndrome' of wei-qi-ying-xue syndrome system, and the 'heat of the lungs and stomach syndromes' of the zangfu-qi-blood-body fluid syndrome system are actually the same syndrome. If similar problems occur in other natural science systems, a unified and more inclusive system needs to be constructed, so that the system's description of this part of the problem is simplified, reasonable, and has a rigorous logical structure. This means that the description of the law of occurrence and development and treatment analysis of external and internal trauma diseases should be solved as a whole, and the similar syndromes involved in several major systems should be structurally adjusted to remove redundancy. Since TCM was not a strict science based on scientific norms and it did not have this concept of scientific development, the situation of coexistence of several major syndrome systems in TCM has continued to this day. Obviously, if the unification of several major dialectical systems is achieved, then the issues of completeness and independence of the state description system involved in these problems will naturally disappear (Yuan Bing and Zhao Zheng 1986).

TCM establishes the human body model and SDS from the holistic level, and the top-level design and overall planning of the model structure and the state variables to be introduced can be carried out. With the continuous discovery of new biomarkers, the introduction of state variables in precision medicine is a process of gradual expansion. With the continuous increase of new clinically significant biomarkers included in the new disease classification system (DCS), the completeness of the SDS formed will be continuously improved. However, if the new DCS of precision medicine is to evolve into an SDS, it is important to solve the problem of independence among the ever-increasing state variables. This means that the redundancy caused by the non-independence of the state variables is removed at any time, so that the state variables can be continuously optimized. Since the discovery of precision medical biomarkers is a horizontal expansion at the micro level, the establishment of a SDS cannot have a top-level design. Therefore, ensuring the independence of state variables can only be done to remove redundancy based on the correlation analysis of biomarkers.

The disease information collection system developed by the 'All of Us Research Program' of the United States for the establishment of a new DCS makes it very easy to realize the relevant analysis of biomarkers. Perform correlation analysis based on such data, remove redundancy on this basis, and optimize the biomarker system to ensure that the remaining biomarkers (state variables) are relatively independent during the development of the DCS. In this way, ensuring the independence of state variables has become a relatively easy technical problem to solve.

7.5.3 Differences in regulation-controlling effects based on state descriptions at different levels

Medicine is an applied science established for the purpose of preventing and treating diseases. Practicability and clinical value are the first considerations for establishing a human body model and SDS. Precision medicine establishes the state description of the human body from the molecular level, and its reference system for research on medications and treatments is largely based on abnormal state variables at the molecular level. As mentioned earlier, the number of variables at the molecular level of the human body that correlate with health status is enormous. In the event of a disease, even if only one in ten million variables are abnormal, the number will be hundreds or thousands. Medications act on the human body, usually on more than one part or element. Their effects on untargeted parts or elements are likely to be side effects. This has been confirmed by the current status of targeted medication research. Even for precision medications developed in the future, it cannot be expected that the parts and properties of the action on the human body will reach the control accuracy like a missile. Research on precision medicine will undoubtedly continue to find more precise medications. But the precision here can only be relative to the specific state variable of the target, and the point of action of its side effects cannot be said to be precise. Moreover, when multiple state variables of the body are abnormal, the precision of the medication for a specific state variable does not mean that other abnormal state variables are also precise or have no adverse effects. When we administer multiple precision medications with their own side effects to a patient at the same time, can we ensure that the overall effect will only pull the abnormality of the target to normal without causing multiple side effects?

Today's modern medicine is still in an era of inaccurate grasp of the actions of medications. For most single state variable abnormalities at the molecular level, there is currently no precision medication that can be applied to the clinic. With the progress of precision medicine research, the number of medications discovered that can be called 'precision' will continue to grow. According to the current research methods of precision medicine, the situation where dozens or even hundreds of precision medications need to be applied at the same time is something that medicine has to face in future. This is actually the current situation in which modern medicine often needs 'combined medication', but with the development of precision medicine, the scale of combined medication will become larger and larger. If multiple medications that target a single abnormal state variable but have more or fewer side effects are used in a complex disease situation, the short-term and long-term actions are uncertain in many cases, and this uncertainty will also show the trend of individualization. To clarify these problems, we will face the research workload of combinatorial explosion based on hundreds of millions of abnormal state variables, and it may take the unremitting efforts of many generations. And even if these problems are clarified, it does not mean that we have a feasible solution for 'personalized' medical treatment.

As we mentioned earlier, the human body model and SDS of TCM are established based on the behavioral functions of the human body at the macroscopic holistic level. Compared with microscopic variables as the frame of reference, medications developed with macroscopic state variables as the frame of reference usually have the effect of regulation-controlling the macroscopic characteristics and functions of the body, and therefore have a broader scope of action. Decades of analysis and research on the viscera and syndromes of TCM in China, as well as clinical studies of integration of traditional Chinese and Western medicine (ITCMWM) have shown that one viscera of TCM often involves the functional activities of multiple anatomical tissues and organs, and a syndrome (state variable) of TCM will involve the abnormality of multiple elements and variables on the micro level. Therefore, the researched medications that can correct abnormal syndromes will undoubtedly exert a therapeutic action on the corresponding abnormal elements and variables at the micro level. And if you start with lower-level micro-variables, to achieve a macro-level control effect, you need to separately study medications that target multiple micro-variables. Due to the non-unique actions and side effects of precision medications on the human body, even if we find all the micro-level variables related to this macro-variable, and separately develop the precision medications for them and give them to the human body at the same time, it does not mean that they are able to pull the abnormal macroscopic variables (syndromes) to the normal state. There are two reasons for this: On the one hand, abnormal state variables at the micro-level may not be the cause of the abnormal state variables at the macro-level, but may only be the appearance or result of the abnormal macro-level variables. Without correcting the cause of the abnormality of the macroscopic variables, it is impossible to eliminate the abnormality of the macroscopic variables, and ultimately it is difficult to completely correct the abnormality of the microscopic variables. Once the action of the medication disappears, there will be a rebound, driving the microscopic variables to return to their original abnormal state. On the other hand, the non-uniqueness of actions of precision medication and its side effects makes it difficult to control the combined force of all its direct and indirect actions, and the patient's personalized health status will also have a personalized impact on the effect of the combined application of precision medications. All of these will make the final result of the combination of precision medications uncertain. In this regard, the research of complexity science has already given a clear theoretical conclusion: for a complex system, the control of high-level elements is not equal to the simple addition of the control of the low-level elements that make it up.

This point is also proved by modern research of CHM. For example, in TCM, the main functions of ginseng and astragalus are to invigorate qi. Modern pharmacological studies on qi-invigorating CHMs have shown that qi is usually reflected in the following actions at the microscopic level: regulating sugar metabolism and lipid metabolism; promoting the synthesis of protein, DNA, RNA; increasing albumin and γ-globulin content; increasing peripheral white blood cells, increase the phagocytic function of the reticuloendothelial system; enhance cellular immunity and humoral immunity, etc. So, if we take a variety of medications at the same time that can respectively regulate sugar metabolism and lipid metabolism, promote protein, DNA, and RNA biosynthesis, increase the phagocytic function of the reticuloendothelial system, and enhance cellular and humoral immunity, will their combined actions show the long-term effect of TCM for invigorating qi? To this question, whether it is based on the analysis and research of pharmacologists or the clinical practice of medical scientists, the answer is very clear: not necessarily. Lowering down the level of controlled amounts of medication research will greatly increase the complexity of the research and increase the uncertainty of medication effects in observing from a macro perspective. In terms of finding precision medications, compared with the bottom-up path of modern medicine, the top-down path of TCM is simpler and more effective. The correspondence between the developed medication and the overall state of the human body will have higher precision.

In fact, in TCM, the research of precision medication for a certain state variable (syndrome) is not limited to the research of single CHMs. Modern pharmacological studies of CHM suggest that CHMs of the same kind with similar actions in TCM may have different mechanisms of action in the human body. Therefore, due to the synergistic effect of medications acting on different parts and different links, in many cases, several medications for the same pathological state are used together, which will show a better therapeutic effect than increasing the dose of a single medication. On the other hand, each CHM (or a class of CHMs) may have its own side effects or biased nature often referred to in TCM. The research of precision medications not only requires the accuracy of targeted treatment, but also minimizes side effects to avoid 'injury to the innocent'. But, in fact, it is not easy to find such a medication. Choosing a medication or medication combination that has better therapeutic effect on the target, and effectively restricting its side effects is usually an easier way to find precise medications, which is the basic idea of the unique prescription compatibility in TCM. A prescription for a specific pathological state, due to combined CHMs that target different links and different mechanisms, is usually more effective than a single CHM. However, since adjuvants that restrict the side effects (or partiality) of the main CHMs and auxiliary CHMs are added to the prescription, a good CHM prescription will often have better accuracy, whether it is for the target or the systemic state.

Based on the concept of complexity science, after medication is composed to a prescription, each of the medications does not work alone. There are chemical reactions among the medications composed to a prescription in the process of processing or taking, and there also be direct and secondary reactions of each medication acting on the human body. The intertwining of the chain of causality caused by these direct or secondary reactions makes the reaction we observe after taking the medication often not a simple addition of the direct actions of the medications that make up the prescription. The interaction among the medications will cause the whole prescription to lose some actions that some medication components have when used alone, and mutate some new actions that none of the medication components have. The actions of prescriptions that people have observed are often the final effect of the resultant force of various direct and indirect reactions, synergy and antagonism that occur in the human body. Such a process cannot be clarified through analysis and decomposition. Therefore, it is also necessary to analyze and study the prescription (herbal formula) as a whole in the same way as the study of a single CHM, from the state change caused by it relative to the SDS (Yuan Bing 2016d).

The core of precision medicine is individualized medicine, which is expected to eventually move towards conditional treatment based on the SDS. The treatment system based on syndrome is precisely the state treatment based on the state description, which is basically consistent with the long-term goal to be achieved by precision medicine. It's just that it hasn't been refined down to the point where it can introduce necessary molecular-level anomalies into its SDS. Precision medicine establishes the description system of the human body state in a bottom-up manner, and it is impossible to have a top-level design in advance. With the horizontal expansion of human understanding of the human body at the micro level and the upward integration of the model structure, it will be found that some variables at the micro level are unnecessary for regulation-control, and the structure of some parts of the model is unreasonable. Therefore, it is necessary to partially reconstruct the model according to a more reasonable structure, and adjust the setting of state variables and the pattern of interrelationships. This can be compared to the construction of various underground pipe networks in cities. If there is no unified planning and design in advance, and various facilities are laid and installed independently, there will often be situations where the pipe network that has been laid needs to be excavated and re-laid according to the new plan. This means much greater workload and social costs. The most important thing is that in the long process before the microscopic research reaches integrity, the inaccuracy and side effects of the precision medications on the overall state, and the uncertainty of the clinical effect of the combination of multiple precision medications, are always the problem that medicine has to face and cannot properly solve.

Then, is it possible to use the human body model and SDS of TCM as the top-level design to achieve the long-term goal of precision medicine? On this basis, with the help of modern scientific detection and analysis techniques, build a more accurate human body model and SDS that integrates traditional Chinese and Western medicine from top to bottom. Is there any feasibility to realize this kind of thinking? This may become an important issue worth pondering in the development of precision medicine.

Chapter 8

The Inevitability of Medical Integration and the Difficulties of Integration

Research in biology and medicine has revealed more and more profoundly the integrity of the human body itself and the close relationship between man and nature. However, so far, whether it is modern medicine based on experiments and analysis, or precision medicine that has emerged in recent years, their expansion to the breadth is carried out independently under the condition of severing the interconnection with other parts. Disease-centered clinical medicine still mainly adopts the method of treating respectively various diseases that occur in the same body on the basis of classifying diseases. Precision medicine also strives to attribute the essence of the disease to individual biomarkers.

With the advancement of science and technology, the total amount of medical knowledge is increasing exponentially. However, due to the limitation of human's physical factors on intelligence, the proportion of knowledge that a person can understand and master throughout his life relative to the total knowledge in this field is getting smaller and smaller. In recent decades, the in-depth research of medicine has reached the molecular level. However, what puzzles many biologists and medical scientists is that the vibrant life phenomenon that cannot be understood at the macro level, after reaching the micro level, loses the whole picture, and the understanding of its essence becomes more and more vague.

In the face of reality, scientists have gradually become sober. Maybe, life cannot be attributed to the interaction between the most basic elements of life, and the control of the human body as a whole cannot also be attributed to the control of the various levels and parts of the whole. To fundamentally understand and control the human body, it is not only necessary to understand the various parts and even the details, but also to integrate them as a whole. As a result, the concept of integrated medicine came into being.

8.1 The inevitability of integration becoming the trend of medical development

In the late 1980s, the American medical community took the lead in putting forward the concept of 'integrated medicine', hoping to integrate the essence of traditional medicine into the modern mainstream medical system, so as to break through the bottleneck of medical development and achieve the goal of effectively preventing and treating various chronic diseases. The Federation of Integrated Medicine Academic Research Health Centers was established in 1999 and has more than 60 member organizations as of 2021, including Johns Hopkins University School of Medicine, Duke University School of Medicine, George University School of Medicine, Mayo Medical Center, etc.

According to the definition of integrated medicine by the Federation of Integrated Medicine Academic Research Health Centers: "Integrative medicine and health reaffirms the importance of the relationship between practitioner and patient, focuses on the whole person, is informed by evidence, and makes use of all appropriate therapeutic and lifestyle approaches, healthcare professionals and disciplines to achieve optimal health and healing" (The Academic Consortium for Integrative Medicine & Health 2015). Proponents believe that integrated medicine is not equal to complementary and alternative medicine, nor is it just a simple combination of traditional medicine and complementary and alternative medicine (Snyderman and Weil 2002, Rees and Weil 2001). Integrative medicine emphasizes the health and rehabilitation of the individual as a whole (including physical, psychological, social, and spiritual), and at the same time, it takes into account the treatment methods of traditional medicine and alternative medicine under the doctor-patient relationship of mutual cooperation and effective communication (Bell et al. 2002). Since integrated medicine does not propose new things beyond 'complementary and alternative medicine', some critics believe that integrated medicine is actually the new name of alternative medicine that is increasingly questioned (McLachlan 2010), and it is also only an alternative supplement to mainstream treatment methods (Whorton 2004).

However, in China, where traditional medicine is highly recognized by the medical community and the public, integrated medicine seems to have a different situation. In November 2009, the 'Medical Development Summit Forum' initiated by 21 medical universities and the magazine 'Medicine and Philosophy', and hosted by 6 national societies reached a consensus on the theme of Medical Integration (Dong Wei 2010). In 2012, Academician Fan Daiming (1953), then vice president of the Chinese Academy of Engineering and president of the Fourth Military Medical University, took the lead in proposing the concept of 'holistic integrative medicine' (HIM). Academician Fan has successively published a series of articles such as 'Introduction to Integrative Medicine' (Fan Daiming 2012), 'Revisiting of Integrative Medicine' (Fan Xing et al. 2013), 'On Integrated Medicine' (Fan Daiming 2014a), 'Holistic Integrative Medicine' (Fan Daiming 2014b), 'Integrated Medicine—New Era of Medical Development' (Fan Daiming 2016), 'HIM, the Inevitable Direction of Medical Development' (Fan Daiming 2017a), 'Holistic Integrative Medicine: Toward a New Era of Medical Advancement' (Fan Daiming 2017b), 'HIM, the only way to a new era of medical development' (Fan Daiming 2017c).

Under the strong promotion of Academician Fan Daiming and the Chinese Medical Practitioner Association, the first China Integrated Medicine Conference was held in Xi'an in October 2016. The meeting focused on the integration of medical-related clinical specialties and appropriate technology applications. The meeting invited 15 academicians and three famous experts to give special reports, and 4000 representatives attended the meeting (Tian Jin 2016). The 2017 China Integrated Medicine Conference was held in April 2017, attended by 52 academicians, more than 150 presidents of medical universities, more than 1,000 deans of hospitals at all levels, and 14,000 representatives. This unprecedented grand gathering of 'China's Integrated Medical Thoughts' was a declaration by the Chinese medical community: The Chinese medical community has soberly recognized the limitations of modern medical analysis methods and has begun to lead medicine towards holistic integration (Xiao Yang 2017).

Unlike the integrated medicine proposed by the American medical community in the early years, the current Chinese version of integrated medicine emphasizes the unity of the whole and the part, and emphasizes the synthesis and comprehensiveness from part integration to the whole. In this regard, Academician Fan Daiming made an interpretation: Integrated medicine is to integrate the most advanced knowledge theories in various fields of medicine and the most effective practical experience in various clinical specialties, and according to the social, environmental, and psychological realities, to modify and adjust it, so as to make it a new medical system that is more in line with and more suitable for human health and disease treatment (Fan Daiming 2017d).

The declaration of the concept of integration and introduction of "integration" containing the concept of synthesis into medicine, undoubtedly represents the direction of modern medicine. In modern medicine, which is still dominated by analytical methods, especially at the level of clinical medicine, hospital management, and medical policy formulation, compared with the previous concept of subdivision and refinement, the integrated concept may bring about significant progress of clinical treatment and medical management systems. However, in modern medicine, which is seriously out of touch with natural science in terms of methods and ideas, medical scientists may not realize what integration means to science. Even if it is only integration in a biological sense, its complexity and workload are unimaginable, let alone the integration of psychological factors, social factors, and environmental factors on this basis.

In the past 400 years, modern medicine based on the concepts of decomposition and single factor analysis, like traditional biology, is basically a static and metaphysical knowledge system. Based on such a theoretical system, it is impossible to achieve an overall grasp of the human body and diseases. Under the existing medical knowledge system, it is obviously too naive to think that integration is a job that clinicians can complete with established concepts, which shows a lack of adequate awareness of the complexity of integration. Integration requires the establishment of a holistic model that reflects the overarched characteristics of the human body and the association of various parts of the human body, and a systematic and comprehensive study of the law of integration based on such a model. Different treatment methods and medications act on the human body at the same time, and their actions are by no means equal to the simple addition of the actions of various treatment methods and medications on the human body. The interaction among different medications/treatment methods and among their direct actions on the human body are highly complex. The huge modern medical system built by medical scientists through unremitting efforts over the past centuries can only be regarded as isolated islands, and the medical system that needs to be established to support the Chinese version of integrated medicine is the vast ocean that can submerge these isolated islands.

In fact, so far, much of the work involved in integrated medicine interpreted by Academician Fan Daiming has been carried out by our medical workers, such as comprehensive medication in clinical treatment, the combination of medication therapy with psychological counseling and nutritional conditioning, the combination of traditional Chinese and Western medicine treatment methods, and the emergence of health management in recent years, etc. All of these are attempts to integrate on the basis of analysis, but they are not under the name of 'integrated medicine'. The Chinese version of integrated medicine so far only involves the popularization of the concept of the integration of treatment methods in the application field, and has not yet put forward a clear idea about the integration of modern medical knowledge systems.

Integrated medicine cannot stay at the slogan stage forever. To truly advance the integration of medicine, it is necessary to build the theoretical system of integrated medicine. However, what method is to use for integration? How does integrated medicine construct its theoretical system? In this regard, the medical community is full of expectations. The article 'Towards the Era of Integration: Integration of Modern Medicine and Integration of Traditional Chinese and Western Medicine' published in the 'Guiding Journal of Traditional Chinese Medicine and Pharmacy' in the 14–15th issue, 2018, discussed these issues with Academician Fan Daiming (Yuan Bing 2018a). There is an old Chinese saying, "The true face of Lushan mountain is lost to my sight, for it is right in this mountain that I reside." To truly understand this issue, it is necessary to broaden the perspective of observation and place the integration of medicine in the larger environment of biology and even the entire natural sciences to discuss. For this reason, we will examine biology, which is more closely related to natural sciences and at the same time is the basis of medicine, and see how it conducted integrated research on life and the problems faced by this research.

8.2 Systems biology: Integrated research on life and the methodological crisis it faces

In the field of biology, the analytical research of cell biology and molecular biology has enabled human beings to have a deep understanding of organisms at the cellular and molecular levels. However, based on these understandings, a systematic and satisfactory explanation cannot be given for the overall behavior of the organism. For a complex biological system, the study of genes and proteins is very important. However, because this kind of research ignores the interaction, support, and integration of various levels in the system, the research results are only limited to explaining the microscopic or partial phenomena of biological systems, and cannot derive information related to the overall behavior and functions of the system. In the late 20th century, with the development of systems science, control theory, and information theory, systems biology came into being.

The definition of systems biology: Systems biology is an academic field that uses holism (rather than reductionism) to integrate information from different disciplines and levels to study, analyze, and understand (i.e., multi-omics integration analysis) how biological systems function. By studying the molecular-level interrelationships and interactions among all the components within each biological system (e.g., gene and protein networks related to cell signaling, metabolic pathways, organelles, cells, physiological systems and organisms), systems biology hopes to eventually build comprehensible models of the entire system. Systems biology uses computer simulations and methods of mathematical analysis to build models of complex biological systems.

Systems biology mainly studies the composition of all components in complex biological systems and all the interrelationships among these components, and analyzes the dynamic process of the system in a certain period of time. Its research goal is to obtain a theoretical model as close as possible to a real complex biological system from a large amount of biological data. Furthermore, the prediction based on the model is summarized as an experiment, and the model is constantly modified and improved through experimental data, so that the theoretical prediction can reflect the authenticity of the biological system. One of the goals of systems biology is to model and discover emerging properties of cells, tissues and organisms (Wikipedia contributors 2023c).

Systems biology includes the integration of all levels and parts of the organism, and ultimately must be integrated into the organism level. This is a bottom-up integration, naturally starting from the most basic elements that make up life. Therefore, the research of systems biology is based on cell biology and molecular biology, and is carried out from the molecular level. One of the initiators of the Human Genome Project, the American scientist Lenoy Hood's (1938) definition of systems biology reflects the characteristics of the early stages of systems biology: Systems biology is a discipline that studies the composition of all components (genes, mRNAs, protein, etc.) in a biological system and their relationship among these components under specific conditions. That is to say, systems biology is different from molecular biology that only cared about individual genes and proteins in the past, it studies all genes, all proteins, and all the relationships among other components (Ideker et al. 2001).

Systems biology mainly studies the modeling and simulation of entity systems (such as individual organisms, organs, tissues and cells), dynamic analysis of biochemical metabolic pathways, the interaction of various signal transduction pathways, gene regulation-control networks, and disease mechanisms, etc. The task of systems biology is first to describe the state and structure of the system, including the definition of the elements of the system and the environment in which the system is located, as well as the in-depth analysis of the interaction between the elements of the system and the interaction between the environment and the system; the second is to make the dynamic analysis of the evolution of the system, including the steady-state characteristics, bifurcation behavior, phase diagram, etc. In addition, the description of the state of biological systems in systems science is hierarchical, and the descriptions on different levels may be completely different. The analysis of system evolution mechanism in system science emphasizes the relationship between the whole and the part. It analyzes how the interaction between subsystems forms the attributes and function of the system as a whole, and strives to find out the connection with the micro level for every behavior of the system as a whole (Sun Lanfang and Jiang Lu 2005).

At present, the research methods of systems biology in the world can be divided into two categories according to the different research tools used: one is experimental methods, and the other is mathematical modeling methods. The experimental method is mainly to understand the system through controlled repeated experiments. The mathematical modeling method is to establish a dynamic model of the system based on the internal mechanism of the system to quantitatively describe the interaction between the various elements of the system, and then predict the dynamic evolution result of the system. In recent years, there has been a trend of combining experimental methods with mathematical modeling methods.

Systems biology research uses artificially controlled conditions to reveal the dynamic characteristics of specific living systems under different conditions and at different times. The research content is mainly the confirmation of the system structure, the analysis of the system behavior, the induction of the system control law and the design of the system. The technology platform of systems biology is mainly a variety of high-throughput omics experiments, including genomics, transcriptomics, proteomics, metabolomics, interactomics, and phenomics (Chang Chang 2006). Based on these studies, the data required to build the model are provided and the structure of the system is identified. Computational biology can provide a strong foundation for the analysis and quantitative prediction of biological systems through modeling and model-based exploration. Computational biology includes data mining and simulation analysis. Data mining is to extract hidden internal laws from a large amount of data and information generated by various experimental platforms and form hypotheses. Simulation analysis is the use of computers to verify the hypotheses formed, to predict the results of biological experiments, and finally to form a virtual system that can be used for various biological research and predictions, namely models.

In recent years, network biology, which studies the interactions between biomolecules, has become the core of systems biology. This includes gene regulatory networks, protein interaction networks, metabolic pathway networks, disease networks, etc. These networks describe the interactions between biological molecules (such as genes, proteins, and metabolites) and how these interactions affect the physiological and pathological processes of organisms. These networks can be used to understand the interactions between molecules, cells, tissues and organs within an organism, as well as the interaction between the organism and its environment. Systems biology aims to understand the overall behavior of biological systems, and network biology provides tools to describe and analyze these systemic interactions. Network biology helps people gain a deeper understanding of the complexity of life by constructing, analyzing and interpreting biological networks (WenJun Zhang 2018).

The basic workflow of systems biology is divided into four stages: the first is to delineate the structure of the selected biological system, including gene interaction networks and metabolic pathways, as well as intracellular and intercellular mechanisms of action, based on an understanding of all components of the system, thereby constructing a preliminary system model. The second step is to change the internal composition (such as gene mutation) or external growth conditions of the object under study by interfering with the system, and then investigate how the system components or structures change under the influence of these disturbances, and integrate the relevant information obtained. The third step is to compare the data obtained through the experiment with the situation predicted according to the model, and modify and improve the preliminary model based on the comparison results. The fourth stage is to set up and implement new experiments to change the state of the system through disturbances according to the predictions or assumptions of the revised model, repeat the second and third steps, and continuously revise and improve the model through experimental data (Baidu Encyclopedia 2023b).

The biggest feature of systems biology is integration. The integration here mainly includes three meanings. First, integrate the components of different properties (DNA, mRNA, protein, small biological molecules, etc.) in the system together for research; secondly, for multicellular organisms, systems biology needs to realize the integration of all levels from genes to cells, organs, tissues, and even individuals; third, the integration of research ideas and methods. Classical molecular biology research is a vertical type of research that uses a variety of methods to study individual genes and proteins. Genomics, proteomics, and various other omics are horizontal research, that is, studying thousands of genes or proteins simultaneously with a single method. The characteristic of systems biology is to integrate horizontal and vertical research into 'three-dimensional' research (Baidu Encyclopedia 2023b).

Genome Medicine is a new concept of medical research proposed by more than 600 famous scientists in the world after the completion of the human genome map. It is a cross-discipline formed by extending the life sciences based on the human genome to clinical medicine. It quickly and efficiently transforms the research results of the human genome into clinical practice, so it is one of the most important research directions in the post-genome era. Genomic medicine will greatly improve the scientific understanding of health and disease states at the molecular level, and enhance the ability to develop effective intervention methods. Therefore, it is a great revolution in life science and clinical medicine in human history (Baidu Encyclopedia 2023c).

In natural sciences, biology has always been the basis of medicine and has been intertwined and developed with medicine, and their research methods and even technical means are in the same line. When we see the complexity of systems biology research dedicated to biological integration, will we believe that medical integration can be completed by clinicians and managers as long as there is an integrated concept? In fact, using the integration concept for medical research has been discussed by scientists in the late 20th century. These include systems biomedicine, systems pharmacology and systems bioengineering proposed as applied disciplines of systems biology, as well as systems medicine (Jin Guantao et al. 2017) which applies the ideas of system science to the medical field. Conceptually, they are exactly the same as today's academician Fan Daiming's 'integration' concept. From this, we can clearly see the attempts of modern medicine to integrate in two different directions:

1. Based on the static knowledge system constructed by analytical medicine, the bottom-up integration through 'analysis-reconstruction'.
2. Based on the achievements of biological integration, it extends to medicine as its application field.

So, should integrated medicine use the ideas of systems biology and the relatively mature methods and technologies to carry out integrated research in the field of medicine? Does the emergence of genomic medicine, systems biomedicine, systems medicine, and systems bioengineering indicate that the integration of medicine can also be achieved through extension and penetration into the medical field based on the results of biological integration?

8.3 The inoperability of 'integration on the basis of analysis'

However, after more than 30 years of research on the integration of organisms, systems biology has shown people not the wonderful prospect of integration, but the helplessness of scientists facing the complexity of life. The cause of this situation comes from the two most basic characteristics of biological systems: complexity and adaptability.

As mentioned above, there are hundreds of millions of variations in protein-coding genes in the human body. Scientists today also clearly know that the occurrence and development of human diseases are far from being completely determined by genes. Proteins, metabolites, functional activities of organs and tissues, and human daily behaviors also play a huge role in the occurrence and development of diseases. Compared with hundreds of millions of genetic mutations, the abnormality of proteins and metabolites will be a larger system that people have to face. Biomarkers at the molecular level alone are dizzying. From the molecular level to the whole, there are intricate interrelationships between the various levels and parts. The complexity of organisms has been clearly shown, so that systems biologists have issued a pessimistic sigh: The complexity of biological systems is far beyond people's imagination. At this stage, it is not appropriate to study the entire biological system, and we can only start with the study of relatively independent 'small systems' with certain functions. However, it is also far from easy to analyze a small system correctly (Sun Lanfang and Jiang Lu 2005).

From the macro level to the micro level through analysis, and then based on the understanding of the micro level, to integrate them into the whole, this is the understanding method of modern science 'analysis-reconstruction'. Analysis and reconstruction are two different stages of the cognitive process. Today's systems biology research mainly focuses on the reconstruction stage of this method. This method was more widely used in the early development of system science after the 1970s, and it is still applicable to simple systems or even simple giant systems. When modern science has thoroughly studied the problems of simple systems and gradually marched into complex systems, it has been found that it is becoming less and less useful (Sun Dongchuan and Lin Fuyong 2004).

The complexity of organisms has already made people frustrated, and what makes systems biologists even more helpless is the inherent adaptability of organisms. In experimental science, the key to a new discovery being recognized is that it can be repeated by other scientists in the same way. If there is no verification by repeated experiments, no matter how amazing the discovery is, it is a worthless 'accident'.

Repeatability is the basic principle of natural science research. It is based on a basic assumption of natural science, that is, there are universal laws under the surface facts, and everything runs under the control of universal laws, and does not vary from person, from time and from place. Scientists discover causal facts through controlled experiments that exclude subjective factors, and then derive universal laws. Therefore, in theory, any successful scientific experiment should be able to be repeatedly tested among different research subjects, which is the objectivity of science. Science does not care about individual phenomena, what it cares about is the universal law applicable to all people in all time and space. They will reappear repeatedly, never disappear or vary from person, time, and place. It is precisely because of this that scientific laws have objectivity and predictability.

Adaptability refers to the phenomenon that the system's response to the stimulus changes due to the continuous or repeated action of the stimulus, which is a major feature that distinguishes a living system from an inanimate system. However, due to the characteristics of organisms, when we carry out repeated experiments based on the concept of 'repeatability', the reproducibility of the experimental results in the strict sense often becomes a luxury requirement. In the experimental methods of systems biology, the most important research method is interference. The development of systems biology is driven by the continuous advancement of interference methods to biological systems. The adaptability of organisms greatly weakens the reproducibility of the experimental subject's response to interference, and the quantitative analysis and calculation of interference and response are an important basis for the construction of models of system biology and the analysis and integration of system behavior. Obviously, the scientific principles established in the course of hundreds of years of development of the science of simplicity seem to have encountered trouble when faced with the adaptive of complex life systems. If we adhere to the 'repeatability' of traditional science, it will be difficult to find natural laws of scientific value in an adaptive biological system; but abandoning repeatability, how can we confirm the regularity of scientific discoveries? "Since You (the Heaven) made me, Zhou Yu, why did You make Zhuge Liang too?" This ancient Chinese saying vividly reflects the frustration of systems biologists in the face of the adaptability of living organisms.

The adaptability of an organism is a reality, and it is an innate characteristic of an organism given by 'God'. Therefore, the problem can only lie in the methods our current science uses to study it. That is to say, the research methods currently used in systems biology may be inherently unsuitable for studying life with complex adaptability.

Perhaps, systems biology's integration based on the past model will still achieve significant scientific discoveries and application results, just as traditional molecular biology, cell biology, and analytical medicine have not yet come to an end. However, in order to achieve its original goal of integrating organisms into the whole, and to confirm the regularity of scientific discoveries of complex adaptive systems, it may have to introduce new methods and evaluation principles, and find another way. However, in traditional Chinese medicine (TCM), medical practitioners from ancient times to the present have always used a very simple way to effectively regulate/control the disease process in the complex human body. Here, complex adaptability is not an obstacle that increases the difficulty of solving problems, but rather a boost factor to make humans simpler and more effective in regulation-controlling complex systems (Yuan Bing 2016c).

What kind of enlightenment do the difficulties and crises encountered by systems biology in the integration process of organisms provide for today's medical integration? In addition to the integration based on analysis, is there any other way for medicine to construct an integrated theoretical system? Applying the integrated method of systems biology to the field of medicine, how will medical scientists deal with the problems caused by the complexity and adaptability of the human body and disease? How to deal with the complex comprehensive medication problem of medications with multiple side effects? How to integrate based on the treatment methods of modern medicine, including the personalized target medications of precision medicine so as to take medicine's regulation-control of diseases towards the holistic level? These are all questions worthy of serious consideration by the advocates of integrated medicine and the medical community.

In fact, in the field of natural science, the discovery of the 'emerging' phenomenon of complex systems is itself a death sentence for the integration based on the analysis method used in the study of complex systems. The birth of complexity science and complex system research methods is the latest progress in natural science's response to complex system research on this basis. Systems biology decomposes the whole organism into relatively independent parts or chains and then conducts research. Although this kind of limited research relative to the whole adopts the concepts and methods of complexity science, this kind of research is still reductionist in nature for the organism as a whole. Obviously, modern biology and medical research not only needs to keep up with the advancement of natural sciences at the technical level, but also must follow up at the methodological level.

Chapter 9

Towards the Medical Era with Holism and Personalized Idea as the Mainstream

As mentioned earlier, precision medicine is essentially based on the concept of reductionism. With the establishment of the new disease classification system (DCS) and target medication system, comprehensive treatment of diseases is still a problem that precision medicine has to face and is more difficult to solve. Since the prospect of achieving integration on the basis of analysis is slim, modern medicine will have to find another way to move toward holistic integration.

After the 1970s, as the dream of 'towards integration on the basis of analysis' was shattered, modern science has gradually developed the complexity science of focusing on the behavioral laws, modeling, and regulation-control of complex systems in the process of exploring the coping methods of complex systems. So, how will modern medicine, which has been closely following the development of modern science in the course of its development in the past few hundred years, conform to the general trend of scientific development and realize the transformation of the medical development model?

9.1 The return of modern medicine toward TCM under the guidance of complexity science

In the past few hundred years, the advancement of medicine and the advancement of natural sciences have been almost synchronized. Every step of the advancement of natural science methods and technology will push medical research in related fields forward. However, the introduction of new scientific concepts is different from the introduction of technical methods and means. It is sometimes accompanied by a certain degree of reconstruction of the existing knowledge system of the subject according to the new concept. In the process of accepting the new scientific concept to reconstruct the knowledge system of the application field, the scientist gradually solidified this scientific concept into a new way of thinking in this field. The introduction of technical methods and means can be done anytime and anywhere, and the more the better. However, the scientific ideas belonging to the methodological level often involve the structure of the theory and require relative stability. It takes time to reconstruct a discipline or field according to a new concept. It is impossible for scientists in the application field to reconstruct the knowledge system according to one concept today, and to overthrow it and rebuild again according to another concept tomorrow.

The current knowledge system of modern medicine is basically constructed under the guidance of the reductionist concept that predominated the natural sciences before the 1970s. As the frontier of modern biology, systems biology is a reconstruction of traditional biology by introducing the concept of systems science since the 1980s. After more than 30 years of research, systems biology, which is ahead of medicine, has greatly deepened human's understanding of the dynamic characteristics of organisms. However, for an open and complex giant system such as the human body, the fatal impact of the 'uncertainty principle' in organisms on the reliability of analytical data, and the technical difficulty of integration caused by the huge amount of data and overly complex interrelationships are all insurmountable obstacles of integration from the micro-level to the macro-level. No matter how advanced the technical means, the reductionist approach of 'reconstructing on the basis of analysis' will ultimately not work for the overall integration of complex life forms. In recent years, systems biology has succeeded in reorganizing individual parts or chains through simulation methods after decomposing the whole into parts. However, it is unimaginable for systems biologists to extend this local or single chain success to the whole that is millions of times more complex. The difficulty in the holistic study of organisms caused by complexity and the fatal impact of adaptability on the repeatability of experiments have left scientists at a loss. Today, systems biologists seem to have seen the end of this discipline: it may be time for modern biology to have to introduce new scientific concepts and rebuild it again.

With the development of complexity science today, it has abandoned the early "analysis-reconstruction" method in dealing with complex systems and began to face complexity directly. Based on the investigation of system input and output, through metaphors and analogs, the system's behavior model is directly constructed from the holistic level. From a methodological point of view, this is a return to ancient holistic science. The difference is that today's metaphors and analogies are not limited to the human thinking level, but can be realized based on big data analysis and machine learning with the help of computer simulation. The state description based on the model will not stay at the intuitive level of experience, and can be established on the empirical basis of rigorous statistical analysis. Today, in the process of studying complex systems such as ecosystems, geological systems, and social systems, complexity science usually uses this method to build simulation models.

Today, in the era when the concept of complexity science has gradually become the mainstream scientific concept, modern medicine that has always followed the concept of mainstream science and used progressive scientific methods and technical means to promote its own development, can only follow the path pioneered by today's complexity science to move towards integration: Leaving aside the entity structure of the human body, metaphor and analogy methods are used to construct the simplified model of the human body from the holistic level. Then using empirical-based qualitative and quantitative analysis methods and computer simulation technology, through the verification and continuous improvement of the model, a theoretical model that can be used to grasp the human body as a whole and comprehensively control human diseases is finally reached. Therefore, the holistic model of the future of modern medicine should obviously have the following characteristics:

1. *Simplify*: For models to be easy to understand and easy to construct, they must be as simple as possible. Simplification is to make the levels involved simpler first, that is, reduction of levels. As a holistic model, only the micro-levels that are relatively far from the whole can be reduced, because a dynamic model containing 60 trillion cells is impossible to understand or construct; the second is to reduce the number of subsystems and ignore the interrelationships that have little impact on the overall performance. Under the premise of ensuring the structure necessary to describe the overall characteristics and behavior, the number of introduced subsystems is as small as possible and the interrelationships are as simple as possible.
2. *There is no one-to-one correspondence with the entity structure of the human body*: The research of complexity science shows that the same function can often be realized by more than one structure. Therefore, it is impossible to draw conclusions about the true internal structure and actual operating mechanism of the system only by investigating the external behavior of the system. Obviously, after ignoring the micro-level structure of the human body, we cannot expect the holistic model to correspond to the entity model at the level of modern medical organs. That is to say, based on the structure of the organ and system level,

the human body model of modern medicine can no longer fully explain human physiology and pathological processes. This means that the sub-systems of the modern medical holistic model cannot be divided according to the anatomical organs and systems, but only according to the functions of the human body at the holistic level. That is to say, like traditional Chinese medicine (TCM), the sub-systems of the holistic model cannot find anatomically corresponding solid organs. Such an internal structure is no longer a physical structure of the human body in an anatomical sense, but more like a functional module in computer software design, that is, a functional subsystem.

3. *From macro to micro, it can be verified by observation and experiment*: As a branch of modern science, the establishment of modern medical theory is inseparable from observation and empirical evidence. But such theoretical models can no longer establish experimental or statistical description methods from micro to macro. It can only establish a certain model structure first, like theoretical physics, and establish the correspondence between human body observable variables and model state variables through corresponding rules. As a result, the expectations made in accordance with the model and corresponding rules can be verified through observation and experimentation.
4. *Combine qualitative and quantitative, and introduce computer simulation technology in the modeling process*: In recent years, at the technical level, every step of progress in modern science and technology will be applied to the medical field in the shortest possible time. Quantitative technology and computer simulation technology have been widely used in medical testing, diagnosis, and treatment since their debut, but they have not been used in the construction of holistic models. The complexity of the holistic model and the seamless connection between modern medicine and modern science and technology make the application of quantification and computer simulation technology a natural and inevitable process when modern medicine is moving towards integration.

When we look at the human body model formed in TCM for thousands of years in such a form of holistic model and the modeling procedure, we notice that the first three points mentioned above are precisely the most important features of the TCM human model. The verifiability and quantification of the model are the first problems that must be solved in the modernization of TCM, and computer simulation technology is a technical means that can be used in the modernization of TCM.

Science is leading mankind's understanding of the world to return to holism, and TCM which originated in ancient times is based precisely on the concepts and methods of ancient holism. Today, life sciences and medicine adopt the method of complexity science to deal with complex systems to establish human models and state descriptions, which is undoubtedly a return to TCM. The difference is that with the introduction of computer simulation technology and empirical concepts, the holistic human body model and SDS will be laid on a rigorous empirical basis. In addition, the clinically significant detection indicators developed by modern medicine (including precision medicine) provide new evidence for the overall model to deepen the description of human physiological and pathological processes; and the effective treatment methods of disease medicine and the targeted drugs with personalized characteristics of precision medicine have laid a solid clinical application foundation for the disease treatment of holistic medicine. The latter will be discussed in the following chapters.

The theoretical system of TCM was gradually formed and completed by medical scholars in their long-term struggle against various diseases that harm humans. It not only has a complete model, but also has disease information collection methods, syndrome diagnosis systems and corresponding treatment systems that are compatible with the model. The 'different paths and the same goal' of modern medicine toward integrated and TCM toward scientific provides a possibility: The human body model established and gradually perfected in the development of TCM for thousands of years, after scientific and standardized transformation, directly serves as the prototype of the holistic model to be established in modern medicine. On this basis, the model system will be developed and improved through the following measures:

1. Based on the principle of 'the simplest and most applicable',
2. Follow the norms of modern science for establishing models and the mechanism for developing and improving models,
3. Introduce quantitative technology, modern detection technology and computer simulation technology,
4. Integrate the syndrome differentiation and treatment of TCM with disease medicine, detection indicators of precision medicine and the corresponding regulation-control system.

From this, it is expected to gradually develop a human body model and corresponding state description system that not only embodies the overall synthesis but also goes deep into the micro level. Based on this, we can achieve precise descriptions and optimal control of the human body's personalized state.

Such a medical system based on the holistic approach to establishing human body models and state descriptions will maximize the integration of knowledge, experience and technology from TCM and modern medicine. It has transcended the narrow geographical and cultural concepts of Eastern and Western medicine and will lead to an unprecedented revolution in concepts and thinking in the medical field. In today's era when the concept of complexity science has gradually become the mainstream scientific concept, we call this brand-new medical system holistic medicine (Yuan Bing 2010).

The term Holistic Medicine has long existed in the West, but the Holistic Medicine we are proposing today has been given a new meaning in the scientific environment of the 21st century. It is a brand-new medical system based on the concepts and methods of complexity science and embodies the holistic concept of TCM and the empirical methods of modern science. It establishes a functional model of the human body at the holistic level, and realizes the personalized description and regulation-control of the patient's pathological state based on this model.

Holistic medicine will still follow the methods of TCM and complexity science, using metaphors and analogies to build functional models that reflect human physiological and pathological activities. But with the help of modern scientific structural analysis concepts and methods, it will give the model a rigorous logical structure. And, with the help of today's big data analysis technology and machine learning, the model can be realized through computer simulation with empirical support. Holistic medicine will still use TCM treatment based on syndrome, but by introducing modern scientific concepts and methods, the SDS based on TCM syndromes can be scientifically improved in terms of the completeness and independence of state variables. Holistic medicine's grasp and regulation-control of the human body's state is holistic and comprehensive. However, based on the principle of 'the simplest and most applicable', by introducing detection methods developed in disease medicine and precision medicine and valuable detection indicators for its state description, the SDS can be refined, and the effective treatment methods of modern medicine can be included in it, so as to achieve the integration of traditional Chinese and Western medicine at the application level.

In the field of biology, after the emergence of systems biology, biology is mainly divided into two major areas: traditional biology and systems biology. Systems biology research does not cover all aspects of biological research, but on the basis of traditional biology, it conducts overall research through modeling. It combines experimental methods with calculation methods, focusing on revealing the dynamic relationship of interactions and mutual influences among various parts and elements. Similarly, holistic medicine will not cover all aspects of the traditional research fields of modern medicine and the personalized research of precision medicine. However, as a medical system that looks at the holistic level and strives to fully grasp the law of disease development and evolution and regulation-control, it will adhere to the principle of 'the simplest and applicable', and is committed to incorporating the law of diseases discovered in modern medicine and effective regulation-control methods into the holistic medicine's treatment system based on disease differentiation. In other words, Holistic Medicine will integrate the achievements of modern medicine in the research field related to the regulation-control of human diseases, and gradually build a new system.

On the other hand, TCM is not a scientific system constructed strictly in accordance with certain definite research methods, but a 'container' that encompasses all aspects related to the human body, health, and disease. When modern medicine emerged from ancient traditional medicine, it abandoned many valuable things in traditional medicine. Of course, holistic medicine cannot cover all aspects of TCM, but its emergence is significantly different from the emergence of modern medicine from traditional medicine. It is a new medical system constructed based on scientific concepts on the basis of inheriting the core methodology and theoretical system of TCM.

As a new medical system based on TCM and modern medicine, it will involve many aspects of the human body and disease research, such as

1. Construction and improvement of human body model, etiology model, medication model, and herbal formula model;
2. The construction of the state description system (SDS) and the improvement of state identification methods;
3. Construction of DCS and improvement of disease diagnosis methods;
4. Research on the evolution of body state under the influence of various factors;
5. Study the general rules of state regulation-control and regulation-control methods for specific state variables and specific body states;
6. The construction of environmental models and spatio-temporal models related to human health, and exploring the impact of the environment on human health based on the model.

In the following content of this chapter, several of these aspects will be discussed to a limited extent.

9.2 The view of the human body and the construction of the human body model in holistic medicine

Like TCM and modern medicine, the study of holistic medicine is centered on the human body. The upper level may involve the natural environment and social environment where people live; the lower level is the various components that make up the human body. The research of holistic medicine adopts the holistic approach, that is, the research is conducted without separating the dynamic relationship between the part and the whole, in order to establish a functional model that reflects the holistic characteristics and behavior of the human body. It can directly introduce the holistic model of TCM as the prototype of holistic medicine, but this prototype is by no means the final model of holistic medicine, rather, it is a transitional model that can be continuously modified and improved based on the results of empirical verification.

As mentioned earlier, the models built according to the holistic approach also have parts and internal structures, but they are definitely not parts and internal solid structures in an anatomical sense. Modeling from the overall level focuses more on the functions of each part and the connections among them. Therefore, each part of the model is actually a functional subsystem, just like functional modules in the 'object-oriented programming' of computer software. This modeling method was not only applied in ancient TCM, but also in the construction of theoretical models in many fields of contemporary natural sciences. In the 1990s, the theoretical model of a complex adaptive system established by the Santa Fe Institute in the United States also used this method.

Also like TCM, in order to make the model easy for people to understand, the division of its functional subsystems cannot be completely separated from the entity organs and morphological structure of the human body, nor can it be completely separated from the experience of people's daily lives. Just as the heart in anatomy, which is responsible for promoting blood circulation throughout the body, has been endowed with the function of 'housing the spirit' and presiding over the mental consciousness and thinking activities in TCM; the spleen which in anatomy is an immune organ, is endowed with the function of being responsible for the absorption of human food and the transport of the absorbed essence in the TCM model.

In this sense, the human body model established by holistic medicine is similar to the conceptual model of the atomic structure constructed by theoretical physics entirely based on the imagination of scientists. With the aid of this atomic model, people can explain the characteristics of atoms and make more accurate predictions about their behavior. As for the actual structure of the atom under a high-power electron microscope, it doesn't matter much. Similarly, with the help of such a model, holistic medicine can form a relatively complete explanation of the basic physiological and pathological processes of the human body, and can also be used for disease analysis and disease development, and outcome prediction, and effectively guide the clinical treatment. Using the method of establishing the system function model from the holistic level, even if we do not understand the actual internal structure of the real system, we can still grasp the characteristics of the system, accurately predict the behavior of the system, and carry-on effective control to it. This is the reason why TCM has not declined after thousands of years, and still radiates dazzling brilliance even with the rapid development of modern science.

The research of modern cybernetics shows that in the holistic model established by investigating only the system's response under the influence of external factors, due to the different influencing factors investigated, the observed system behavior is different, and the internal structure obtained will be different. Even if the influencing factors investigated are the same and the observed system behavior is the same, different researchers may introduce different structures and build different models. Therefore, in the development process of many fields of natural science, there is a stage in which more than one model coexists. For example, the optical wave model and particle model coexisted in the optical field from the beginning of the 17th century to the beginning of the 20th century. Another example is the coexistence of quantum mechanics and relativity in theoretical physics from the beginning of the 20th century to the present. And in TCM, the viscera-qi-blood-body fluid model, the six meridians model still coexists with the wei-qi-ying-xue model, and the triple-jiao model, each of which is independently used for disease analysis in different scopes.

Generally speaking, integrating several models with a small coverage into a model with a larger coverage expands people's horizons, and gives people a deeper understanding of objective objects, making their control more effective. In physics, the dispute between the wave theory of light and the particle theory ended in the 'wave-particle duality' of light in the early twentieth century. The establishment of the 'wave-particle duality' model of light led to the birth of quantum mechanics, a major achievement in physics in the 20th century. It greatly deepened people's understanding of the microscopic world, and also contributed to the rapid development of science and technology in the 20th century. In the larger view of physics, the theory of relativity shows people the cosmoscopic picture of the real world, and quantum mechanics shows people the microscopic picture of the real world. But the real world is a unity of micro and macro. Therefore, the establishment of a unified field theory that can unify the cosmic world and the micro world is the dream of modern physicists. It will trigger an unprecedented breakthrough in the history of physics, greatly advancing the human understanding of nature in terms of depth and breadth.

Similarly, if several existing syndrome model systems in TCM are combined into one, it can undoubtedly deepen people's understanding of the laws of human physiology and pathology, and improve the comprehensive analysis and control of human diseases in TCM to a higher level. Therefore, holistic medicine can introduce the human body model of TCM as the starting point for model research. A better choice is to construct a more inclusive holistic model that can integrate several existing syndrome model systems in TCM into one, as the basis for the study of the holistic medical human body model.

Holistic medicine is a theoretical medical system that studies the laws of the human body's physiological activities and the laws of interaction with the environment from a holistic perspective, and studies the laws of the occurrence and development of diseases, and the methods of adjustment and control. Different from modern medicine, it does not focus on revealing the entity structure of the human body and the actual mechanism of the physiological and pathological process. Instead, it reveals the intrinsic laws of the human body and diseases from a functional perspective through the dynamic investigation of physiological and pathological activities of the human body, thereby accurately predicting and comprehensively regulating and controlling human life processes and disease processes.

In the medical field, the so-called differences in disciplines are essentially the differences in the human body outlook, and all subsequent research and control methods are based on the human body outlook. The human body view of holistic medicine is:

The human body is a living organism with a high degree of self-organization, self-adaptation, and self-adjustment, and an open and complex giant system with nonlinear, time-varying, random, and fuzzy nature. The various parts of the human body, the human body and the external environment (including nature and society) are interconnected and restricted to each other, forming an indivisible whole. The human body system is composed of a batch of subsystems with different functions, and the subsystems are connected and restricted in many different ways to maintain the life process of the human body. If we stipulate that the functional state of each subsystem is represented by a set of state variables, then the set of all state variables of all subsystems constitutes the state space of the human body. The state of the human body is determined by its initial state, the characteristics of the human body itself, and external input. And as the output of the human body, any changes in the shape, function, subjective feeling and objective detection indicators of the human body are determined by the combination of the human body's state and input.

Under normal circumstances, the various subsystems of the human body are interconnected and restricted, and the human body's adjustment mechanism responds to changes in the external environment, maintaining a relatively stable dynamic balance within the human body, between the human body and the external environment. As a result, the human body system completes the life process in accordance with specific birth, growth, maturity, and aging procedures. Disease means that the balance between the various subsystems, the human body and the external environment is broken, some state variables of the system deviate from the normal range, and the movement of the system deviates from the normal life process. If $X_1^0(t)$, $X_2^0(t)$,…$Xn^0(t)$ are used to represent the normal state of the body, the disease can be described by $|X_1(t)–X_1^0(t)|$, $|X_2(t)–X_2^0(t)|$,… $|Xn(t)–Xn^0(t)|$. The normal life program and pathological life program can be respectively described as $f(X_1^0(t), X_2^0(t),...Xn^0(t))$ and $g(X_1(t), X_2(t),...Xn(t))$. Obviously, the task of regulation and control is to intervene by appropriately changing the input of the human body system (such as medication, acupuncture, injection, physical therapy, psychotherapy, etc.).

$$let\ |Xi(t)–Xi^0(t)| < \varepsilon\ (i = 1,2,\dots n)$$

$$let\ |f(X_1^0(t), X_2^0(t),\dots X_n^0(t))–g(X_1(t), X_2(t),\dots X_n(t))| < \varepsilon$$

That is to restore the steady-state balance between the human body's various subsystems, between the human body and the external environment, and restore the human body's normal life program (Yuan Bing and Zhao Zheng 1986).

For complex systems with self-organization, self-adaptation, and self-regulation capabilities, in order to simplify the identification and regulation control of system states, we can introduce a batch of state variables that represent the functional state of the system as a whole. Using state variables at the overall level to express the effects of medications makes it easier to grasp their macroscopic, indirect and long-term effects produced through the human body's self-organization, self-adaptation and self-regulation capabilities. What should be explained here is that the state variables at the two levels have a certain correlation, but the state variables at the overall level are by no means a simple addition of the state variables at the subsystem level. The effects of therapeutic drugs targeting state variables at the overall level often cover related state variables at the subsystem level. For example, in TCM, the effect of qi-tonifying medicines that target qi deficiency at the overall level often covers qi deficiency in different organs at the subsystem level.

9.3 State description method of the holistic medicine

We collectively refer to the energy, material and information that people ingest and obtain from the external environment in an active or passive way in the process of life as the input of the human body, including the impact of nature and society on the human body, the energy and substances that the human body takes from the outside world for survival, and the artificial intervention in order to regulate and control the state of the human body and the direction of disease. We collectively refer to the various energies, substances, and information that people release to the environment in the course of their lives as the output of the human body, including the excretions and pathological products of a person's normal metabolism, the subjective feelings (symptoms) that a person can express, and the various physical signs and detection indicators that can be observed from his or her external appearance and behavior. Here, we collectively refer to the various sensible and observable quantities in the output of the human body as the clinical findings of the internal state of the human body.

The holistic medicine links the state variables in the human body model with clinical findings that people can feel and observe according to certain rules. Thus, according to the model, the output of different body states under the action of various inputs can be predicted, and the model can be verified and perfected according to the corresponding relationship between the input and output.

9.3.1 Description of human body state: state variables and state space

Generally, it is conducted according to certain rules that expectations based on the overall model are attributed to empirical observations and experiments. After dividing the human body into different sub-systems, corresponding state variables should be introduced to characterize the functional state of each sub-system. Then establish the corresponding relationship between the state variables and the human body sensible/observable/detectable clinical finding variables, so that the state of the body can be described through the model. For example, after observing the clinical finding group that appears on the human body at time t, based on the correspondence between the clinical finding system and the state variable system, the state interval Xi(t) where the corresponding state variable Xi can be determined, thereby determining the body state X(t):

$$X(t) = \{X_1(t), X_2(t), \ldots\ldots X_n(t)\}$$

[Among them, Xi(t)(i = 1,2,...n) represents the state interval of state variable Xi at time t] (Yuan Bing and Zhao Zheng 1986).

From the relationship between the state variables and the influence of various external inputs on the state variable system of the human body, the changing trend of its state can be inferred. From this, we can analyze, understand and grasp the physiological and pathological process of the human body based on the model, and guide its effective regulation and control.

The effective treatment system based on syndrome that has been used in TCM for thousands of years was established in this way. A single basic syndrome in TCM, such as heart-qi deficiency, usually refers to the pathological state interval of a state variable, as described in Part 1, Chapter 4, Section 4.4. A compound syndrome composed of more than one basic syndrome combination, such as deficiency of both qi and blood, and deficiency of both yin and yang, refers to the state subspace formed by the combination of pathological state intervals of more than one state variable. When the state interval of each state variable at a certain moment is determined, a set of values composed of the state intervals of all state variables at that moment uniquely determines the state subspace of the human body at that moment.

From the holistic level to establish the human body model and state description, the model usually has only two levels, the holistic system level and the subsystem level. Correspondingly, there are naturally only two levels of state variables. For example, the state variables describing the holistic level include qi deficiency, blood deficiency, and qi stagnation; the state variables describing the subsystem level include lung-qi deficiency, heart-blood deficiency, and liver-qi stagnation. The sum of state variables at the holistic level constitutes the state space at the holistic level of the human body, and the sum of state variables at the subsystem level constitutes the state space of the human body at the subsystem level. In the process of constructing the model in TCM, due to the introduction of the sub-systems of the viscera and organs, a two-level description of the state is formed. In other traditional medicine, there is usually only one level to describe the state of the human body, such as blood, mucus, yellow bile, and black bile in the 'four body fluids' of ancient Greek medicine; earth, water, fire, and wind in ancient Indian medicine; and Taiyin, Taiyang, Shaoyin and Shaoyang in Korean medicine.

Today, in modern medicine, only understanding the cause, location, and pathology of the disease is far from being able to determine the occurrence, development, and transformation trends of the disease in the human body, and it is far from being able to effectively treat diseases with a certain degree of individualized characteristics. In physiology research, differences in the response and adaptation of human individuals to external stimuli under physiological conditions are attributed to differences in constitution; in pathogenesis and pathology, differences in the susceptibility of human individuals to certain pathogenic factors and the tendency of disease development are attributed to differences in constitution. In the treatment of diseases and the analysis of curative effects, the differences in the response and curative effect of human individuals to medications and treatment methods are also considered to be caused by constitution. Precision medicine shows the prospect of human beings being able to effectively grasp and control individualized differences from a micro level. And at the holistic level, how to describe a person's constitution, how to distinguish the differences in constitution, and how to conduct personalized treatment in response to different constitution?

Since the 1970s, some visionary scholars in TCM, represented by Professor Wang Qi (1943), have realized that it is possible to establish a description and regulation-control system of human constitution based on the treatment system based on syndrome. With the unremitting efforts of Professor Wang Qi and his research team for more than 30 years, the research of TCM constitution continues to advance the understanding of human constitution classification and the improvement of its regulation-control methods. In 2009, under the auspices of the State Administration of TCM, the Chinese Society of TCM formulated the 'Classification and Judgment of TCM Constitution' standards (Wang Qian 2009). Since then, the TCM constitution theory began to become standardized and has been recognized by the TCM community, and has also attracted the attention of the medical community around the world.

The standard of Classification and Judgment of Constitution in TCM divides the constitution into nine basic types: calmness, qi deficiency, yin deficiency, yang deficiency, qi stagnation, blood stasis, phlegm-dampness, damp-heat, and idiosyncratic characteristics (China Association of TCM 2009). Obviously, this constitution classification standard is basically based on the TCM syndrome classification. Among them, calmness constitution is a normal and unbiased state of people; qi deficiency, yin deficiency, yang deficiency, qi stagnation, blood stasis, phlegm dampness, and damp-heat constitution were introduced based on the syndromes of the TCM treatment system based on syndrome; only idiosyncratic constitution is more special, which is based on describing the allergic state in modern medicine (Yuan Bing 2015b).

According to scientific concepts, to classify things at the same level, usually only one principle can be used at a time. If two or more principles are adopted at the same time, things which belong to a category according to one principle should be classified to a different category according to the other principle. Such a classification system will inevitably have contradictions and repetitions, and it is difficult to have rigorous logic.

In modern medicine, blood type is a method of classifying blood, usually referring to the classification of red blood cells. At present, there are 30 blood group systems that have been discovered in broad human populations and recognized by the International Association of Blood Transfusion. Among them, the ABO blood group system and the Rh blood group system (Rhesus factor) are the most important. The ABO blood group system is based on the presence or absence of different combinations of type A and/or type B antigens on the surface of human red blood cells, divided into four blood types: A, B, O, and AB; the Rh blood group system is based on the presence or absence of Rh antigens on human red blood cells (Rhesus factor), and divided into two blood types, Rh+ and Rh–. Obviously, the ABO blood group and the Rh blood group are independent blood group classification systems to each other and are based on two different classification reference points. Just imagine, if someone combines the blood types measured by the two classification methods into a blood type system that includes five blood types: A, B, O, AB, and Rh+, it will definitely make people with common medical knowledge feel inexplicable. The constitution classification standard currently promoted by the Chinese Society of Chinese Medicine is similar to such a classification system (Yuan Bing 2016b).

In this standard, the eight constitutions of calmness, qi deficiency, yin deficiency, yang deficiency, qi stagnation, blood stasis, phlegm-dampness, and damp-heat are relatively independent, but the idiosyncratic constitution does not have such independence. The experience of TCM treatment of allergic diseases shows that for allergic patients, it is still necessary to formulate treatment plans from person to person on the basis of syndrome differentiation in order to receive good results. In TCM, syndromes at the holistic level can clearly classify and describe the individual differences of patients at the holistic level, which can obviously be used to classify and describe human health-related constitutions. The treatment methods for these syndromes developed in TCM obviously can effectively regulate/control the corresponding constitution types and their combinations. But the introduction of idiosyncratic constitution, 'a piece of rat feces ruined a pot of porridge', made the system used to describe human physique nondescript and lost its rigorous scientific nature. Realizing the major flaws of the 'standard' and the confusion it caused, in July 2013, the 10 constitution types proposed in the article 'The Development Trend of Constitutional Medicine Based on Holistic Medicine' abandoned the special constitutions in the 'Standard', and added two types of constitutions: blood deficiency and fiery heat (Yuan Bing 2016b, Yuan Bing 2013a). Of course, how many and what batch of basic syndromes are used to describe the human body constitution is more appropriate, depends on the completeness and independence of the state variable (syndrome) system at the holistic level, and must satisfy the principle of simplest applicability. Among the state variables (syndromes) at the holistic level that satisfy the above conditions, only state variables with a certain degree of stability can be used to describe the human constitution. Syndromes describing exogenous diseases or critical illnesses are obviously not suitable for describing the human constitution because they do not have sufficient stability.

The holistic level of the human body and the collection of all state variables involved in all subsystems constitute the description system of the state of the human body, in which there are currently less than 100 state variables. During clinical diagnosis, as long as the abnormal state variables are identified through the symptoms and signs of the patient, the patient's health status can be comprehensively understood.

It is very meaningful to divide the description of the human body state into layers. The recognition of the state of the holistic level (constitution recognition) helps to grasp the overall health of the patient, and the recognition of the state of the subsystem level allows us to further refine the patient's disease. The important thing is that the state at the holistic level and the state variables at the subsystem level are related to each other. For example, the qi deficiency at the holistic level is related to the lung qi deficiency, spleen qi deficiency, heart qi deficiency, and kidney qi deficiency at the subsystem level. The state recognition at the two levels shows the state deviations at the holistic level, and also shows which subsystem functions these deviations mainly affect. Treating state variables at the holistic level reduces the complexity of treating multiple subsystem abnormalities at the subsystem level. And knowing which subsystem the abnormality mainly occurs can make the treatment plan more targeted to the subsystem where the disease occurs. This kind of treatment plan not only highlights the key points, but also takes care of all aspects.

In natural sciences, for some relatively simple systems, state variables are usually variables that can be directly detected or observed. But in the human body, a complex living organism with multiple levels and closely related parts, the relationship between state variables that describe the functions and attributes of the whole or subsystems at a macro level and the sensible or observable clinical finding system often present a many-to-many network relationship. That is, a state variable is associated with multiple clinical finding variables, and a clinical finding is usually associated with multiple different state variables. Therefore, the association between state variables and clinical findings often does not have a one-to-one specificity. Each state interval of each state variable corresponds to a set of clinical findings consisting of symptoms, signs and detection indicators. The association between each clinical finding and the state interval of the associated state variable presents a certain statistical certainty, and the specificity of one-to-one correspondence has become a rare exception.

The most basic clinical findings that define each state interval of the state variables are determined when the model is conceived, and the state variables are introduced. For example, in TCM, the deficiency of heart yang is defined based on the most basic symptoms of palpitations, chills, and cold limbs. Heart-blood deficiency is defined by the basic symptoms of insomnia, dreaminess, forgetfulness, heart palpitations, and pale complexion. With such a batch of state variables, and when the corresponding relationship between the state interval of each state variable and the observable appearance of the human body has been established, clinically, based on these clinical manifestations, it can be determined which state variable deviates from the normal state, and the direction and extent of the deviation, thereby determining the state of the body.

9.3.2 The establishment of the corresponding relationship between state variables and clinical findings

State variables are introduced based on preliminary investigations of human physiological and pathological phenomena. At the beginning of the introduction, a state interval of a state variable was only defined by a few basic clinical findings. Science always strives to find certainty, here it is to find the observable variables of the human body that have a linear correspondence with the state variables. However, the experience of TCM for thousands of years tells us that in the human body, a whole in which the functions of various parts are closely related to each other, it is very difficult to find the observable variables that have a qualitatively one-to-one correspondence with the state variables, let alone a quantitative linear relationship. In this way, it is difficult to accurately determine the state variables that deviate from the normal state and the state interval in which they are based only on basic clinical findings of the state variable originally defined. That is, it is not easy to accurately identify the patient's syndromes. In recent decades, the bottleneck that has hindered the modernization of TCM theory, Chinese herbal medicine (CHM), and TCM testing technology is the standardization and objectivity of the TCM syndrome system. To this end, it is necessary to use statistical methods or a combination of qualitative and quantitative methods to find observable variables of the human body that are meaningful for state variable identification through a wider range of clinical observations.

Modern medicine has introduced advanced testing instruments and adopted methods that combine qualitative and quantitative methods. It can be said that the diagnosis of diseases has relatively accurate and objective standards. Holistic medicine constructs a holistic model of the human body, and the standardization and objectification of the corresponding relationship between the state interval (syndrome) of state variables and observable variables of the human body is also a problem that has to be faced and solved. In recent decades, the development of system science, complexity science, and the study of TCM expert systems since the 1980s have provided a theoretical and practical basis for the solution of this problem: On the basis of expert experience, a combination of qualitative and quantitative methods are used to conduct repeated iterative verification and practical verification using computer simulation technology or human-machine integration. From this, a mathematical model that reflects the corresponding relationship between each syndrome and the observable variables of the human body can be established. The specific steps are as follows:

1. *Expansion of the observational variable set required to determine the state interval (syndrome) of state variables, and introduce a batch of objective indicators with higher specificity.*

   To find a batch of clinical findings corresponding to a certain state interval of a certain state variable, it is not an easy task for the medical community, and a lot of clinical statistics and scientific experiments are required. For example, if we want to determine the clinical findings closely related to the state interval (syndrome) H21-Deficiency of Heart qi in Table 4.5 in Chapter 4, we design the experiment as follows: Since heart-qi deficiency is defined by the basic clinical findings of palpitations, shortness of breath, fatigue, especially after activities, and pale complexion, we will count a large number of cases with the above signs and select appropriate methods to test these patients. Measuring heart rate, blood pressure, electrocardiogram, extremity rewarming time, cardiac imaging, etc., and statistically processing the test results, we can calculate the conditional probability P(ω/H21) of each test index when the diagnosis of heart-qi deficiency is confirmed. It can be limited: Among these patients, only those who are treated effectively with the method of replenishing the heart and energy have statistical significance. If it is found that the heart rate is slowed down, tachycardia is caused by slight activity, the blood pressure is low, the pressure difference is small, the electrocardiogram is low voltage, the imaging shows that the heart enlargement and other indicators are more likely to appear in this batch of patients, then these indicators can be used as observable variables to determine the deficiency of heart-qi.

2. *Through statistical analysis and experiments, determine the diagnostic significance of these observable indicators to the state interval (syndrome) of the corresponding state variables.*

   After screening out a batch of observable variables and corresponding detection methods that are meaningful for determining the state of the state variables, these methods can be tested on a large number of patients. Based on the test results, the conditional probability $P(H_{21}/\omega)$ of being diagnosed with heart-qi deficiency when each of the above clinical findings appears can be calculated. Among them, the detection indicator of larger P(ω/H21) is often meaningful in explain the pathology, and the detection indicator of larger $P(H_{21}/\omega)$ is of greater significance in determining the appearance of the syndrome of heart-qi deficiency H_{21}. A batch of larger $P(\omega/H_{21})$ and a batch of large $P(H_{21}/\omega)$ detection indexes can be screened out from the two rounds of the above-mentioned one positive and one negative test, which can be added to the set of clinical findings that determine the heart-qi deficiency H_{21}. The two rounds of trials described here, one positive and one negative, are reflected in the actual research process as two different statistical algorithms for the same batch of clinical cases. The detection indicators that are screened out by the two algorithms and have a greater degree of relevance to a specific state variable can be added to the clinical finding set corresponding to the state variable as a basis for state variable diagnosis. In this way, the correspondence between state variables and the clinical finding system is more accurate and objective than before.

The degree of correlation between the state interval of state variables and the observable indicators of the human body can sometimes be quantified, such as blood pressure, heart rate, transaminase, alpha-fetoprotein content and other testing indicators of modern medicine. Holistic medicine will also introduce some quantifiable instrument detection indicators that are meaningful for state recognition in the establishment of the human body holistic model. As a result, the collection and identification of clinical findings not only have certainty of whether they appear, but also the degree of quantification to measure the degree of clinical findings.

Some observable indicators are necessary conditions for the establishment of corresponding syndromes, such as sore throat for exogenous wind-heat. Some observable indicators correspond to corresponding syndromes in the form of probability, such as palpitations and insomnia for heart-qi deficiency. For quantitative detection indicators, a linear correlation (if any) between the measurement values of these indicators and the degree of state variables can be established; or they can be divided into several intervals to convert quantitative problems into qualitative problems for processing. For a qualitative indicator, the diagnostic significance of the state variable (syndrome) can be quantified based on the conditional probability of the corresponding state when it appears. In the actual process of clinical findings analysis, logical algorithms and quantitative algorithms are usually selected or used in combination according to the specific conditions of the relationship between the clinical findings and specific syndromes.

As a result, the basic rules and algorithms for determining the state interval (syndrome) of state variables based on the sensible/observable variables of the human body can be determined. Furthermore, according to these rules and algorithms, with the help of artificial intelligence, machine learning or human (expert judgment) combined with machine, through the iterative deduction of computer simulation technology, the diagnosis model of syndrome can be verified. And in the verification process, by constantly modifying the rules and algorithms, let the accuracy of the syndrome identification based on the model reach a satisfactory level.

The perceived/observable variables used in holistic medicine include symptoms and signs, and various detection methods developed by modern science (including modern medicine) will also be available in the future. It will try its best to find some observable variables that have a direct correspondence with the state variables, have greater significance in determining a certain state of the body, and are relatively more objective than symptoms and signs. For example, whether the cardiac hypertrophy detected by imaging and the low voltage detected by electrocardiography can be used as observable variables of heart-qi deficiency and heart-yang deficiency; whether blood viscosity and embolism of cardiovascular angiography can be used as observable variables of heart blood stasis, etc., all of which will be questions worthy of study when holistic medicine introduces observable variables. This means that holistic medicine will integrate all available detection methods of TCM and modern medicine, and establish a more objective, more accurate measurable variable system that includes qualitative and quantitative analysis based on the TCM syndrome system. It will use logical, statistical and fuzzy mathematics methods to make the correspondence between observable variables and model state variables more accurate. On this basis, scientific empirical principles are adopted, computer simulation and artificial intelligence technology are introduced, and the model is constantly revised and improved in the verification based on experience and practice. The aim is, in the near future, to make this holistic model a rigorous and scientific theoretical model that can be verified by controlled trials (Yuan Bing and Zhao Zheng 1986).

Extending the information collected during the disease course from the external information of the human body such as symptoms and signs to the internal information of the human body that needs to be detected by instruments, including the analysis of blood and excrement, bioelectricity, and image information inside the human body, its significance is far beyond the objectification and precision of disease information detection. This means that the human body model of holistic medicine will be built on the basis of human entity structure and function. Just like in complexity science, scientists use the 'sand pile model' to describe the formation and characteristics of self-organized critical states to reveal the process of formation and collapse of sand piles. Furthermore, through metaphors and analogies of this process, it scientifically reproduces the power law that appears in the process of earthquakes, forest fires, biological extinctions, and even traffic jams in urban traffic. Holistic medicine uses the functional model constructed through metaphors and analogies to reveal the law of the evolution of the human body's state in the course of disease, and can predict the trend of the disease, and guide the regulation-control process of the pathological state.

Obviously, under the state medical framework of holistic medicine, anatomy, histology, physiology, biochemistry, cytology, and molecular biology, which are the basis of modern medical theory, have no theoretical significance. The human body is only the carrier of the symptoms, signs and detection indicators that need to be collected in the process of identifying the pathological state of the human body for holistic medicine. The role of the knowledge system constructed by these basic disciplines only exists in the process of discovering these detection indicators. Of course, holistic medicine's understanding of human health and disease processes is not limited to state medicine. As we will see later, in the disease medicine of holistic medicine, the knowledge system constructed by these basic disciplines of modern medicine will still exist and play a role. Under the framework of holistic medicine, it is only a supplement to the insufficiency of state medicine. Its positioning and the way it functions will be very different from today.

9.3.3 Standardization of the clinical findings system

To establish the corresponding relationship between the state variable system and the clinical finding system on the basis of qualitative and quantitative, an indispensable prerequisite is that the clinical finding system must be standardized. As mentioned earlier, the clinical findings include various subjective feelings of the patient, various physical signs that the practitioners can observe, and various indicators and images that can be detected by scientific instruments. Therefore, the standardization of the clinical finding system includes the accurate descriptions and degree classification of the patient's various subjective sensory responses, the qualitative definition and quantitative classification of the information observed and perceived by the clinicians, and the introduction of various objective instrumental detection methods and indicators. In this regard, modern medicine has gradually formed a set of detailed descriptions of the clinical manifestations of various diseases at various stages and has developed a set of standardized diagnostic criteria/procedure systems.

Although it occurs in the same human body at the same time, the clinical finding information needed to diagnose the disease and identify the syndrome is usually different or not exactly the same. In other words, the symptoms, signs, and detection indicators that are meaningful for the diagnosis of the disease are not necessarily meaningful for determining the state (syndrome) of the state variables relative to the model, and vice versa. For example, to distinguish between chronic gastritis and gastric ulcer, people are concerned about the nature of stomach pain whether it is painful after eating or being hungry. In order to identify the nature of the spleen and stomach's cold, heat, deficiency and excess and qi stagnation and blood stasis, medical practitioners are more concerned about the nature of stomach pain whether it is distension or tingling, whether it likes to be warm and be pressed, or distended and refuses to be pressed. To this end, it is also necessary to develop a set of clinical finding systems that are meaningful for syndrome identification, including symptoms, signs, and detection indicators. And such a system can be established with a combination of qualitative description and quantitative analysis by referring to the process of establishing diagnostic criteria in modern medicine.

The clinical findings information that patients can feel, the medical practitioner can observe or perceive, and may be detected by the instrument can be divided into three categories in terms of recognizability: One type is deterministic information that can be quantified to a certain degree of accuracy, such as blood pressure, heart rate, body temperature, etc.; the other type appears in the logical form of presence and absence, and its degree cannot be accurately quantified, such as symptoms such as headache, dizziness, nausea, etc. In the third category, not only can degree not be accurately quantified, but also the presence or absence cannot be accurately determined, such as tongue condition, pulse condition, complexion, etc. We call the latter two types fuzzy information.

Standardization of the clinical findings system means giving a unified name and definition of each clinical finding, and to determine its qualitative characteristics and quantitative range. The deterministic information is self-explanatory, the difficulty is how to grasp the vague information.

Although it is impossible to accurately quantify the former type of fuzzy information, it can be approximately regarded as deterministic information processing through the logical grading of its degree through the provisions of its characteristics. For example, heart murmurs can be divided into different types such as 'rumbling' and 'blow-like' according to their nature, and can be roughly divided into six levels according to their degree:

Grade I: The tone is extremely weak and short, requiring careful auscultation to hear;

Grade II: The sound is soft, the time limit is not too long, and it is easier to hear;

Grade III: Medium intensity, easy to hear;

Grade IV: The sound is loud and very easy to hear;

Grade V: The sound is very loud, rough, and has a long time, but it cannot be heard when the entire chest end of the stethoscope is slightly away from the chest wall;

Grade VI: The sound is extremely loud, and can still be heard even when the whole chest end of the stethoscope is slightly separated from the chest wall.

More roughly, the degree of stomach pain can be divided into three levels:

Grade I: Dull pain, discomfort, but does not affect general activities;

Grade II: More severe pain, restricted body position, affecting general activities;

Grade III: The pain is severe, and the patient cried, howled, and rolled, which was unbearable.

The latter type of fuzzy information can be grasped by decomposing it into a set of more basic variables. For example, the pulse condition can be characterized by a batch of variables such as pulse rate, pulsation amplitude, waveform inflection point position, and amplitude of each harmonic. The totality of these basic variables constitutes the state space of pulse conditions, and each pulse condition is a fuzzy subset defined in the pulse condition space. The qualitative grasp of the pulse condition can be done by roughly grading each basic variable to determine the degree of relevance of each basic variable to each pulse condition subset that is related to it, and then use fuzzy mathematics to solve it. The quantitative description of the pulse condition can be solved by finding a set of characteristic variables for each pulse condition subset from these basic variables, and then establishing the functional relationship between the degree of each pulse condition and these characteristic variables. This is exactly the method that people usually use in the research of pulse detection and analysis instruments in TCM. Using this method or the instrument developed by this method, the fuzzy information after processing can be described qualitatively and quantitatively like deterministic information to a certain extent (Yuan Bing and Zhao Zheng 1986).

However, using basic variables to describe the characteristics of pulse conditions also has the problem of completeness and independence of basic variables. As long as the basic variables that can be detected by science and technology are not enough to describe all the characteristics of the pulse condition, this basic variable set is incomplete. That is to say, with the support of these technical means, it is not enough to fully identify the pulse condition. In addition, the same pulse condition in different people may be reflected in the similarity of 'imagery' that relies on human perception. This similarity is sometimes difficult to extract through objective detection methods. In other words, it is technically difficult to use objective detection methods to reproduce TCM pulse conditions through analytical features and feature synthesis.

9.4 The etiology model of holistic medicine and its integration with modern medical etiology

As mentioned earlier, TCM's understanding of the cause is based on the body's response state caused by the cause. Based on the changes in the human body state caused by pathogenic factors acting on the human body, metaphor and analogy methods are used to establish a functional model describing the characteristics and pathogenic actions of pathogenic factors. This is how the etiology model of the 'six evils' of wind, cold, heat, dampness, dryness, and fire were established. The use of modeling methods of complexity science to establish models of pathogenic factors is the mainstream method for holistic medicine to grasp pathogenic factors. As mentioned earlier, holistic medicine's grasp of the human body's state is achieved by introducing state variables that describe its functions and attributes on the basis of the model. The grasp of the state of the state variables (syndrome) is achieved by identifying clinical findings of the patient after establishing the relationship between the state variable and clinical findings. Similarly, the understanding of pathogenic factors is based on the response state of the human body, and it is also necessary to describe and distinguish the response state of the human body. Therefore, it is also necessary to introduce state variables and establish the relationship between state variables and related clinical findings.

Based on the body's response state to grasp the cause, the corresponding medication research is naturally based on the body's response state caused by the pathogenic factor as the frame of reference. The effective medications for the pathogens thus found can undoubtedly correct the corresponding response state, but usually also have direct or indirect killing or inactivation actions on the pathogens. As mentioned above, using this method, even if we don't know anything about the actual pathogen that causes the disease, and we have not selected medications that can kill the pathogenic factors, we can still improve the pathological state caused by the pathogenic factors, so as to achieve the purpose of curing diseases (Yuan Bing and Zhao Zheng 1986).

Among the treatment methods for pathogens that can survive in the human body, antibiotics are the most mature in modern medical treatment methods. So, how can these effective treatments developed by modern medicine be incorporated into the clinical medical system of the holistic medicine? There are two ways: one is to refine the etiology model, and the other is to incorporate these pathogenic factors into the disease medical system of holistic medicine.

The refinement of the etiology model can be carried out in the following ways. If a certain pathogen invades the human body, the various initial clinical finding groups appearing in the population are completely included in the clinical finding group of a certain pathogenic factor described by the etiology model, then this pathogen can be regarded as a subtype of the pathogenic factor. For example, if the initial response state (clinical finding groups) caused by the SARS virus invading the human body is completely covered by the clinical finding group corresponding to exogenous wind-heat, in other words, the clinical finding set that appears at the initial stage of the SARS virus invading the human body is a subset of the clinical finding group corresponding to the exogenous wind-heat, the SARS virus can be regarded as a subtype of exogenous wind-heat. As a result, the refinement of the etiology model established by 'identify the symptoms to find the etiology' can be realized.

But the usual situation is that when the SARS virus invades the human body, the response status of different individuals will be different. Some manifested as exogenous wind-heat, only fever, headache, sore throat, cough and other symptoms; some manifested as cold on the outside and hot on the inside, with fever and sore throat, intolerance to cold, chills, body pain, etc.; some people also have symptoms such as vomiting, diarrhea, etc., of damp-heat in the spleen and stomach, and damp-heat in the large intestine. In other words, the body's response state (syndrome) caused by the SARS virus is not unique. Therefore, in this case, the method of refining the exogenous wind-heat cannot be used.

As mentioned earlier, the clinical practice of TCM has been based on the combination of syndrome differentiation with disease differentiation since ancient times. In the modern 4-axis classification system of diseases, many diseases, such as tuberculosis and typhoid fever are specified based on pathogenic factors. The content of these diseases covers the analysis and research of the pathogenic factors themselves, the research on the clinical manifestations and the law of occurrence and development of the initial disease, and the research on the treatment methods of the cause, which has formed a detailed knowledge system. What needs to be done to incorporate the knowledge and treatment of these diseases into the holistic medicine system is to incorporate such diseases into the DCS of holistic medicine. In this way, under the holistic medicine framework, the organic combination of personalized treatment for pathological states and treatment for diseases caused by specific etiologies can be achieved. The issue of incorporating modern medical diseases into the holistic medicine disease system will be discussed further later.

The aforementioned pathogenic factors that can survive in the human body are not only micro-organisms, but also parasites, which are collectively referred to as biological pathogenic factors. The characteristics of micro-organisms are: they can spread, cause immune response, and can be killed by heat or chemical disinfectants. Common micro-organisms include bacteria, viruses, mycoplasma, rickettsiae, spirochetes, chlamydia, and fungi. For example, common human tuberculosis, dysentery, tetanus, carbuncle and furuncle are caused by various bacteria. More infectious diseases are caused by viruses than bacteria, such as viral hepatitis, measles, influenza, etc. Typhus and tsutsugamushi are caused by Rickettsia; diseases caused by spirochetes include relapsing fever, syphilis, and Leptospira illness, etc.

Parasites can be divided into two types: protozoa and helminths. Protozoa are single-celled animals, larger than bacteria, and can be seen clearly with an optical microscope. The diseases caused by protozoa include malaria, kala-azar, amoebic dysentery, trichomonal vaginitis, etc. Worms are multicellular animals with organ differentiation and are generally visible to the naked eye, such as roundworms, pinworms, ginger worms, schistosomiasis, liver flukes, and tapeworms.

In addition to biological pathogenic factors, modern medical research on pathogenic factors also includes (Wen Liyang et al. 2001):

1. *Physical pathogenic factors*

   Knife cuts, spikes, crushing, contusions, gunshot wounds, insect bites, burns, frostbite, ionizing radiation, noise, vibration, electric shock, and radiation damage are all physical pathogenic factors.

2. *Chemical pathogenic factors*

   Food poisoning, snake venom and gas poisoning are all chemical factors at work. Some substances, such as benzene, lead and organophosphorus pesticides, which are exposed to the surrounding environment in the form of gas, dust or liquid in industrial and agricultural production work, can poison people and cause illness under certain conditions.

3. *Nutritional deficiency*

   The nutrients needed by the human body include water, sugar, protein, fat, various vitamins and minerals (such as calcium, iron), etc. If these substances are lacking or insufficient, they can be the cause of disease. Dehydration, malnutrition and vitamin A, B and C deficiency can cause corresponding keratitis, angular cheilitis and scurvy, while calcium and iron deficiency can cause rickets and anemia.

4. *Allergies and immune deficiencies*

   Allergens in allergic diseases (such as shrimp, crab, pollen, lacquer, penicillin, etc.) can cause rubella, rhinitis, dermatitis, and severe bronchial asthma and anaphylactic shock. In addition, due to congenital hypoplasia or acquired secondary reasons (such as long-term immuno-suppressive therapy, protein wasting enteropathy, etc.), immuno-deficiency diseases that cause cellular and humoral immune dysfunctions also fall into this category.

5. *Genetic factors*

   The genetic material (DNA) and chromosome changes in the reproductive cells (sperm and egg) of the parents of the previous generation often cause diseases in the offspring, such as hemophilia, color blindness, congenital stupidity, and hermaphroditism. Marriage of close relatives is more likely to produce such genetic diseases.

In addition to the exogenous 'six evils', the research of TCM pathogenic factors also includes the following aspects (Beijing College of TCM 1974):

1. *The seven emotions of internal injury include happiness, anger, worry, thinking, sadness, fear, and shock.*

   The effects of seven emotions on the five internal organs: happiness damages the heart, anger damages the liver, anxiety and thinking damage the spleen, panic and fear damage the kidneys, and sadness damages the lungs. The seven emotions affect the qi maintenance: happiness causes the qi to slow down, anger causes the qi to rise, sadness causes the qi to disappear, fear causes the qi to move downwards, fright causes chaos in the movement of qi, and anxiety/thinking causes the qi to coalesce.
2. *Eating disorders include improper eating, eating preference, and unclean eating.*

   Improper eating mainly refers to excessive hunger or overeating, and irregular eating. It will cause food accumulation over time and damage the spleen and stomach.

   The diet preference includes preference for cold and heat, preference for fat and sweet, preference for drinking, preference for five flavors and so on. The preference for food types can cause diseases such as rickets, simple goiter, night blindness, etc., due to lack of certain nutrients. Overeating raw and cold foods can damage the spleen and stomach, leading to deficiency and coldness of the spleen and stomach; overeating pungent dry foods can easily hurt the stomach yin and cause stomach heat. A preference for fat or sweet foods tends to produce dampness and phlegm; a preference for drinking can damage the liver and damage the spleen. Unclean diet can easily damage the intestines and stomach.
3. *Exertion and rest disorders*

   Excessive labor includes excessive physical labor, excessive mental labor, and excessive sexual life. Excessive physical work can easily consume qi and damage the essence; overuse of the brain can easily damage the heart and spleen, resulting in the deficiency of the heart and spleen; excessive sex life can easily damage the kidney essence.

   Excessive comfort will also exhaust qi and hurt the spleen over time.

In addition, both modern medicine and TCM contain pathogenic factors related to the accumulation of metabolites, such as hyperlipidemia and hypercholesterolemia in modern medicine. Regarding similar pathogenic factors, TCM also establishes etiology models based on the body's response state, such as the phlegm model and blood stasis model.

On the basis of integrating the pathogenic factors of modern medicine and TCM, from the perspective of treatment-related, the pathogenic factors of holistic medicine can be divided into the following categories:

1. After invading the human body and causing diseases, the biological pathogenic factors that still exist in the human body include the exogenous 'six evils' of TCM and the pathogenic microorganisms of modern medicine.

   For such diseases, the treatment of modern medicine usually considers killing or eliminating the pathogenic factors; TCM focuses on controlling and improving the response state caused by these factors, and of course, it does not rule out the direct action on the pathogenic factors. Although, for diseases caused by bacteria, modern medicine or TCM alone can achieve good results, such as antibiotics against bacteria, and CHM directly against the body's pathological conditions induced by bacteria or virus. However, since the two therapies usually act on different links related to the disease to a large extent, the comprehensive application can usually play a synergistic effect.

2. The pathogenic factors that still exist in the human body after causing the disease, in TCM, include food accumulation caused by improper diet, excessive accumulation of certain nutrients caused by diet preference, and phlegm, blood stasis, etc.; in modern medicine, include the accumulation of metabolites, such as blood lipids, blood sugar, and cholesterol, etc. Modern medical treatment of such diseases usually focuses on reducing the intake of substances that cause the disease, or promoting the patient's metabolism. TCM focuses on adjusting and improving the response state caused by these factors. Because the cause of such diseases is sometimes not due to excessive intake, but a problem with the metabolic mechanism, treatments for the pathological state of the human body may improve the metabolic mechanism and thus exert long-term actions. The methods of modern medicine have positive significance for quickly correcting the accumulation of pathological products and reducing secondary damage. At this time, the treatment methods of modern medicine and TCM are actually the relationship between 'treatment of symptoms' and 'treatment of root causes', and the combined application of the two usually produces a certain synergistic effect.
3. Regarding the lack of certain nutrients necessary for human life, whether in TCM or modern medicine, the adjustment of the diet structure is achieved by changing the insufficient intake or correcting the dietary preference. In addition, modern medical therapies are superior in supplementing the lack of nutrients, while TCM may improve patients' problems in the absorption of these nutrients. If it is not for malnutrition caused by hunger or dietary structure problems caused by diet preference, people's usual intake is sufficient to provide the necessary nutrients for human life activities.
4. Some pathogenic factors that cause the body's state to deviate or disease are not entities, and they cannot be directly treated except to avoid contact with them to reduce the occurrence of diseases. For example, the 'seven emotions' of TCM, the abnormality of diet and living rules, and some physical factors, emotional factors of modern medicine. In this case, the treatment has to focus on the deviation of the body state caused by the cause of the disease, that is, completely following the method of TCM syndrome differentiation.

The different ways of understanding the cause of disease in TCM and modern medicine and the resulting differences in the way of controlling the cause have formed two different treatment systems for pathogenic factors. However, in most cases, the inconsistency of the treatment focus of the two may form a therapeutic synergy, thereby enhancing the healing power of human beings against pathogenic factors and the damage caused by them. For those diseases whose causes cannot be found in time (such as influenza caused by a frequently mutating virus), or diseases caused by unknown causes, the treatment method based on correcting the body's response state caused by the cause has shown great advantages.

Integrating the treatment methods developed by modern medicine and TCM into the framework of holistic medicine can more effectively carry out clinical research on the synergy of the two types of treatment methods. On this basis, according to the actual situation of the patient, the two methods are organically combined and used in clinical treatment to exert a synergistic effect, which will undoubtedly enhance the medicine's ability to control diseases and human health.

9.5 Research on medications and treatment methods of holistic medicine

Diseases are caused by pathogenic factors that cause certain damage to the body, and the body's state deviates from normal homeostasis. In order to treat diseases, and restore and maintain physical health, holistic medicine must have corresponding means to eliminate pathogenic factors, repair disease damage, and correct deviations in state variables. In other words, holistic medicine must also establish its own treatment methods and drug system.

9.5.1 Medications, formulations, and research methods for them in holistic medicine

Corresponding to the SDS of the human body and diseases of TCM and modern medicine, two major medication systems of CHM and Western medication have been formed. The human body model of TCM, which is the reference frame of CHM research, is based on the overall human level. Its status variables are relatively macroscopic syndromes that can be attributed to perceivable/observable symptoms and signs. Therefore, the medication actions found are relatively macroscopic actions described by symptoms and signs, and some relatively long-term actions can even be found. Modern medicine regards the entity etiology, pathological changes, and specific detection indicators as the controlled amounts of medication research, therefore, the first discovery in the research process is naturally effective drugs targeting these controlled amounts. Their macroscopic and long-term actions (including certain side effects) are usually discovered gradually after a long period of time after being put into clinical use and after clinical observation and statistical research on the use of a large number of patients.

For example, Aspirin (acetylsalicylic acid) has been used clinically for more than 100 years. In the early days, people mainly used it for antipyretic, analgesic and anti-inflammatory purposes. It has a good effect on alleviating mild or moderate pain, such as toothache, headache, neuralgia, muscle aches and dysmenorrhea. It is also used to reduce fever in febrile diseases such as colds, and flu and treat rheumatism and other such pains. About 40 years ago, people noticed its inhibitory action on platelet aggregation, which can prevent thrombosis. So, it is clinically used to prevent transient ischemic attack, myocardial infarction, artificial heart valve and venous fistula or other thrombosis after surgery. However, statistical studies in recent years have found that prophylactic use of aspirin has no benefit in low-risk groups (e.g., those without a history of heart disease and stroke, and without underlying medical conditions), and may instead increase the risk of major bleeding and cancer death (Richman and Owens 2017, McNeil et al. 2018). Obviously, the actions of medications discovered early are usually far from the full actions of medications, and the discovery of their long-term actions is a long process after the beginning of clinical application. For example, Tamiflu (oseltamivir phosphate), which claims to be able to treat influenza A and B, was only found to have serious adverse reactions after it was proven effective in clinical trials and put into clinical application (Bai Zhiyang 2012).

It can be seen that the different common characteristics of CHM and Western medication are not mainly determined by the nature of the medication itself. The difference in the controlled amount as the research reference system led to the discovery of the actions of different characteristics of the medication, thus forming two different systems of CHM and Western medication. If we take a CHM (or its purified extract) into a clinical trial with the reference system of modern medicine, we will naturally find some of its therapeutic actions that can be described in modern medical terms, making it a Western medication, such as Artemisinin. And if a Western medication can be fully described based on the reference system of TCM, then this medication can naturally be used as a CHM for clinical treatment of TCM, such as aspirin.

To establish a medication system for holistic medicine, the first thing to face is to incorporate the existing CHM-formula system of TCM and medications of modern medicine with good curative effects and less side effects into the medication system of the holistic medicine. The clinical diagnosis and treatment of diseases in TCM has always been a combination of syndrome differentiation with disease differentiation. The description of the action of CHM in TCM is mainly target to syndromes, but there are also descriptions of actions to disease or symptoms, such as hemostasis, relieving cough, and relieving asthma. The changes in the human body model, SDS, and DCS brought about by holistic medicine will inevitably affect the description of the actions of CHM. The medications of modern medicine are basically developed with disease as the reference frame. When a certain disease is included in the category of holistic medicine, as a means of treating the disease, the corresponding medication is naturally included in the treatment system of holistic medicine. As for whether the medications of modern medicine can be used for the treatment of syndromes, firstly, it depends on whether it shows the therapeutic effect for a certain syndrome in the statistical observation based on the syndromes as the frame of reference; secondly, it depends on its efficacy and safety, that is, whether its efficacy is superior compared with CHM for this syndrome.

Although from a methodological point of view, the relationship between the personalized DCS established by precision medicine and conventional DCS is similar to the relationship between the TCM syndrome system and the disease system, this does not mean that it can be incorporated into the personalized SDS of holistic medicine. The research on isomorphism and homomorphism models of system science shows that based on the same reality system, in principle, an infinite number of isomorphism models can be constructed, and the number of homomorphic models that can be constructed is even greater. There are endless possibilities for the relationship between the model constructed from the holistic level and the model constructed from the micro level. As long as their structures are incompatible, it is impossible to use the micro-level SDS as a refinement of the holistic SDS, and integrate the two into a unified system. Therefore, whether the conventional DCS or the new personalized DCS of modern medicine has to be included in the DCS in holistic medicine, the corresponding treatment methods will also be included in the holistic medicine treatment system target to diseases.

Ancient TCM books about the knowledge of CHM usually record the properties of CHMs in all directions, including the therapeutic actions, action trends and side effects. Since modern medicine characterized by disease medicine has not established an SDS that can fully describe the state of the body, its revelation of the actions of medications mainly focuses on the actions on pathogenic factors, pathological changes, and the structure and function of various levels of the human body, as well as side effects. Holistic medicine is the medicine that includes two systems of state-based regulation-control and disease-based treatment, the description of the properties and actions of medications must naturally include the actions on the state and the actions on the disease. In other words, all medications included in the holistic medicine system must have a comprehensive description of their functions (including therapeutic actions and side effects) corresponding to the SDS and DCS of the holistic medicine.

As mentioned earlier, whether it is the state variables (syndrome) under the holistic medical SDS or the disease (symptom) under the DCS, it will ultimately be attributed to a specific set of clinical findings. Therefore, to incorporate a medication into the description system to which it did not belong, all that needs to be done is to summarize its therapeutic actions (or side effects) relative to the corresponding description system based on past observations and statistics on the clinical application of the medication. For example, according to the adverse reactions identified during the use of Tamiflu after it was approved for marketing (Roche Registration Co. Ltd. 2023):

1. *Whole body*: swelling of face or tongue, allergic reaction, hypothermia.
2. *Skin*: rash, dermatitis, urticaria, eczema, toxic epidermal necrolysis, Stevens-Johnson syndrome, erythema multiforme.
3. *Digestive system*: hepatitis, abnormal liver function test.
4. *Heart*: Arrhythmia.
5. *Gastrointestinal tract*: gastrointestinal bleeding, hemorrhagic colitis.
6. *Nerve*: epileptic seizures.
7. *Metabolism*: worsening diabetes.
8. *Mental*: abnormal behavior, delirium, including the following symptoms, such as hallucinations, irritability, changes in level of consciousness, confusion, nightmares, delusions.

From the above clinical manifestations, the actions of Tamiflu relative to the SDS of TCM can be concluded: attribute heat, easy to generate heat and fire, easy to cause bleeding, mania and mental disorder, etc. According to scientists from the patent holder Gilead Sciences, Inc., Tamiflu uses shikimic acid as the starting point for synthesis, and shikimic acid can only be obtained from the extract of Chinese star anise. The adverse reaction of Tamiflu is basically consistent with the TCM understanding of the properties of the original plant of shikimic acid-star anise. The description of star anise on its properties and actions in Chinese Materia Medica: spicy and sweet in taste, warm in nature. It has the actions of warming yang, dispelling cold and regulating qi, and treats middle cold, vomiting, cold colic in the abdomen, low back pain caused by kidney deficiency, and dry and wet beriberi. In the view of TCM practitioners, using it to treat influenza A, which is invaded by wind-heat and usually has severe heat symptoms, is tantamount to 'adding fuel to fire', which is incredible.

Holistic medicine mainly uses the state represented by state variables to describe the disease process. The state variables at the holistic level are usually relatively macroscopic and abstract, and usually correspond to many and a variety of abnormalities of structure and function at the entity or micro level of the human body. Therefore, medications discovered with these state variables as controlled quantities may act in different links in the human entity and have different mechanisms of action. Experience tells us that to pull a system out of a certain stable state and act on multiple state variables that support this steady state at the same time, will have a better effect than using a greater force to act on one of the state variables. In addition, in the course of the disease, patients often have abnormalities in more than one state variable, and the mutual influence between abnormal state variables is sometimes manifested as the transmission of pathological actions, and sometimes manifested as the influence of physiological functions. In the treatment process, simultaneous intervention of multiple state variables is often better than intervention of one state variable alone, because the synergistic effect among the medications often results in a better overall effect. Therefore, among the herbal formulas of TCM, some are aimed at a single state variable abnormality, such as Yinqiao powder for exogenous wind-heat and Bazheng powder for bladder damp-heat. Some are for more than one related state variable abnormality, such as Danggui Buxue Decoction for replenishing qi and blood, and Guilu Erxian glue for strengthening kidney yin and kidney yang. Some are for specific diseases and a combination of related state variables, such as Chuanxiong tea for wind-cold headache and Sishen pills for treating spleen-kidney yang deficiency and early morning diarrhea. To incorporate these herbal formulas into the holistic medicine system, it is only necessary to summarize their performance and therapeutic actions relative to the SDS and DCS of holistic medicine based on the changes in clinical findings system caused by their action on the human body.

9.5.2 Clinical trials and efficacy evaluation methods of holistic medicine medications

Today, in the field of modern medicine, with the rise of precision medicine, randomized controlled trials based on conventional DCS are gradually giving way to new clinical trial models such as basket trials and umbrella trials. These trials simultaneously observe a variety of different diseases corresponding to one (or a set of) biomarker abnormalities, as well as a variety of personalized classifications covered by one disease, which usually means more sample size and longer clinical trial cycles. Large-scale clinical trials featuring randomized, double-blind, and controlled trials are still the gold standard for testing the safety and effectiveness of innovative treatment methods. However, in the past, medications developed based on this seemingly rigorous scientific trial often found some long-term side effects after being used in the clinic for a period of time, and in some cases, the side effect was so serious that it even led to the medication being eliminated or banned. As a result, more and more people began to question the reliability of the results of randomized controlled trials. In addition, the economics and rationality of such clinical trials are being criticized more and more. Patient groups claim that large-scale clinical trials are time-consuming and labor-intensive, and will delay the time for breakthrough medications that may save lives to enter the market. Promoters of radical marketization policies advocated replacing the bloated FDA with a mechanical 'market decision' model. They claim that as long as the safety of the medication is guaranteed, whether it is effective or not, it should be licensed for sale on the market. Because market competition and rational consumers will allow truly cost-effective medications to naturally win (Zhou Jifang 2017).

Against this background, Real-World Evidence (RWE) has naturally been respected by people. Real-world Research (RWR) refers to research in which the research data comes from the real medical environment and reflects the actual diagnosis and treatment process and the patient's health status under real conditions. Real-world Data (RWD) is very wide, and it can be massive data generated by patients through various channels such as outpatient clinics, hospitalizations, examinations, surgeries, pharmacies, wearable devices, and social media. The data type can be research data, such as data from patient surveys based on specific research purposes, patient registration studies, electronic medical records, and intervention studies based on real medical conditions (such as actual effectiveness randomized controlled trials); or non-research data, such as health-related data routinely monitored, recorded, and stored by various institutions (such as hospitals, medical insurance departments, civil affairs departments, and public health departments), hospital electronic medical records, medical insurance claims databases, public health surveys, and public health monitoring (such as monitoring of adverse drug events), birth/death registration items, etc. (Lily 2019).

In order to better manage the uncertainty in reimbursement decision-making and effectively monitor the safety of medications after marketing, a large number of research results close to the actual clinical medical treatment and epidemiological data closer to the natural environment are required, which also makes the application of real-world research more extensive. With the development of big data and machine learning technology, especially the widespread application of EDC (Electronic Medical Record Report), the evidence strength and importance of observational studies with large sample sizes have gradually increased, which even poses a challenge to randomized controlled trials in health policy decision-making (Figure 9.1).

The traditional randomized controlled trial (RCT) as the highest level of clinical evidence standard requires strict control of test conditions, standardized treatment in potentially effective target populations, small sample size and short follow-up time. Real-world research can include patients with complex and multiple diseases; treatment can be allocated in a non-random manner; multiple measures can be used in the treatment according to the needs of the patient and the treatment strategy of the clinician; the clinician can be allowed to determine the dose based on the disease and the patient's general condition instead of determining the dose according to the plan, so as to accurately meet the needs of the patient; a longer study period can be set to measure the long-term benefits and risks of interventions; broader indicators, such as functional indicators, patient report outcomes, cost-benefit indicators, etc., can be included to make research evidence more extrapolable and more clinically practical (Lily 2019).

Traditional clinical trials are often prospective, while real-world studies are mostly retrospective cohort studies and case-control studies. Traditional clinical trials are time-consuming and laborious, with cumbersome procedures, and involve huge expenditures on medications and personnel. In contrast, real-world research can obtain information with the same value as clinical trials at a relatively low cost (Zhou Jifang 2017). Traditional clinical trials are centered on medical practitioners and the objective indicators of laboratories, which represents a cold old-fashioned 'biomedical model;' while real-world research is a patient-centered, 'bio-psycho-social' model, which considers patients' subjective feelings as important as objective indicators. Currently, real-world research has applications in all stages of medication development and life cycle. In the medication development stage, real-world data can be used to understand the current treatment of related diseases, disease burden, and to help design clinical trials and patient recruitment. After the medication is marketed, prospective/retrospective studies can be used to track the long-term safety, effectiveness, and

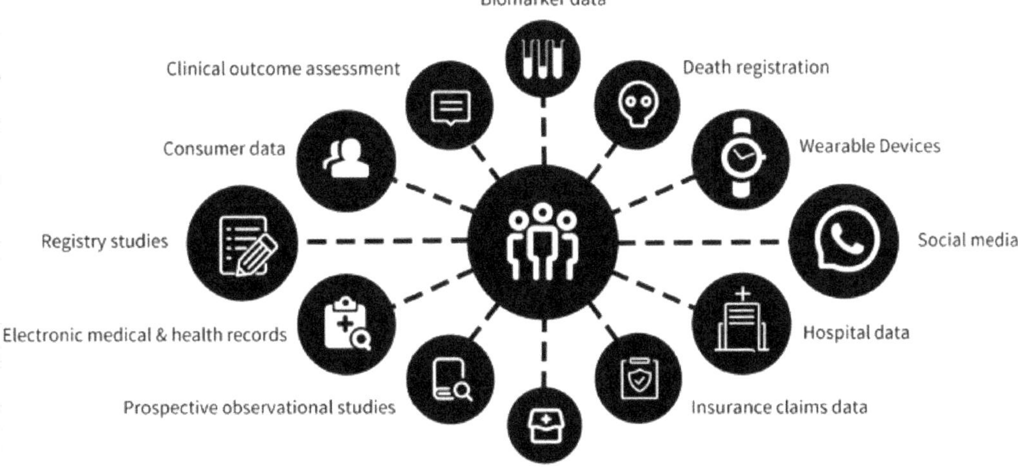

Figure 9.1. Diverse real-world data drives real-world evidence generation.

patient compliance of the medication in the real world. More complex subgroups of patients with multiple diseases can be included to study the effect differences of different subgroups (Lily 2019). In December 2016, the U.S. Congress published the 21st Century Cures Act on the official website, proposing approval to use real world evidence to replace traditional clinical trials for expanded indications (Office of the Commissioner 2020). On December 6, 2018, the FDA announced the Real-World Evidence Program Framework, which provides a clear route map for achieving RWE's goal of supporting medication approval decisions (U.S. Food & Drug Administration 2018a).

Real-world research (RWS) is the systematic collection and analysis of real-world data (RWD) routinely generated in clinical practice. It is a complementary not antagonistic relationship to randomised controlled clinical trials (RCTs). Both require scientifically sound research designs, experimental protocols, and statistical methods. The criterion for distinguishing between real-world studies and randomized controlled clinical trials is the context in which the study is conducted. Real-world research data comes from medical institutions, families, communities, etc., rather than an ideal environment with many strict constraints (The New World of Health 2020). A comparison of the two is shown in Table 9.1.

At present, real-world research and clinical trials have a tendency to achieve the same goal by different paths. On the one hand, clinical trials emphasize the inclusion of diverse populations and improve the accuracy of trial results in real scenarios. Although the accuracy of the test results may be reduced due to the diversity of patients, the extrapolation of the test results is guaranteed, that is, external validity. On the other hand, is the rise of pragmatic trials. Clinical trials simulate real treatment scenarios as much as possible, and observe the efficacy and safety of certain treatment methods with a large sample of patients with fewer constraints. Obtaining clinical research data under this reduced intervention has the advantages of strong acceptability, low cost, flexible operation, etc., and has been more and more widely used.

At the same time, real-world research is also making various efforts to reduce bias. One of the most criticized aspects of retrospective research is the existence of various biases. Compared with the ability to balance the heterogeneity of patients through randomization, retrospective studies cannot fundamentally solve the problem of the lack of comparability between the experimental group and the control group. Nowadays, researchers in real-world studies will, like traditional clinical trials, determine the analysis plan before analysis, to minimize the bias caused by selective reporting. Secondly, real-world research is also undergoing methodological innovations to make better use of the massive amounts of data continuously generated in real life. Big data, machine

Table 9.1. Comparison of randomized controlled clinical trials and real-world studies. (The New World of Health 2020, Wu Jieping Medical Foundation 2018.)

	RCT	**RWS**
Research purposes	Focused on efficacy research	Diverse research purposes, including efficacy studies
Research design	Randomized controlled, prospective study	Random or non-random sampling, can also be observed; can be forward-looking, can also be retrospective
Research population	The population is relatively single, with many and strict inclusion/exclusion criteria	The population is diverse, and the inclusion/exclusion criteria are relatively loose
Implementation scenario	Ideal world: Highly standardized environment	Real world: Healthcare facility, community, home
Sample size	Calculated according to statistical formulas, the sample size is small	According to the real data environment or statistical formula, the sample size can be large or small
Data sources	Standardization, the collection process is more strictly regulated	Diverse sources, which can be prospective or retrospective, can be specially collected, or can be based on existing databases
Research time	Short	Short or long
Research endpoint	Most of the research endpoints are based on the assessment of outcome indicators	To obtain all treatments and long-term outcomes as study endpoints
Main advantage	The internal validity of the research results is high	The research results have strong applicability
	The current standard scheme for pre-market evaluation of new products	Great for evaluating long-term outcomes
	Can prevent selection bias	Test conditions and environments for everyday clinical practice
	Good comparability between groups and simple statistical analysis methods	Expenses dedicated for testing not high
Shortcoming	Requires maximum selection of homogeneous patients	
	Can only be used to evaluate short-term effects	The test conditions are not strict
	The experimental environment is quite different from the daily clinical environment	The experimental design is relatively simple
	Requires substantial dedicated trial funding	Test results are prone to bias

learning, artificial intelligence and other technologies have been widely used in the methodological innovation of retrospective research (Zhou Jifang 2017).

At the 2018 ASCO meeting, the organizers of the conference announced in a high profile that CancerLinQ would cooperate with the FDA to examine the reliability of various newly approved anti-tumor medications in real populations. This company collects real patient data from different types of cancer clinics across the United States, and then uses innovative big data to guide personalized new medication development after anonymization, helping the FDA to make new medication approval decisions. On the other hand, the drug regulatory authority will not relax strict requirements on medication safety while accelerating medication review. Massive data can quickly verify rare adverse medication reactions and help real-time supervision of medication safety issues. The integration of clinical trials and authenticity research done by CancerLinQ may indicate the main development direction of the medication evaluation system in the future (American Society of Clinical Oncology 2017).

Over time, real-world research and clinical trials continue to learn useful parts from each other and reduce their own shortcomings. The boundary between the two will become more obscure. As Dr. Sean Khozin, senior expert on oncology of the U.S. FDA, said at the ASCO annual meeting:

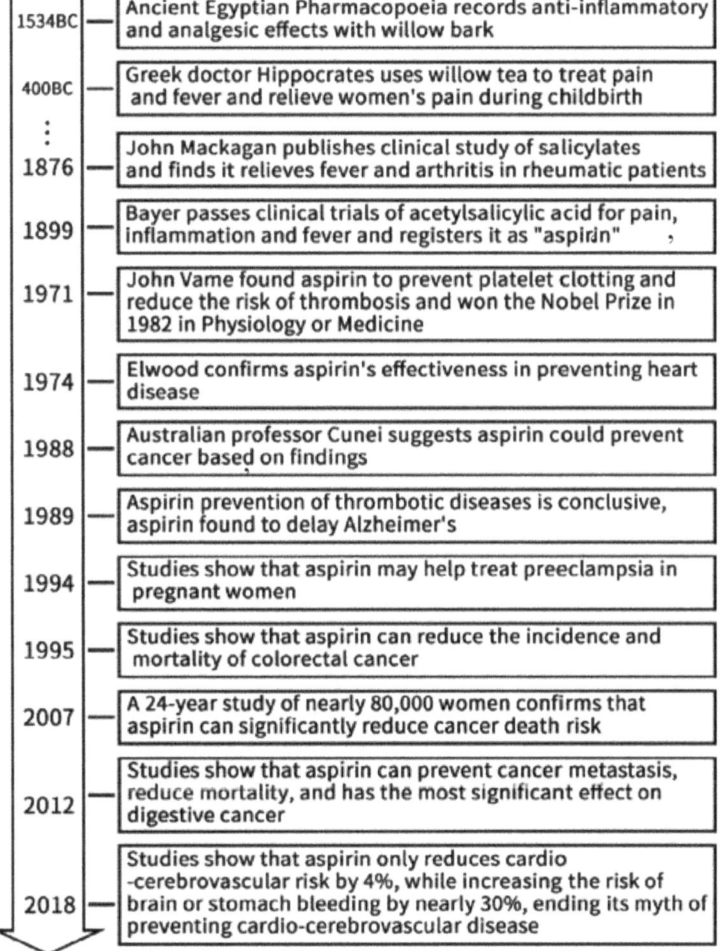

Figure 9.2. Past and present life of aspirin.

"We need to overcome the limitations of randomized controlled trials and use real-world data to guide the reform of the medical system and the formulation of health policies" (Zhou Jifang 2017).

Figure 9.2 shows the process from the emergence of aspirin, the 'miracle medicine for a century' to the deepening of people's understanding of its actions. Today, from the perspective of real-world research, isn't the 'past life and present life' of aspirin just real-world research that is ongoing and will continue to be done? (Yuan Bing 2021).

On Nov. 6, 2018, the State Drug and Food Administration of China issued the 'Technical Guidelines for the Clinical Research of Syndrome-type TCM New Medications' (hereinafter referred to as the Guiding Principles), which aims to provide basic guidance for the development, effectiveness and safety evaluation of clinical trials of new CHMs based on syndromes (State Food and Drug Administration 2018). The reference frame for clinical research of new TCM medications with syndromes can be a simple TCM syndrome, a certain syndrome type under a certain TCM disease, or at least three different Western medicine disease types under a certain TCM syndrome. The issuance of the Guiding Principles means that TCM syndromes will begin to be included in the category that can be evaluated through traditional prospective clinical trials. Obviously, after the US FDA opened the basket test and umbrella test models to promote the development of innovative medications in precision medicine, this is clearly the 'unbinding' of the Chinese CFDA's evaluation methods for similar CHM. Previously, under the current medication evaluation system, medications for TCM syndromes could not undergo industry-recognized clinical trials and efficacy evaluations, and it was naturally impossible to have recognized effectiveness.

Throughout the ages, TCM's understanding and evaluation of the effects of thousands of CHMs and innumerable herbal formulas are based on long-term real-world research, just like modern medicine for aspirin, the miracle medicine for a century. Only, due to the limitation of scientific methods and technical level, the research of TCM on them cannot reach the rigor of clinical research on aspirin based on empirical evidence in modern medicine. Today, the medication evaluation system has got rid of the fetters of traditional clinical trials and evaluation methods, and has begun to move towards the real world. In this context, it is worth pondering whether restricting CHMs and herbal formulas, which have been based on real-world research in the past, to the evaluation system of randomized controlled trials has positive significance for the development of TCM.

However, whether it is to use traditional clinical trials or real-world research, safety evaluation is essential for the research of medications and formulas in holistic medicine. Randomized controlled clinical trials and their corresponding clinical efficacy evaluation system are designed based on the approach of 'changes only one variable at a time' against the background of reductionism as the mainstream scientific concept. This method is obviously not suitable for observing the all-round effects of medications on various parts of the human body under various conditions. Due to the relatively clear criteria for disease diagnosis, as the efficacy evaluation of medications of modern medicine gradually develops toward real world research, medication research target to disease in holistic medicine will gradually move towards an evaluation system that combines traditional clinical trials and real-world research. In holistic medicine, there will still be situations where the definition of state variables (syndrome) is not clear enough, and some curative effects are long-term effects that require long-term follow-up observation. Moreover, in the ill population, the coexistence of multiple syndromes is common. Therefore, for medication research on state variables (syndrome), it seems more appropriate to adopt real-world research that can include complex patients with multiple diseases, can allocate treatment in a non-random manner, and requires a large number of cases. It's just that today's real-world research can no longer be the same as in ancient times, but should be based on rigorous empirical and statistical analysis. To realize this kind of large-scale real-world research, standardization of the SDS is an indispensable prerequisite. Therefore, the standardized symptoms, signs, and detection index systems established under the framework of holistic medicine should be a unified system suitable for the SDS and DCS of holistic medicine. On this basis, it is necessary to adopt big data and computer artificial intelligence technology to establish a standardized information collection system.

With a standardized human body model, SDS and DCS, with a unified, standardized system of clinical findings, with a standardized information collection system, and with a data mining and statistical analysis system based on big data analysis and machine learning technology, it is possible to extract information about changes in the clinical finding system that is valuable for understanding the properties of medications and formulas from real-world studies on a large scale and sustainably. Based on this information, we can continue to deepen our understanding of the properties of medications and formulas, so that holistic medicine's grasp of medications, formulas and other treatment methods is based on a rigorous empirical basis.

9.6 Holistic medical modeling and war game deduction

Wargaming is a method of predicting battle results and tactical strategy effects by simulating battle scenarios and strategic decisions. It is usually conducted in a virtual environment, using various models and rules to simulate combat entities, weapons systems, geographical environments and decision-making processes.

The origins of wargaming can be traced back to ancient times, when board games were used to simulate combat and strategic situations. This form of simulation helps develop tactical and strategic thinking while also being a form of entertainment. Over time, wargaming has gradually evolved into a more complex and systematic approach, incorporating elements from tactics, strategy, intelligence, command, decision-making, and other aspects. In modern times, with the development of computer technology, wargames have gradually shifted from paper simulations to computer simulations. Computer technology makes simulations more accurate and complex, capable of simulating large-scale combat scenarios and changing tactical decisions. At the same time, the application of artificial intelligence and big data analysis technology has also brought a higher level of intelligence and automation to war games (Lin Deli et al. 2013).

Today, in addition to the military field, the application of war game deduction has expanded to other fields, such as:

1. *Emergency management*: In emergencies such as natural disasters and epidemics, war games can be used to evaluate different emergency plans.
2. *Urban planning*: Used to simulate urban development and planning, and evaluate the impact of different urban construction plans.
3. *Traffic management*: Used to simulate traffic flow, traffic signal optimization, etc., to optimize urban traffic management.
4. *Environmental protection*: Used to simulate environmental pollution, resource utilization, etc., and formulate environmental protection policies.

In treatment decisions, doctors are often faced with the choice of multiple treatment options, such as different drug combinations, surgical methods, radiotherapy, or chemotherapy options. Each treatment option may have a different impact on a patient's disease and health status. Through war games, doctors can simulate the execution and results of each treatment plan, including treatment effects, patient responses, and potential risks, and predict the patient's treatment process. Such simulations can help doctors better understand the advantages and limitations of different options and expected outcomes, so they can choose the most appropriate treatment strategy.

In addition, war games can be used to optimize treatment options. By continuously adjusting the parameters and variables in the model, doctors can try different treatment strategies and find the best treatment plan. This optimization process allows doctors to more comprehensively consider patients' individual differences and treatment needs and provide personalized treatment plans to achieve the best treatment results.

Overall, wargaming is a powerful tool in treatment decision-making, which can help doctors predict and evaluate the effects and risks of different treatment options, optimize treatment options, and maximize the success rate of treatment and the quality of life of patients. It provides doctors with more comprehensive, systematic and scientific decision-making support, helping to achieve the goals of precision medicine and personalized treatment.

The purpose of building a model is for application. Holistic medicine establishes models of the human body, models of causes, medications, and treatments. Through model-based war games, we can accurately grasp the status of the system and analyze the evolutionary process of disease occurrence and development. On this basis, we can make the regulation-control of diseases and body status more targeted and effective.

Chapter 10

Holistic Medicine

A Medical System that Combines State Medicine and Disease Medicine

Today, under the impact of the three major trends of the times, the rise of complexity science, the transformation of modern medicine to personalized medicine, and the comprehensive revival of traditional Chinese medicines (TCM), medicine is on the eve of major changes. Holistic medicine will merge TCM and modern medicine into a medical system that integrates state medicine and disease medicine. So, is it necessary and feasible for holistic medicine to realize this kind of integration? In the future theoretical system of holistic medicine, what kind of structural relationship will be based on the holistic model of TCM, the DCS of modern medicine, and the personalized DCS of precision medicine? What is the relationship between holistic medicine and modern medicine about the knowledge system of the structure and function of the human body at various levels?

10.1 The necessity and significance of 'go forward separately and strike together' of state medicine and disease medicine

The current theoretical model of TCM has about 20 factors (heart, liver, spleen, lung, kidney, Qi, blood, body fluid, essence, etc.), and its SDS is close to 100 state variables. Biological research shows that the total number of human cells is approximately 60 trillion. At the molecular level, there are more than 100 million variants of genes alone, and there are much larger proteomes, transcriptomes, metabolomes, and so on. Obviously, there is a long distance from the holistic level of the organism to the molecular level, and the road from analysis to synthesis is therefore out of reach. The research of modern biology and medicine on the human body further reveals that within the human body system, the connections between various parts and elements are intricate, and there is no strict hierarchical relationship. In many cases, the interaction at the molecular level directly affects changes in behavior at the level of organs and tissues. For example, adrenal cortex hormone acts on the heart, directly causing the rhythm of the heart to be faster; serotonin acts on the peripheral tissues, directly causing the contraction of blood vessels.

Although the new DCS being constructed by precision medicine may gradually develop into the state medical system of modern medicine, it is still basically a static model based on the concept of reductionism. As mentioned earlier, it is impossible to achieve a comprehensive and dynamic description of the human body using a bottom-up 'analysis-integration' approach, nor is it possible to achieve an overall synthesis of the body's state and disease regulation control. This has been proved whether it is based on the conclusions of complexity science or the experience of integrating life in systems biology.

The use of a top-down approach to refine the description of the human body's state will not follow the hierarchical structure of the human body revealed by modern medicine, but can only refine the human body from the perspective of behavior and function. The description and definition of these behavioral functions are achieved through external symptoms, signs and testing indicators. Of course, these detection indicators can belong to the organ and tissue level, as well as the cell and molecular level. Just like the All of Us Research Program being launched in the United States, it not only includes indicators of the genome, proteome, and metabolome, but also behavioral data such as the clinical manifestations of the human body at the holistic level.

From the description of the overall state of the human body in TCM to the revelation of life activities such as genes, proteins, and metabolites at the molecular level, there is a vast unknown space. Science is not omnipotent. Today's science does not expect to be able to completely open the black box of the human body, exposing thoroughly the dynamic processes of the functional activities and relationships of all structures from the molecular level to the holistic level. Scientists are only trying to speculate on the internal laws of this black box by investigating the input and output of the black box, and gradually verify and perfect the understanding of these laws, so as to achieve the purpose of caring for health and effectively treating diseases.

In theory, there are models that not only conform to the entity structure of the human body but also reflect the functional characteristics of the human body as a whole. As mentioned earlier, it is impossible to integrate from the micro to the whole by completely opening the black box of the human body and integrating it on the basis of analysis. So, is it possible to achieve this model by connecting the holistic level model with the micro level model? We know that a model built from the holistic level based on the investigation of the basic physiological and pathological activities of the human body cannot be consistent with the entity structure of the human body. While building a model from the holistic level, the model structure usually does not exceed two levels. Because if there are more than two levels, it is difficult to attribute model-based predictions to definite expectations and test them with practice. Since an infinite number of homomorphic models can be constructed theoretically based on the same input and output, the probability that the functional model constructed from the holistic level and the entity model constructed from the micro level can exactly match in structure is almost zero. In other words, it is basically impossible to integrate the state description of the holistic level and the personalized DCS of precision medicine into a system through refinement. As a result, the practically valuable part of the personalized DCS constructed by precision medicine will still exist in the disease medicine system in the holistic medicine.

The revealing of various possible states in the human life process and the understanding of diseases and their laws are two effective paths for medicine to explore the vast unknown space of the human body in different directions. Along these two roads, medical science has made great achievements so far. In TCM, the 'treatment based on syndrome' based on state description and the 'disease differentiation and treatment' based on disease diagnosis have always been combined in clinical practice. The development of modern medicine to this day has also realized the limitations of traditional disease medicine, and has begun to introduce a 'personalized medicine' mechanism to build its state medicine system. Disease medicine and state medicine are to 'go forward separately and strike together' to the unknown space in the human body from different directions and angles. The individualized pathological process and regulation-control law that cannot be described by disease medicine, the state medicine provides an effective solution; and disease medicine's understanding of the common laws of diseases and the common treatment methods developed for diseases cannot be replaced by state medicine.

Therefore, under the framework of holistic medicine in the 21st century, disease medicine and state medicine will coexist. It is just that the future disease medicine will be the integration and unification of modern DCSs (including personalized DCS constructed by precision medicine) and DCS of TCM at a higher level. The future state medicine will be based on the holistic level but integrates the micro-level detection indicators, so it is a scientific system with a rigorous and logical structure and laid on an empirical basis. In it, modern detection methods, including micro-level detection indicators such as genomes, proteomes, and metabolomes, can be introduced to achieve precise and objective descriptions of the human body state. Integrating the regulation-control methods of the state medicine and the disease medicine means the organic combination of TCM treatment based on syndrome with the specific methods of TCM and Western medicine for diseases. As a result, a more holistic and more precise human disease and health regulation-control system will be achieved (Yuan Bing 2017b).

10.2 The disease medicine system of holistic medicine

As mentioned earlier, TCM has gradually established a DCS from ancient times to the present based on the cause, location, disease characteristics and clinical manifestations of the disease. TCM's understanding of certain entity causes that continue to survive in the human body after the disease and metabolites such as 'phlegm' and 'blood stasis' are based on the state deviation caused by them. For non-entity causes that are 'no trace to be found' after acting on the human body, treatment can only be carried out for the state deviation caused by it. Therefore, treatment for the cause of the disease is basically attributed to the treatment of the syndrome of the body's specific response state caused by the cause. The treatment for diseases defined by syndromes in modern medicine also mostly attributes these syndromes to specific TCM syndrome or combination of TCM syndromes, and then adopts the TCM treatment method based on syndrome. In the true sense, disease differentiation and treatment mainly target the types of diseases that can be attributed to symptoms, such as headache, dizziness, nausea, and vomiting. The treatment of this type of disease has corresponding symptomatic treatment medications to choose from, such as Chuanxiong Rhizoma and Viticis Fructus for headache, Gastrodiae Rhizoma and Uncariae Ramulus cum Uncis for dizziness, Pinelliae Rhizoma and ginger for nausea and vomiting, and so on.

The coding system related to injuries, diseases and causes of death, ICD-11, which took effect in 2022, is also a multi-axis classification system based on the cause, location, disease characteristics, and clinical manifestations. It contains 150 diseases and 196 syndromes from TCM. The items belonging to syndromes overlap with the SDS of TCM, so obviously they should not be included in the DCS of holistic medicine. Holistic medicine is a medical system with the purpose of treating diseases and regulation-controlling the health of the human body. In order to reduce the complexity of treatment, the classification of diseases should naturally follow the principle of 'the simplest and applicable'. Based on this principle, the DCS of holistic medicine can simplify and optimize the ICD-11 international DCS from the following aspects:

1. In the DCS, delete as far as possible diseases that are combined by elements corresponding to different treatment methods, and retain only the diseases defined by independent elements corresponding to different treatment methods. As a result, the DCS can be greatly simplified.

   For example, bronchitis caused by streptococcus is a type of disease defined by the combination of etiology, disease location, and pathological changes. Streptococcal infection may cause infectious diseases in other parts of the body, such as tonsillitis and pneumonia, but the treatment methods for the cause are the same. The specific pathological changes of bronchial infection can also be caused by other reasons, but the symptomatic treatment methods for the same local pathological changes are basically the same. Decomposing this compound disease type into disease types defined by relatively independent elements will greatly reduce the compound disease types formed by the refinement of diseases based on multiple axes, and the correspondence between diseases and treatment methods will be clearer.

2. If the treatment of a disease formed by a combination of different elements cannot be attributed to a combination of treatments for the elements, the compound disease should be retained.
3. For different pathological changes in the same disease site, or different pathogenic microorganisms of the same category, etc., if there is no difference in the treatment methods, there is no need to subdivide them into different disease types. This means that if broader-spectrum antibiotics or medications with wider applicability for pathological changes are developed, the corresponding classification of diseases that have been refined can be simplified and no longer refined.

4. When the same pathological changes that occur in different disease locations correspond to the same treatment method, the upper disease location can be used to merge multiple different disease locations, thereby simplifying the DCS. This simplification also applies to compound disease types defined based on specific etiology plus specific pathological changes.
5. For the same disease that occurs in different groups of people or at different times in the life process, if it does not lead to differences in treatment methods, there is no need to refine the classification of the disease. These include menopausal arthritis, age-related macular degeneration, etc.
6. If the refinement of the disease does not bring about the refinement of the treatment methods for the disease, or this refinement can be completed through the classification of the state, then the disease should not be classified into further classifications, such as menopausal syndrome, menopausal hot flashes, excessive menopausal bleeding and so on. However, menopausal syndrome is still meaningful as a definition of the scope of the disease.
7. For diseases defined based on symptoms and signs, if the different types of diseases formed after refinement correspond to different treatment methods, the upper-level diseases do not have to be retained. For example, stomach pain is named in TCM, and it also exists in the International Classification of Diseases ICD-10. Since gastritis, gastric ulcer, abnormal gastric motility and other detailed disease types are present, and there is no unified treatment method for it, there is no need to retain such disease names. This disease name is no longer included in ICD-11.

Therefore, when the DCS of holistic medicine is established based on the DCS of TCM and the international DCS of ICD-11, it is not ruled out that with the differentiation of disease-specific treatment options, the disease classification may be further refined, but the DCS will not be subdivided without limit. Perhaps the personalized DCS based on precision medicine will gradually expand, but the main change trend of the disease classification based on the conventional DCS is to simplify and optimize the existing system. The simplification of the disease system will help reduce complexity, and facilitate the overall grasp of the patient's disease state and comprehensive treatment. Obviously, no matter the refinement or simplification of the DCS, it is oriented to the progress of treatment methods. That is, under the premise of effective classification of treatment methods, the simpler the DCS and the fewer disease types, the better (Yuan Bing 2017b).

As mentioned above, the personalized DCS being constructed by precision medicine is a different DCS from ICD-11. The currently proposed personalized disease classification is based on a single biomarker abnormality, and there will also be a combination of biomarker abnormalities suitable for the same target medication. However, this kind of disease classification and conventional disease classification are usually a many-to-many network relationship, which cannot be included in the conventional DCS. As mentioned earlier, although this new DCS and the SDS based on the holistic level (the TCM treatment system based on syndrome) are both personalized DCSs, due to technical reasons, this new DCS cannot be integrated as a refinement into the SDS of holistic medicine. Therefore, the DCS of holistic medicine will include two parts, the conventional DCS and the new personalized DCS, corresponding to the matching disease differentiation and treatment system. As a result, a more scientific and reasonable DCS has been formed that integrates eastern and Western medicine, and integrates the experience and wisdom of all mankind in fighting diseases for thousands of years.

10.3 The highest state of medicine in the 21st century: The organic combination of overall regulation-control and specific treatment

The formation of the medical model of dividing diseases into categories for diagnosis and treatment is not based on the law of occurrence and development of diseases, but more on the way humans learn and understand the medical knowledge system. With the rapid growth of the total amount of medical knowledge, the proportion of the knowledge that a person can master throughout his life is smaller and smaller than the total knowledge in this field, which also makes the scope familiar to modern medical workers more and more narrow. However, the clinical practice of traditional Chinese and Western medicine has shown that the coexistence of multiple syndromes and/or multiple diseases is common in a patient.

As mentioned earlier, the treatment of diseases in modern medicine lacks integrity, and the specific methods for diseases of disease medicine are difficult to incorporate into the SDS. The TCM treatment based on syndrome fundamentally solves the holistic integration and individualization of disease treatment, but this does not mean that all problems in the course of disease treatment have been solved. In addition to the treatment based on syndrome, TCM also has adjustments in adding and deducing Chinese herbal medicines (CHM) to suit various diseases and symptoms. Many effective treatments for pathogen or specific pathological changes developed in modern medicine have quick effects in eliminating the cause, improving and relieving specific pathological changes, and sometimes they cannot be achieved by simple TCM treatment based on syndrome. In TCM, the concept of 'treating the symptoms if urgent, and treating the root cause if chronic' also emphasizes the importance of rapid remission or improvement of certain diseases or symptoms in the treatment of diseases. And many treatments for pathogens or symptoms developed in modern medicine are exactly effective methods to 'treat the symptom'.

For example, in the treatment of hypertension, although the antihypertensive medications of modern medicine cannot eliminate the cause of the continuous increase in blood pressure, timely reduction of excessive blood pressure to a safe level can minimize the risk of cardiovascular and cerebrovascular diseases. Holistic medicine solves the cause of the increase in blood pressure caused by the abnormal state of the body through treatment based on syndrome to cure the root, and the combined use of antihypertensive medications to treat the symptoms reduces the risk of slow onset of the root cure. And gradually reducing the dosage of the medications targeting symptoms with the gradual effectiveness of the root-treatment medications is a safe and fundamental treatment method to cope with hypertension.

Another example is the treatment of intractable insomnia. Simply taking sleeping pills will cause dependence, and the dosage needs gradually increase as the condition gets worse. In addition, the sleeping quality maintained by sleeping pills is not good. From the perspective of TCM treatment based on syndrome, insomnia is related to Qi deficiency, blood deficiency, yin deficiency or yin deficiency and fire prosperity, and internal heat. However, the effect of simple treatment based on syndromes, even with the addition of CHMs to calm the nerves, is sometimes not ideal for improving insomnia in a short period of time. In this case, TCM treatment based on syndrome combined with sleeping pills can achieve better sleep quality than sleeping pills alone. Combining treatment based on syndrome with sleeping pills can help patients maintain a certain amount of sleep a day, and at the same time can gradually improve their sleep quality. As the patient's condition improves, the dosage of sleeping pills can be gradually reduced until they are completely stopped. Using this method, it is expected that the sleep time and sleep quality of patients with refractory insomnia can be gradually improved in a step-by-step manner.

Other examples include the treatment of malignant tumors, which currently account for approximately one-sixth of the global causes of death. We know that the main reason for the death of cancer patients is not the proliferation of cancer cells. In the process of cancer development and changes, the proliferation and spread of cancer cells, the condition of the various physiological functions of the human body to maintain life, and the patient's physical fitness and disease resistance determine the direction of the disease. The proliferation and spread of cancer cells are restricted by the body's immunity, and at the same time, cancer destroys the body's normal physiological functions and consumes the patient's physical and disease resistance. The obstacles of human physiological function provide suitable 'soil' for the occurrence and spread of cancer, and at the same time restrict the recovery of the human body's physical fitness and disease resistance. The weakness of the body's physical fitness and disease resistance makes the proliferation and spread of cancer cells out of control, and at the same time, it is also difficult for a weakened body to maintain normal life activities. The condition of cancer patients develops and changes in the interaction of the above three aspects.

People has always been looking forward to finding substances that directly kill cancer cells, just like dealing with bacteria, to completely eliminate the cancer cells produced by the human body, so as to fundamentally cure cancer. However, because cancer cells and normal cells of the same tissue origin only differ in the degree of differentiation, it is very difficult to find a substance that only targets mutated cancer cells and has no or less impact on normal cells, whether medication therapy or immunotherapy. Often the stronger the killing action on cancer cells a therapy can have, the worse the damage the damage to normal cells. The most important aspect of modern medical anti-cancer research is to find therapies and medications that can kill cancer cells as much as possible while damaging normal cells as little as possible.

In recent decades, with the advancement of methods and technologies for the treatment of cancer in modern medicine and the advancement of clinical research on the treatment of cancer with ITCMWM, cancer is no longer an incurable disease that is tantamount to the death penalty. The ideas of 'Cancer is just a chronic disease' and coexisting with cancer is accepted by more and more people. As a result, cancer treatment has changed from simply killing cancer cells to comprehensive treatment, including killing cancer cells, controlling the proliferation and spread of cancer cells, preventing, reducing and eliminating complications caused by cancer, improving and maintaining the stomach, the digestion and absorption function of the intestines, the improvement of human physical fitness and disease resistance, and, as much as possible, ensuring the quality of life of patients, and so on.

TCM treatment based on syndrome under the holistic concept, that is, the state regulation-control method of holistic medicine, has significant effects in the following aspects of cancer treatment:

1. Treat ulcers, nodules, polyps, chronic inflammatory and benign tumors and other pre-cancerous lesions in time to prevent the occurrence of cancer and the malignant transformation of pre-cancerous diseases.
2. Control the growth of tumors, reduce or eliminate lesions.
3. Improve or eliminate complications caused by cancer, such as pleural effusion and ascites.
4. Improve and restore gastrointestinal function, enhance human immunity and physical fitness, reduce and eliminate cachexia, and improve the quality of life of patients.

Simple TCM treatment has shown better effects for cancer patients with low malignancy and slower growth. But only relying on TCM treatment based on syndrome, its direct killing effect on cancer cells is limited. Therefore, it is often seen in clinical practice that some patients have gradually improved their physical condition as the complications are eliminated and their resistance strengthened, but the Western medical examination of the lesions did not shrink, and some increased instead. Recognizing the limitations of using conventional treatment based on syndrome for such malignant chronic diseases, TCM scholars have been trying to find a way to 'contend poison with poison' in order to treat malignant tumors as effectively as treating ordinary chronic diseases. However, due to the limitation of the level of science and technology, the 'poison' that TCM practitioners try to use often do not correspond well to the 'poisons' of malignant tumors, so the results are not ideal. Modern medicine has always focused on eliminating tumors and killing cancer cells, which is precisely the 'poison' that TCM practitioners dream of to eliminate. Compared with the TCM method of attacking poison with poison, modern medical surgery, chemotherapy, radiotherapy, especially targeted therapy, immunotherapy, etc., have obvious advantages in treating cancer lesions.

In terms of cancer diagnosis, modern medical imaging examinations, such as CT, ultrasound, MRI, PET and other techniques, can accurately determine the location of the lesion; cytology and pathological diagnosis methods can determine the benign and malignant tumors and cell types, the degree of differentiation; as well as biochemical examination, immunological examination, tumor marker examination and other detection methods. These methods and technologies provide an objective basis for cancer diagnosis and monitoring and evaluation of treatment effects.

Obviously, in the treatment of cancer, the advantages of TCM in overall regulation-controlling and the advantages of modern medicine for local lesions, are complementary to each other, while each one has its own strengths. Combining the treatment methods of traditional Chinese and Western medicine on the basis of modern Chinese and Western medicine disease information monitoring will undoubtedly greatly improve the level of cancer treatment. Therefore, holistic medicine treating cancer should focus on the overall condition of cancer patients, with TCM treatment based on syndrome as the core for holistic regulation-controlling, and integrate the effective methods of modern medicine for cancer lesions. Here, the methods of removing blood stasis, eliminating tumors, laxative and water-removing methods in TCM, and various treatment methods for cancer lesions in modern medicine are all methods of 'treating the symptoms when urgently needed', which can be combined and used in a timely and appropriate manner.

In short, holistic medicine in the 21st century will still include the two systems, the state medicine and disease medicine, while clinical treatment is the organic combination of the treatment methods based on two systems. But this combination is different from the integration of traditional Chinese and Western medicine (ITCMWM) based on modern medical concepts in Mainland China, it is an integration led by a holistic concept. The two major medical systems that originated in the East and the West have very different methodologies and theoretical frameworks, but science belongs to all mankind and should not be defined by region. In modern medicine in the 21st century, with the establishment of holistic medicine led by state medicine, and the return of modern medicine to Eastern traditional medicine in terms of concepts and methods, future medical integration and structural reorganization are inevitable. The result of medicine integration led by the holistic concept will be the organic combination of holistic regulation-control and specific treatment, which will be the highest state of medicine in the 21st century.

References

Alastair Bruce Scott-Hill. 2013. The Paranormal is Normal: The Scientific Validation of Reincarnation, the Paranormal, and Your Immortality. Xcell Books.

American Society of Clinical Oncology. 2017. CancerLinQ Partners with FDA to Study Real-World Use of Newly Approved Cancer Treatments. Retrieved from https://www.asco.org/about-asco.

Anhui Cultural History Editorial Committee. 2000. Anhui Cultural History, Vol. 2, Nanjing University Press, p. 1585 pages (Kim Jong-hee).

Association of Accredited Naturopathic Medical Colleges. 2020. What is Naturopathic Medicine? Learn more Now with AANMC. AANMC. https://aanmc.org/naturopathic-medicine/.

Bacon, Francis. 1878. Novum Organum, sive Indicia Vera de Interpretatione Naturae. Clarendon Press.

Bai Chunqing. 2011. Thirty Years of Traditional Chinese Medicine Expert System. Medical Information, 24(2): 550–552.

Baidu Encyclopedia. 2022. Bioholographic theory. Baidu Encyclopedia, https://baike.baidu.com/item/生物全息论/10462770.

Baidu Encyclopedia. 2023a. Integrated Chinese and Western Medicine, https:// baike.baidu.com/item/中西医结合/657978.

Baidu Encyclopedia. 2023b. Systems Biology, Baidu Encyclopedia, https://baike.baidu.com/item/系统生物学/665855.

Baidu Encyclopedia. 2023c. Genomic medicine, Baidu Encyclopedia, https://baike.baidu.com/item/基因组医学/8232755.

Bai Zhiyang. 2012. Anti-flu star drug "Tamiflu" be exposed to strong side effects or can cause death, Sohu Health, https://health.sohu.com/20120104/n331143220.shtml.

Baselga, J., Tripathy, D., Mendelsohn, J. et al. 1999. Phase II study of weekly intravenous trastuzumab (Herceptin) in patients with HER2/neu-overexpressing metastatic breast cancer. Seminars in Oncology, 26(4 Suppl 12): 78–83.

Bayer. 2018. FDA Approves Vitrakvi® (larotrectinib), the First Ever TRK Inhibitor, for Patients with Advanced Solid Tumors Harboring an NTRK Gene Fusion (1,2). CISION. https://www.prnewswire.com/news-releases/fda-approves-vitrakvi-larotrectinib-the-first-ever-trk-inhibitor-for-patients-with-advanced-solid-tumors-harboring-an-ntrk-gene-fusion1-2-300755551.html.

Beijing College of TCM. 1974. Foundation of Traditional Chinese Medicine, Shanghai People's Publishing House.

Bell, I. R. et al. 2002. Integrative medicine and systemic outcomes research: Issues in the emergence of a new model for primary health care. Arch. Intern. Med. January 2002, 162(2): 133–40. PMID 11802746. doi:10.1001/archinte.162.2.133.

Brahmer, J. R., Tykodi, S. S., Chow, L. Q. et al. 2012. Safety and activity of anti-PD-L1 antibody in patients with advanced cancer. New England Journal of Medicine, 366(26): 2455–2465.

Cai Jingfeng et al. 2000. General history of chinese medicine (Modern Volume). People's Medical Publishing House, Beijing.

Capra, Fritjof. 1983. The Turning Point: Science, Society, and the Rising Culture. Bantam Press.

Castiglioni, Arturo, translated by Edward Bell Krumbhaar. 2019. A History of Medicine. Routledge.

Chang Chang. 2006. Research progress in systems biology. Life Science Research, 10(2): 1–6.

Chen Dashun, et al. 1988. Various Theories of Traditional Chinese Medicine—Correspondence Teaching Materials in National Higher Education Colleges of Traditional Chinese Medicine. Hunan Science and Technology Press.

Chen Keji and Song Jun. 1999. Inspiration from evidence-based medicine to Integrative Medicine. Chinese Journal of Integrative Medicine, 19(11): 643–644.

Chick, S. E. and Sanchez, S. M. (Eds.). 2008. Simulation Modeling Handbook: A Practical Approach. CRC Press.

China Association of Traditional Chinese Medicine. 2009. Classification and Determination of TCM Constitution, China Traditional Chinese Medicine Publishing House.

China Traditional Chinese Medicine News. 2019. 70 years of brilliant achievements in the revitalization and development of traditional Chinese medicine, China Traditional Chinese Medicine Network. https://www.cntcm.com.cn/2019-06/24/content_62165.htm.

Cohen, I. Bernard. 1987. Revolution in Science. Belknap Press.

CPC Central Literature Research Office. 1987–1998. Manuscripts of Mao Zedong since the founding of the PRC, Vol. 7. Central Party Literature Press. p. 451.

Dai Zhaoyu et al. 2000a. Responses and Reflections on the Side Effects of Xiao Chai Hu Tang, 2000 Current Situation and Trends of Japanese Traditional Medicine. Hong Kong, Asian Medicine Press, pp. 31–36.

Dai Zhaoyu et al. 2000b. An Overview of the Current Situation of Kampo Medicine in Japan, 2000 Current Situation and Trends of Japanese Traditional Medicine. Hong Kong, Asian Medicine Press, pp. 14–23.

Dawes, M., Summerskill, W., Glasziou, P. et al. 2005. Sicily statement on evidence-based practice. BMC Med Educ. 5(1): 1. doi:10.1186/1472-6920-5-1. PMC 544887. PMID 15634359.

Descartes, René. 1637. Discourse on the Method of Rightly Conducting One's Reason and of Seeking Truth in the Sciences. Penguin Publishing Group, 2000.

Ding Zhi Wei, Shi Yu, Zhang Yongqiang. 2016. Perforators, the Underlying Anatomy of Acupuncture Points. Alternative Therapies, May/June, 2016, 22(3): 25–30.

Dong Wei. 2010. Expert Appeal: It is time to change the medical model, China Youth Online - China Youth Daily, Jan. 4 2010, http://zqb.cyol.com/content/2010-01/04/content_3011411.htm.

Druker, B. J., Talpaz, M., Resta, D. J. et al. 2001. Efficacy and safety of a specific inhibitor of the BCR-ABL tyrosine kinase in chronic myeloid leukemia. New England Journal of Medicine, 344(14): 1031–1037.

Eddy, D. M. 1990. Practice policies: Where do they come from? JAMA, 263(9): 1265. https://doi.org/10.1001/jama.263.9.1265.

Eddy, D. M. 2005. Evidence-based Medicine: A Unified Approach. Health Affairs, 24(1): 9–17. PMID 15647211. doi:10.1377/hlthaff.24.1.9.

Einstein, A. and Infeld, L. 1938. Evolution of Physics. Cambridge University Press.

Epstein, David 2017. When Evidence Says No, But Doctors Say Yes. ProPublica. Retrieved from https://www.propublica.org/article/when-evidence-says-no-but-doctors-say-yes.

Ervin Laszlo. 1978. The Systems View of the World: A Holistic Vision for Our Time. New York, George Brazille, Inc.

Expert Committee on Implementation Specifications for Real World Research in Clinical Medicine in China. 2017. Implementation regulations for real World Research in Clinical Medicine in China [J/CD]. Chinese Journal of Experimental and Clinical Infectious Diseases (Electronic Edition), 11(6): 521–525. DOI: 10.3877/cma.j.issn.1674-1358. 2017. 06. 001.

Ezkurdia, I. et al. 2014. Multiple evidence strands suggest that there may be as few as 19 000 human protein-coding genes. Human Molecular Genetics, 23(22): 5866–5878. https://doi.org/10.1093/hmg/ddu309.

Falconer, Kenneth. 2003. Fractal Geometry: Mathematical Foundations and Applications. John Wiley & Sons, Ltd.

Fan Dongping. 2005. Types of emergent theory and its theoretical appeal. Science and Technology and Dialectics, 22(4): 49–53.

Fan Dongping. 2006. Emergence and hierarchy of complex systems. Academic Research, 12: 35–39.

Fan Daiming. 2012. A preliminary exploration of integrative medicine. J. Negative, 3(2): 3–11.

Fan Daiming. 2014a. An introduction to integrative medicine. J. Negative, 5(5): 1–13.

Fan Daiming. 2014b. Holistic integrative medicine. Am. J. Digest. Dis., 1(1): 22–36.

Fan Daiming. 2016. Integrative medicine: A new era of medical development. J. Chinese Medical Journal, 96(22): 1713–1718.

Fan Daiming. 2017a. HIM, the inevitable direction of medical development. J. Negative, 8(1): 1–10.

Fan Daiming. 2017b. Holistic integrative medicine: Toward a new era of medical advancement. J. Front Med., 11(1): 152–159.

Fan Daiming. 2017c. HIM, the only way for medical development in a new era. J. Negative, 8(3): 1–19.

Fan Daiming. 2017d. My views on holistic integrative medicine. J. British Medical Journal Chinese Edition, 20(10): 547–548. DOI: 10.3760/cma.j.issn.1007-9742.2017.10.106.

Fan Xing, Yang Zhiping and Fan Daiming. 2013. Revisiting integrative medicine. J. Medicine and Philosophy, 34(3A): 6–11, 27.7.

Gao Chong et al. 2013. Research progress on the pharmacological effects of traditional Chinese medicines for promoting blood circulation and removing blood stasis, Drug Evaluation Research, 36(1): 64–68.

Goldstein, Jeffrey. 1999. Emergence as a construct: History and issues. Emergence, 1(1): 49–72. doi:10.1207/s15327000em0101_4.

GRADE working group. 2006. Welcome to the GRADE working group. Retrieved from www.gradeworkinggroup.org.

Guangzhou College of TCM. 1979. National Medical College Trial Textbook-Herbal Formulas. Shanghai Science and Technology Press.

Guo Changqing. 1998. Handbook of Practical Micro-Needling. China Medical Science and Technology Press.

Guo Feng and Zhao Na. 2007. Research status and prospect of pharmacological effects of Epimedium. J. China Pharmaceutical., 16(24): 70–72.

Guo Xinfeng and Lai Shilong. 2010. Research status and enlightenment of evidence-based medicine in the field of Kampo medicine in Japan. Chinese Journal of Evidence-Based Medicine, 10(5): 625–628.

Hao Guangming. 2004. The current modernization of traditional Chinese medicine is a fake modernization. Chinese Archives Of Traditional Chinese Medicine, 22(1): 149–151 (DOI)10.13193/j.archtcm.2004.01.148.haogm.069.

Hao Jinghua and Lu Peiqi. 2010. Grade 5 Primary School Textbooks. Jiangsu Education Press, Nanjing.

Hawking, Stephen and Mlodinow, Leonard. 2005. A Briefer History of Time: A Special Edition of the Science Classic Classic – Illustrated. Bantam Books.

Henry, N. Pollack. 2003. Uncertain Science and Uncertain World. Cambridge University Press.

Herrera-Perez et al. 2019. A comprehensive review of randomized clinical trials in three medical journals reveals 396 medical reversals. eLife, 8. https://doi.org/10.7554/elife.45183.

He Ziyuan. 2020. Roadmap for evidence-based research in TCM confirmed: "Witness" Decade Action Launched in Beijing, Global Business Information Network, 2020.1.10, http://www.huangxiaobo.org/hqxw/115596.html.

Hochhaus, A., Larson, R. A., Guilhot, F. et al. 2017. Long-term outcomes of Imatinib treatment for chronic myeloid leukemia. New England Journal of Medicine, 376(10): 917–927.

Hua Guofan and Jin Guantao. 1979. Traditional Chinese Medicine: A miracle in the history of science. Dialectics of Nature Newsletter, 1(2): 20–32.

Huang Xinrong. 2007. Complexity Science and Philosophy, Central Compilation Press.

Huang Xinrong. 2012. Complexity Science Method and Its Application, Chongqing University Press.

Ideker, T. et al. 2001. Integrated Genomic and Proteomic Analyses of a Systematically Perturbed Metabolic Network. Science, 292(5518): 929–934.

International Forum. 2019. Deng Tietao: China has not cultivated real Chinese medicine practitioners for decades. World Traditional Medicine Network, 2019-05-07, http://tmbos.com/index/article/view/id/7126.html.

Jin Guantao and Hua Guofan. 2005. Cybernetics and Scientific Methodology, Nova Publishing House, Beijing.

Jin Guantao, Ling Feng, Bao Yuhai and Jin Guanyuan. 2017. Principles of Systems Medicine. China Science and Technology Press.

Jin, W. et al. 2020. Infrared imageries of human body activated by teas indicate the existence of meridian system. Research Square. https://doi.org/10.21203/rs.3.rs-20030/v1.

John Holiand. 2001. Emergence: From Chaos to Order. Copyright by John H. Holiand.

Krauss, A. 2018. Why all randomised controlled trials produce biased results. Annals of Medicine, 50(4): 312–322. https://doi.org/10.1080/07853890.2018.1453233.

Ke Xuan. 2017. "Emotions" captured by filming, scientists reveal the mystery of human energy field—Interview with Russian physicist Konstantin Korotk, Epoch Times Health 1+1, https://www.epochtimes.com/gb/17/11/16/n9848899.htm.

Law, A. M. and Kelton, W. D. 2020. Simulation Modeling and Analysis. McGraw-Hill Education.

Liang Maoxin and Sun Dehua. 1991. The current situation, problems and solutions of standardization research of traditional Chinese medicine. Chinese Journal of Medicine, 6(5): 58–60.

Li Daozhen. The Strange Human Energy Field. Prohibited News. 2020, https://www.bannedbook.org/bnews/comments/20200605/783245.html.

Li Jianshan et al. 2002. An Overview of Scientific Methods, Beijing, Science Press.

Lily, Q. 2019. What is real-world research? How is it different from traditional research? Tencent.com, December 14, 2019, https://new.qq.com/omn/20191214/20191214A06FGL00.html?pc.

Li Meng. 2016. The road of experience: bacon and descartes on the method and philosophical basis of modern science. Journal of Yunnan University (Social Sciences Edition), 15(05): 9–23.

Lin Deli, Yang Shuqing and Xue Xuejun. 2013. War Game Deduction Technology and Application. Military Science Press.

Ling Yikui and Yan Zhenghua. 1984. Textbooks for Medical Colleges - Chinese Pharmacy, Shanghai Science and Technology Press.

Li Qiangou, Zhou Xuewen and Shan Zhaowei. 2001. Practical Chinese Medicine Gastroenterology, Beijing, People's Health Publishing House.

Li, R. Y. M. et al. 2019. Ranking of risks for existing and new building works. Sustainability, 11(10): 2863. https://doi.org/10.3390/su11102863.

Liu Jinyang. 2007. The two forms of reductionism and their thinking essence. Beijing: Journal of Dialectics of Nature. 28(6): 25–31.

Liu Shenlin. 2014. Principles and Paths of Integrative Medicine, Nanjing University of Chinese Medicine Graduate School website, https://gra.njucm.edu.cn/2017/1012/c2892a24400/page.htm.

Li Xiantao et al. 2019. Exploration of research technology and method of TCM syndrome diagnosis scale. J. Tianjin Traditional Chinese Medicine, 36(2): 122–124.

Li Xisuo. 2002. Five Thousand Years of History of Cultural Exchanges Between China and Foreign Countries, Vol. 2, World Knowledge Press.

Li Youping and Liu Ming. 1999. Evidence-based medicine and the modernization of traditional Chinese medicine. Chinese Journal of Information on Traditional Chinese Medicine, 6(12): 14–16, 32.

Li Youping et al. 2007. China Evidence-Based Medicine Center's strategy for promoting the modernization of traditional Chinese medicine. Chinese Journal of Evidence-Based Medicine, 7(3): 159–161.

Li Youping, Li Jing, Sun Xin et al. 2016. The origin and development of evidence-based medicine in China: dedicated to the 20th anniversary of evidence-based medicine in China. Chinese Journal of Evidence-Based Medicine, 16(1): 2–6.

Li Zhijun. 2004. The Eastward Spread of Western Learning and the Real Learning in Ming and Qing Dynasties. Shanghai Ancient Books Publishing House, Bashu Publishing House.

Ludwig von Bertalanffy. 1987. General System Theory - Foundations, Development, Applications. New York, George Brazille, Inc.

Luenberger, D. G. 1979. Introduction to Dynamic Systems: Theory, Models, and Applications. John Wiley & Sons.

Luo Linming et al. 2017. Research progress on the active ingredients and their mechanisms of antitumor effect of ginseng. J. Chinese Herbal Medicine, 48(3): 582–596.

Lu Yongbo. 2006. Textbooks for Higher Education Institutions - Systems Engineering, Tsinghua University Press. Beijing Jiaotong University Press.

Lv Aiping. 2013. Research on animal models of combination of disease and syndrome: From theoretical innovation to technical challenge. Chinese Journal of Integrative Medicine, 33(1): 6–7.

Mandelbrot, B. B. 1967. How long is the coast of Britain? Statistical selfsimilarity and fractional dimension. Science, 155: 636–638.

Mandell, B. F. 2021. Clinical decision-making strategies. MSD Diagnostic and Treatment Manual Medical Professional Edition. Retrieved from https://www.msdmanuals.cn/professional/special-subjects/clinical-decision-making/evidence-based-medicine-and-clinical-guidelines.

McLachlan, J. A. 2010. Integrative medicine and the point of credulity. BMJ. 341: c6979. https://doi.org/10.1136/bmj.c6979.

McNeil, John J. et al. 2018. Effect of aspirin on cardiovascular events and bleeding in the healthy elderly. New England Journal of Medicine, 10–18, 379(16): 1509–1518.

Meng, S., Tang, Z., Zhu, M. et al. 2023. Perforating cutaneous vessels: A key feature of acupoints – Anatomical evidence from five-Shu acupoints in the upper limbs. Clinical Anatomy, 1–10. https://doi.org/10.1002/ca.24077.

Middleton, G. et al. 2020. The National Lung Matrix Trial of personalized therapy in lung cancer. Nature, 583(7818): 807–812. https://doi.org/10.1038/s41586-020-2481-81-8.

Mi Hong, Li Shuzhuo, Hu Ping et al. 1997. Two Household Registration and Population Surveys in the Late Qing Dynasty and the Early Republic of China, Historical Research.

National Administration of TCM. 2010. Basic Standards for Writing Medical Records of TCM. http://www.natcm.gov.cn/yizhengsi/gongzuodongtai/2018-03-24/3072.html.

National Cancer Institute. 2022. Complementary and Alternative Medicine (CAM). https://www.cancer.gov/about-cancer/treatment/cam.

National Institutes of Health (NIH). 2023. National Center for Complementary and Integrative Health (NCCIH). https://www.nih.gov/about-nih/what-we-do/nih-almanac/national-center-complementary-integrative-health-nccih.

National Health Commission. 2020. Statistical Bulletin of my country's Health Development. Central People's Government of the PRC. http://www.gov.cn/guoqing/2021-07/22/content_5626526.htm.

National Ministry of Health. 1989. Measures for the Classification of Hospitals. https://zh.m.wikisource.org/zh-hans/医院分级管理办法（试行草案）.

National Ministry of Health. 1994. Basic Standards for Medical Institutions (Trial), http://hospital.hf.cas.cn/hfzl/hfzl_yyxw/hfzl_tztg/201706/W020170616337930798108.pdf.

National Ministry of Health. 2010. Basic Norms for Writing Medical Records, WYZF. No. 11, January 22, 2010, http://www.nhc.gov.cn/yzygj/s3585u/201002/0517a82e35224ee0912a5d855a9d249f.shtml.

National Research Council. 2011. Toward precision medicine: building a knowledge network for biomedical research and a new taxonomy of disease. https://www.nap.edu/catalog/13284/.

National Standardization Administration. 2008. Discipline Classification and Code. National Standard of the People's Republic of China GB/T13745-2008. June 30, 2008, https://xkb.pku.edu.cn/docs/2018-10/20220328083301969071.pdf.

Neuman, M. I. et al. 2014. Durability of Class I American College of Cardiology/American Heart Association Clinical Practice Guideline Recommendations. JAMA, 311(20): 2092. https://doi.org/10.1001/jama.2014.4949.

NHS sites. 2022. Complementary and alternative medicine, https://www.nhs.uk/conditions/complementary-and-alternative-medicine/.

Nogier, Paul. 1957. Auricular point distribution map shaped like an embryo's reflection. German Acupuncture and Related Technology Magazine.

Office of the Commissioner. 2020. 21st Century Cures Act. U.S. Food and Drug Administration. https://www.fda.gov/regulatory-information/selected-amendments-fdc-act/21st-century-cures-act.

Office for the 7th National Census of the State Council. 2021. Main Data of the Seventh National Census in 2020. China Statistics Press Co., Ltd.

Otsuka Kyoo. 2000. History and Current Situation of Kampo Medicine, 2000 Current Situation and Trends of Japanese Traditional Medicine, Hong Kong, Asian Medicine Press, pp. 8–10.

Ouyang Xuemei. 2020. Protection, Inheritance and Development of Chinese Medicine by the Communist Party of China, Guangming Daily, April 1, 2020. http://dangshi.people.com.cn/n1/2020/0401/c85037-31657020.html.

Pan Feng and Chen Keji. 2005. Intersection of Integrated Chinese and Western Medicine Research and Modern Technology, Scientific Interview, Official Website of Chinese Academy of Sciences. https://www.cas.cn/ft/zxft/200507/t20050728_1689618.shtml.

People's Health Publishing House. 1979. People with Sensitive Meridians—Research Data Collection on Meridian Sensing Phenomenon, People's Health Publishing House.

PMI Working Group. 2015. The Precision medicine initiative cohort program – building a research foundation for 21st century medicine. Retrieved from https://www.nih.gov/sites/default/files/research-training/initiatives/pmi/pmi-working-group-report-20150917-2.pdf.

Porter, D. and Porter, R. 1997. The Cambridge History of Medicine. Cambridge University Press.

Pray, Leslie A. 2008. Eukaryotic genome complexity. Nature Education, 1(1): 96. Retrieved from https://www.nature.com/scitable/topicpage/eukaryotic-genome-complexity-437/.

Prigogine, Ilya and Stengers, Isabelle. 1997. The End of Certainty - Time, Chaos and the New Law of Nature. Free Press.

Qian Xuesen, Yu Jingyuan and Dai Ruwei. 1990. A new field of science: Open complex giant systems and its Methodology. Journal of Nature, 12(1).

Qian Xuesen. 1988. On System Engineering. Hunan Science and Technology Press.

Qian Xuesen. 2001. Systematics of Creation. Shanxi Science and Technology Press.

Qin Dulie and Bao Yiwan. 1989. Introduction to Computer Simulation and Expert System of Traditional Chinese Medicine, People's Health Publishing House.

Rees, L. H. and Weil, A. 2001. Integrated medicine: Imbues orthodox medicine with the values of complementary medicine. BMJ, January 20, 2001, 322: 119–120. doi:10.1136/bmj.322.7279.119.

Ren Zhuang. 2020. Deepening Evidence-Based Research and Promoting "Inheritance and Innovation". China News of Traditional Chinese Medicine, 5th edition, 2020-01-16, https://www.cntcm.com.cn/2020-01/16/content_70371.htm.

Richman, I. B. and Owens, D. K. 2017. Aspirin for primary prevention. The Medical Clinics of North America (Review), 101(4): 713–24. PMID 28577622. doi:10.1016/j.mcna.2017.03.004.

Roche Registration Co., Ltd. 2023. Drug Instructions: Oseltamivir Phosphate Capsules, Yaozhi Data, https://db.yaozh.com/instruct/15618.html.

Sackett, David L. et al. 1992. Evidence-based medicine: A new approach to teaching the practice of medicine. JAMA. November 1992, 268(17): 2420–25.

Sackett, D. L., Rosenberg, W. M., Gray, J. A. et al. 1996. Evidence based medicine: What it is and what it isn't. BMJ, 312(7023): 71–72. doi: 10.1136/bmj.312.7023.71.

Schork, N. J. 2015. Time for one-person trials. Nature, 520(7549): 609–611.

Schneeweiss, A., Chia, S., Hickish, T. et al. 2013. Pertuzumab plus trastuzumab in combination with standard neoadjuvant anthracycline-containing and anthracycline-free chemotherapy regimens in patients with HER2-positive early breast cancer: a randomized phase II cardiac safety study (TRYPHAENA). Annals of Oncology, 24(9): 2278–2284.

Schwaederle, M. et al. 2015. On the road to precision cancer medicine: Analysis of genomic biomarker actionability in 439 patients. Mol. Cancer Ther., 14(6): 1488–1494.

Shanxi Provincial Institute of TCM. 1976. Commentary on "Medical Forest Correction". Beijing, People's Health Publishing House.

She Huimin. 2015. Precision medicine makes treatment more personalized. People's Daily Online. http://politics.people.com.cn/n/2015/0622/c70731-27188674.html.

Shen Ziyin and Wang Wenjian. 2004. Practice proves the importance of integrative medicine. Chinese Journal of Integrative Medicine, 24(12): 1062–1063.

Sherman, M. E. et al. 2011. Multidisciplinary Canadian consensus recommendations for the management and treatment of hepatocellular carcinoma. Current Oncology, 18(5): 228–240. https://doi.org/10.3747/co.v18i5.952.

Slamon, D. J., Clark, G. M., Wong, S. G. et al. 1987. Human breast cancer: Correlation of relapse and survival with amplification of the HER-2/neu oncogene. Science, 235(4785): 177–182.

Smith, M. S. and Logan, A. C. 2002. Naturopathy. Medical Clinics of North America, 86(1): 173–184. https://doi.org/10.1016/s0025-7125(03)00079-8.

Snyderman, R. and Weil, A. 2002. Integrative medicine-bringing medicine back to its roots. Archives of Internal Medicine, 162(4): 395. https://doi.org/10.1001/archinte.162.4.395.

State Food and Drug Administration, 2018. Technical Guiding Principles for Clinical Research of New CHM Medications with Syndrome. State Food and Drug Administration website, Nov. 1 2018, http://www.nmpa.gov.cn/WS04/CL2138/331783.html.

Sterman, J. D. 2000. Business Dynamics: Systems Thinking and Modeling for a Complex World. Irwin/McGraw-Hill.

Sun Dongchuan and Lin Fuyong. 2004. Introduction to Systems Engineering. Beijing: Tsinghua University Press.

Sun Lanfang and Jiang Lu. 2005. Systems biology: Systems science and biological systems. J. Systems Engineering Theory and Practice, 25(10): 67–72.

Sun Xiaoli and Zhang Zengyi. 2004. Ten Relationships in Scientific Method, Xuelin Press.

Swain, S. M., Baselga, J., Kim, S. B. et al. 2015. Pertuzumab, trastuzumab, and docetaxel in HER2-positive metastatic breast cancer. New England Journal of Medicine, 372(8): 724–734.

Tan Lu and Jiang Lu. 2009. Introduction to Systems Science, Beijing Normal University Press, Beijing.

Terry, Sharon F. 2015. Obama's Precision Medicine Initiative. Genet Test Mol. Biomarkers, 19(3): 113–114.

The Academic Consortium for Integrative Medicine & Health 2015. Introduction: Academic Consortium for Integrative Medicine & Health. (n.d.). https://web.archive.org/web/20151212034221/http://imconsortium.org/about/about-us.cfm.

The Medical Education Network. 2005. Perspective on the Modernization of Traditional Chinese Medicine—Thoughts triggered by the article "The current modernization of traditional Chinese medicine is a fake modernization". Medical Education Network. https://www.med66.com/html/ziliao/07/1/509421719867b607d48a0a1b0ae35c89.htm.

The New World of Health. 2020. Concepts and Applications of Real World Evidence (RWE). Personal Library, June 19, 2020, http://www.360doc.com/content/20/0619/21/70270324_919432545.shtml.

The Ministry of Education press conference. 2016. The Ministry of Education introduced the reform and development of Chinese medicine higher education. The Ministry of Education website. http://www.gov.cn/xinwen/2016-10/26/content_5124517.htm.

Tian Jin. 2016. Academicians and experts gathered in Xi'an to discuss a new model for the development of integrated medicine. China News Network. https://www.chinanews.com.cn/m/sh/2016/10-08/8024199.shtml.

Tian Xiaohang. 2019. The world's first evidence-based Chinese medicine center was established at the China Academy of Chinese Medical Sciences, Xinhuanet, 2019.3.12, http://www.xinhuanet.com/politics/2019-03/12/c_1124226955.htm.

Topalian, S. L., Hodi, F. S., Brahmer, J. R. et al. 2012. Safety, activity, and immune correlates of anti-PD-1 antibody in cancer. New England Journal of Medicine, 366(26): 2443–2454.

U.S. Food and Drug Administration. 2003. Press Release: FDA Announces Plans to Prohibit Sales of Dietary Supplements Containing Ephedra. https://ods.od.nih.gov/HealthInformation/Ephedra.aspx.

U.S. Food and Drug Administration. 2007. Complementary and Alternative Medicine Products and their Regulation by the Food and Drug Administration Draft Guidance for Industry. https://www.fda.gov/regulatory-information/search-fda-guidance-documents/complementary-and-alternative-medicine-products-and-their-regulation-food-and-drug-administration.

U.S. Food and Drug Administration. 2017. FDA approves first cancer treatment for any solid tumor with a specific genetic feature. https://www.fda.gov/news-events/press-announcements/fda-approves-first-cancer-treatment-any-solid-tumor-specific-genetic-feature.

U.S. Food and Drug Administration. 2018a. Framework for FDA's Real-world Evidence Program. Retrieved from https://www.fda.gov/media/120060/download.

U.S. Food and Drug Administration. 2018b. FDA approves larotrectinib for solid tumors with NTRK gene fusions. https://www.fda.gov/drugs/fda-approves-larotrectinib-solid- tumors-ntrk-gene-fusions-0.

Wang Junping. 2014. "Worship Famous Teachers" in theoretical study and "Worship Famous Doctors" in clinical practice: Can't train "Chinese medicine practitioners on books". People's Daily, August 1, 2014, p. 19. http://cpc.people.com.cn/n/2014/0801/c83083-25382183.html.

Wang Qian. 2009. China issued the first standard of "Classification and Determination of Constitution of Traditional Chinese Medicine", Central Government Portal, http://www.gov.cn/govweb/fwxx//jk/2009-04/10/content_1282011.htm.

Wang Yongyan, Liu Baoyan and Xie Yanming. 2003. Applying the method of evidence-based medicine to construct TCM clinical evaluation system. Chinese Journal of Basic Medicine in Traditional Chinese Medicine, 9(3): 17–23.

Wang Yongyan and Huang Luqi. 2019. Based on high vision, building a Chinese TCM evidence-based medicine center. Chinese Journal of Evidence-Based Medicine, 19(10): 1131–1137.

Wang Zengtao. 2023. The background and process of studying the physical anatomy of traditional acupoints. Shandong Medicine, 63(9): 1–11.

Wear, A. and French, R. 2011. The Medical Revolution of the Seventeenth Century. Cambridge University Press.

WenJun Zhang. 2018. Fundamentals of Network Biology. World Scientific Publishing Co Pte Ltd, https://doi.org/10.1142/q0149.

Wen Liyang et al. 2001. Introduction to Medicine—Textbooks for National Medical Colleges and Universities, People's Health Publishing House.

Whorton, J. C. 2004. Nature Cures: The History of Alternative Medicine in America. Oxford University Press on Demand.

Wiener, Norbert. 1948. Cybernetics: Or Control and Communication in the Animal and the Machine. Paris (Hermann & Cie) & Camb. Mass. (MIT Press), 2nd revised ed. 1961.

Wikipedia contributors. 2020. List of systems of the human body. In Wikipedia. https://en.wikipedia.org/wiki/List_of_systems_of_the_human_body.

Wikipedia contributors. 2021. Ian Stevenson. In Wikipedia. https://en.wikipedia.org/wiki/Ian_Stevenson.

Wikipedia contributors. 2022. Demographics of China. https://en.wikipedia.org/wiki/Demographics_of_China.

Wikipedia contributors. 2023a. Evidence-based medicine. In Wikipedia. https://en.wikipedia.org/wiki/Evidence-based_medicine.

Wikipedia contributors. 2023b. Medical guideline. In Wikipedia. https://en.wikipedia.org/wiki/Medical_guideline.

Wikipedia contributors. 2023c. Systems Biology, In Wikipedia, https://zh.wikipedia.org/wiki/系统生物学.

Withnall, A. 2014. Study provides evidence that "near-death" experiences are real. The Independent. https://www.independent.co.uk/news/science/life-after-death-largestever-study-provides-evidence-that-out-of-body-and-neardeath-experiences-may-actually-be-real-9780195.html.

Wu Jieping Medical Foundation. 2018. Guide to Real World Research, 2018 Edition, https://www.sohu.com/a/251847383_331432.

World Health Organization. 2013. WHO Traditional Medicine Strategy 2014–2023. WHO website, https://www.who.int/publications/i/item/9789241506096.

World Health Organization. 2018. WHO releases new edition of the International Classification of Diseases (ICD 11). World Health Organization. https://www.who.int/zh/news/item/18-06-2018-who-releases-new-international-classification-of-diseases-(icd-11).

World Health Organization. 2023. WHO Traditional Medicine Strategy 2014–2023. Retrieved from www.who.int, 15.

Wu Yilong. 2015. Precision cancer medicine: Towards road to the future. Evidence-Based Medicine, 15(1): 1–2.

Xiao Sansan (edit and organize). 2020. Changes in hospital grade review, what should the hospital do in 2020? China Net Medical Channel, http://med.china.com.cn/content/pid/182877/tid/1013.

Xiao Yang. 2017. The China Integrative Medicine Conference was held in Xi'an. Xinhuanet. http://www.xinhuanet.com/local/2017-04/30/c_129581846.htm.

Xie Shusheng. 2009. How far is biological reductionism still to go. China Reading News, 2009-2-25.

Xinhuanet. 2021. Strengthening evidence-based research of traditional chinese medicine to promote high-quality development of traditional chinese medicine. http://www.xinhuanet.com/health/2021-04/29/c_1127391350.htm.

Xiong Shili. 1985. New Consciousness-Only Theory, Zhonghua Book Company.

Xu Guozhi. 2000. System Science, Shanghai Science and Technology Education Press, Shanghai.

Yang Jisheng and Jiang Bing. 2018 Talking about System Science. Hefei University of Technology website. http://som.hfut.edu.cn/info/1006/1963.htm.

Yuan Bing. 1985. Establishing a clear theoretical structure—on the scientization of TCM human model. Journal of Beijing University of Traditional Chinese Medicine, 8(2): 32–33.

Yuan Bing and Zhao Zheng. 1986. Modern Methods of Traditional Chinese Medicine. Hubei Science and Technology Press.

Yuan Bing. 2000. Establishing chinese medicines research methods conforming to chinese medicine characteristics. Chinese Journal of Basic Medicine in Traditional Chinese Medicine, 6(11): 34–36.

Yuan Bing. 2010. Holistic Medicine: Theoretical Medicine Integrating Chinese and Western Medicine. Modern Medicine Press.

Yuan Bing, 2013a. The development trend of the body constitution medicine based on holistic medicine. World Chinese Medicine, 8(7): 811–814.

Yuan Bing. 2013b. On the development trend of constitutionology of Chinese medicine and its historical status. China Journal of Traditional Chinese Medicine and Pharmacy, 28 (1): 9–11.

Yuan Bing. 2014. Ebola is a challenge and an opportunity. China Traditional Chinese Medicine News, 3rd edition, October 20, 2014, viewpoint.

Yuan Bing. 2015a. Viral raging: A challenges and opportunities of Modern Medical science. Hong Kong Journal of Traditional Chinese Medicine, 10(2): 11–16.

Yuan Bing. 2015b. Constitutional Medicine: The Historical Opportunities Faced by the Development of Modern Medicine—Also Analyzing the Essence of TCM Constitution Types and the Relationship with TCM Syndrome, China News of Traditional Chinese Medicine, 3rd Edition, August 21, 2015.

Yuan Bing. 2016a. Constructing 'Precision Medicine' Integrating Chinese and Western Medicine—Modern Medicine toward the Age of Complexity Science, World Science and Technology - Modernization of Traditional Chinese Medicine and Materia Medica, 18(4): 563–569.

Yuan Bing, 2016b. Problems and progress strategy of constitutionology development. China Journal of Traditional Chinese Medicine and Pharmacy, 31(3): 917–921.

Yuan Bing. 2016c. The future precision medicine: Medicine towards easy control complex adeptability. Medicine and Philosophy, 37(8A): 13–18.

Yuan Bing. 2016d. Establishing of precise states control system: The route that brings TCM to precision medicine (2). Journal of Beijing University of Traditional Chinese Medicine, 39(4): 273–276.

Yuan Bing. 2017a. Looking at the way out of Japanese Kampo medicine from the general trend of medical development. Chinese Medicine Herald, 23(1): 1–6.

Yuan Bing. 2017b. Towards state medicine: The medical revolution initiated by precision medicine. China Journal of Traditional Chinese Medicine and Pharmacy, 32(4): 1434–1448.

Yuan Bing. 2018a. Towards the Era of Integration: The Integration of Modern Medicine and the Integration of Chinese and Western Medicine—A Discussion with Academician Fan Daiming Guiding Journal of Traditional Chinese Medicine and Pharmacy, 24(14): 1–6, 24(15): 4–8.

Yuan Bing. 2018b. Returning to Tradition of Chinese Medicine, Towards Holistic Medicine—Modern Medicine in the Post-Precision Medicine and Systems Biology Era, Medicine and Philosophy, 39(1A): 15–20.

Yuan Bing and Fan Gang. 2018. Chinese medicine how to enters the artificial intelligence Age. China Journal of Traditional Chinese Medicine and Pharmacy, 33(02): 698–703.

Yuan Bing. 2020. Traditional Chinese medicine, which targets the body's response state, is the more effective way to treat epidemic infectious diseases. Chinese Journal of Integrative Medicine, 2020.10.

Yuan Bing. 2021. Towards a Clinical Efficacy Evaluation System Adapted for Personalized Medicine. Pharmacogenomics and Personalized Medicine, 14: 487–496. https://doi.org/10.2147/pgpm.s304420.

Yuan Zhongyu et al. 2021. Comparison of evidence-based research on traditional medicine in China and Japan. West China Medicine, 36(4): 539–544.

Zhang Boli. 2002. The Treatment based on Syndrome and evidence-based medicine. Chinese Journal of Evidence-Based Medicine, 2(1): 1–3.

Zhang Junhua, Sun Xin, Li Youping et al. 2019. The present and future of evidence-based Chinese medicine. Chinese Journal of Evidence-Based Medicine, 19(5): 515–520.

Zhang Lianming. 2023. System (referring to the whole composed of parts), https://baike.baidu.com/item/系统/479832.

Zhang Xichun. 2009. Medical Application Based on TCM and Combined with Western Medicine. Shanxi Science and Technology Press.

Zhang Yingqing. 1980. Bioholographic Law. Journal of Substance Science, 2: 50–53.

Zheng Chaowei et al. 2013. The confusion and the way out of the research on the essence of spleen deficiency syndrome. Journal of Guangzhou University of Traditional Chinese Medicine, 28(3): 314–316.

Zhou Jifang. 2017. Real world research and clinical trials reach the same goal by different routes. Pharmaceutical Economic news, August 3, 2017, F03 edition.

Zhou Yiqiang. 2006. Practical Chinese Medicine Oncology. Beijing, Traditional Chinese Medicine Classics Press.

Zhu Danian. 1979. Physiology—National Textbook for Higher Education Institutions. People's Medical Publishing House.

Zhu Zongxiang. 1989. Biophysics of Acupuncture and Meridian: Scientific Verification of China's First Great Invention, Beijing Press.

Index

T

For Product Safety Concerns and Information please contact our EU
representative GPSR@taylorandfrancis.com
Taylor & Francis Verlag GmbH, Kaufingerstraße 24, 80331 München, Germany

www.ingramcontent.com/pod-product-compliance
Lightning Source LLC
LaVergne TN
LVHW081316110826
845149LV00006B/1519

* 9 7 8 1 0 3 2 4 8 0 3 5 0 *